Solving the Climate Puzzle

Solving the Climate Puzzle
The Sun's Surprising Role

Javier Vinós

Critical Science Press

Madrid

2023

Published by Critical Science Press
ISBN: 978-84-125867-7-0 (Paperback)

To my siblings, Lide, Ana and Íñigo, because growing up with you was extraordinarily enriching and I am deeply grateful that after so many years our relationship is still as strong and warm as when we were children.

CONTENTS

viii

LIST OF FIGURES

Puzzle Pieces

Figures

Tables

FOREWORD

In his new book, *"Solving the Climate Puzzle, The Sun's Surprising Role,"* Javier Vinós has produced a masterful summary of observational facts about Earth's climate and the theories that have been proposed to explain these facts. This is a long book, 400 pages, but well worth reading for the excellent figures alone. Extensive citations to original papers add to the length, but the references are a valuable resource. I know of no other book that presents so many detailed and interesting facts about Earth's climate, now, in the past, and what may happen in the future. There is a thorough discussion of theories of our current ice age, beginning with the century-old, pioneering work of Milankovich. There are thorough discussions of the various proxies for past climates, including the radioisotope ^{14}C, which indicates a much larger influence of the Sun than current dogma will admit. And there is much, much more, all presented with admirable qualitative clarity. There is less emphasis on quantitative details, which many readers will welcome. The most compelling takeaway message is that the maniacal focus on carbon dioxide (CO_2) as the "control knob" of the Earth's climate is a profound delusion. Vinós calls this the "Enhanced Greenhouse Effect Hypothesis." After many decades of research, and many tens of billions of dollars spent, the quantitative measure of how much changing CO_2 affects the climate through an enhanced greenhouse effect is as poorly known today as it was in 1908 when Svante Arrhenius estimated in his book *"Worlds in the Making"* that Earth's surface would warm by $S = 4°C$ if atmospheric CO_2 concentrations were doubled. A typical estimate by today's climate-alarm establishment is almost the same, $3°C$! *Parturient montes, nascetur ridiculus mus!*[1]

It is very hard to defend a climate sensitivity as large as $3°C$. Most estimates of the direct, "instantaneous" effects of a doubling of the CO_2 concentration, a 100% increase, imply a decrease of radiation to space of only about 1%. Because of the T^4 Stefan-Boltzman law of isothermal blackbody radiation, which remains approximately valid for Earth with its greenhouse gases, a 1% flux decrease can be made up for with a 0.25% increase of the absolute temperature T. An approximate value of T is about 300 K, so the feedback-free temperature increase from doubling CO_2 should be about 0.75 K or $S = 0.75°C$. To get a politically correct sensitivity, say $S = 3°C$, requires that positive feedbacks increase this number by a factor of 4 or 400%. But most natural feedbacks are negative, not positive, in accordance with Le Chatelier's Principle.

The book makes it clear that the modest changes in temperature observed over the past century, as the concentration of atmospheric CO_2 has risen from about 280 parts per million (ppm) in the year 1850 to about 430 ppm today, are comparable to many similar temperature changes that have occurred throughout the interglacial period we are living in today. None of the previous temperature changes could have been caused by human emissions of CO_2. Some of the current warming may be due to human-induced increases in CO_2, but much of it is probably due to natural causes.

---

[1] Mountains are in labor, a ridiculous mouse is born. The Mountains in Labour. Aesop.

There is no credible scientific support for the claim that the current warming is, or will be, an existential threat to humanity. On the contrary, more atmospheric CO_2 will probably turn out to have been a major benefit to life on Earth, since additional CO_2 has such a positive effect on the productivity of agriculture forestry and on photosynthetic life in general.

As the book makes clear, climate is always changing, often more dramatically than the modest changes we have seen since the year 1850. What is causing these changes? Answering this question has been set back at least 50 years by the politically imposed dogma that CO_2 is the control knob of climate. In a kind of scientific Gresham's law, a debased, politically dictated "Enhanced Greenhouse Theory" crowds out competing theories based on the gold standard of sound, observational science.[2] The book describes one plausible theory involving the Sun: "The Winter Gatekeeper Theory," and there are other, equally plausible theories that should be taken seriously.

This book should stiffen the spines of brave policymakers to stand up and resist this latest *"extraordinary popular delusion and madness of crowds"* — to paraphrase the title of Charles MacKay's classic and accurate description of the current "climate emergency."

William Happer
Cyrus Fogg Brackett Professor of Physics, emeritus, at Princeton University
Former director of the Department of Energy's Office of Science
Princeton, NJ, USA
October 22, 2023

---

[2] Gresham's law is a monetary principle stating that "bad money drives out good."

ABBREVIATIONS

- Units -

Δ: Delta, a Greek letter that means "change in" when used with units.
Gt: Gigatonne, one billion tonnes.
hPa: Hectopascal, one hundred pascals. Unit of pressure equal to millibar.
km: Kilometer
nm: Nanometer, one billionth (10^{-9}) of a meter.
mb: Millibar
ms: Millisecond, one thousandth of a second.
µm: Micrometer, one millionth (10^{-6}) of a meter.
ppm: Parts per million.
PW: Petawatt, one quadrillion (10^{15}) watts.
sq: Squared.
TW: Terawatt, one trillion (10^{12}) watts.

- Formulas -

CO_2: Carbon dioxide

- Acronyms -

AD: Anno Domini. Number of years since the beginning of the Christian era in the Gregorian calendar.
AMO: Atlantic Multidecadal Oscillation
AR: Assessment report published by the IPCC.
BC: Before Christ. Label to indicate a number of years before the beginning of the Christian era in the Gregorian calendar.
GHG: Greenhouse gas
HadCRUT: Hadley Climate Research Unit temperature
IPCC: Intergovernmental Panel on Climate Change
ITCZ: Intertropical Convergence Zone
KNMI: Koninklijk Nederlands Meteorologisch Instituut
LOD: Length of day.
NASA: National Aeronautics and Space Administration
NH: Northern Hemisphere
NOAA: National Oceanic and Atmospheric Administration
PDO: Pacific Decadal Oscillation
QBO: Quasi-biennial oscillation
QBOe: Easterly orientation of the quasi-biennial oscillation
QBOw: Westerly orientation of the quasi-biennial oscillation
SH: Southern Hemisphere
SILSO: Sunspot Index and Long-term Solar Observations
UN: United Nations
UV: Ultraviolet

CHAPTER 1
INTRODUCTION

Unequivocal science

In recent decades, an unquestionable dogma has prevailed worldwide, posing a formidable challenge to the diversity of thought and expression that has historically enriched and nourished culture and scientific progress. This dogma states that humans are seriously endangering life on the planet and our own existence with our CO_2 emissions. Recently, leading medical journals and the World Health Organization have identified climate change as *"the greatest threat to global health in the 21st century."*[3] It is truly astonishing to encounter such a characterization, especially in the wake of a pandemic that may have killed 18 million people.[4]

The impassioned plea for immediate action to mitigate rising temperatures from health journal editors around the world underscores a critical point: the scientific consensus is unequivocal. The latest report from the Intergovernmental Panel on Climate Change (IPCC) states emphatically that humans are responsible for global warming. This conclusion is based on the claim that the observed warming is primarily driven by emissions from human activities, with greenhouse gas-induced warming being partially masked by aerosol cooling.[5] The IPCC concludes that human activities have unequivocally caused global warming. As a scientist, however, I am well aware that the science is rarely unequivocal on poorly understood and highly complex scientific issues such as climate change.

Nearly a decade ago, I set out to find the supposedly conclusive evidence that our emissions are the primary driver of observed climate change, not just a contributing factor. As we are all being asked to make sacrifices to reduce emissions, it is critical that we are well informed about this smoking gun. However, my search did not lead to a simple and clear answer. Instead, it led me to the notion of scientific consensus and computer models. This answer was unsatisfactory to me because scientific progress comes from questioning the established consensus, not from passively accepting it. Otherwise, we could continue to believe that the Earth is the center of our solar system. Furthermore, it is widely accepted that climate models are not without flaws. For those who are not aware of this fact, I urge you to read this book, where I show in detail how scientists themselves acknowledge these flaws.

We must recognize that while computer models are valuable tools for generating ideas and expanding knowledge, they have no direct connection to physi-

---

[3] Atwoli, L., et al., 2021. N. Engl. J. Med. 385 pp.1134–1137.
doi.org/10.1056/NEJMe2113200 **The cited articles can be accessed by typing the full string of their digital identifier in the address bar of a web browser.**

[4] Wang, H., et al., 2022. The Lancet, 399 (10334), pp.1513–1536.
doi.org/10.1016/S0140-6736(21)02796-3

[5] IPCC AR6 Climate Change 2023: Synthesis Report, pg.43.
doi.org/10.59327/IPCC/AR6-9789291691647

cal reality because they are a product of the human mind. Their inherent limitations are evident when we consider the possibility of different models producing conflicting results, a clear demonstration that they are not scientific proof. It is highly unlikely that the results of current models will be valid two decades from now, while the scientific evidence gathered by Babylonian astronomers more than two millennia ago is still valid today.

My tireless quest to understand the causes of climate change has taken me nine years and involved a detailed examination of thousands of relevant scientific papers. I have applied the rigorous scientific method with integrity to the evidence presented in the articles, ignoring the opinions of their authors. On many occasions, I have taken the data from these articles and reworked and analyzed it in a variety of ways. The culmination of this painstaking effort is the book you now hold in your hands.

Unlike many books that simply "tell" the science behind climate change, this book attempts to "show" the hard evidence and data that support an alternative interpretation of that science. This approach allows the reader to draw their own conclusions based on the evidence presented, rather than relying solely on the perspectives of others. Admittedly, this book is more complex than others on the subject, and a certain level of scientific literacy helps to fully understand its contents. Nevertheless, I have tried to make it *"as simple as possible, but no simpler."*[6] While not every reader will understand all aspects of the book, everyone will undoubtedly gain a profound understanding of climate science. Even the most eminent climate scientists would discover new insights within its pages, for climate science is an immensely complex field in which no one can claim sufficient knowledge, given the large stream of new research articles published daily.

Climate change as a scientific question

From a scientific perspective, climate change is essentially an energy change. For the global climate on the planet's surface to change, there must be a change in the energy content of the upper layer of the ocean, the surface, and the lower layer of the atmosphere. In particular, global warming depends on an increase in the energy content of this part of the planet. This limits the possible causes of climate change and requires a thorough understanding of the planet's energetics. This book focuses on energy because climate change is inextricably linked to it.

The main focus of climate scientists' research is human-caused climate change because the IPCC was created in 1988 out of *"concern that certain human activities could change global climate patterns constituting a threat to present and future generations."*[7] *"The role of the IPCC is to assess ... the scientific ... information relevant to understanding the scientific basis of risk of human-induced climate change."*[8] In addition to assessing this information, the United Nations' 1988 decision to endorse the IPCC triggered one of the most spectacular explosions in scientific research. Since 1988, the number of articles published each year on climate change has increased by a factor of 50 (fig. 1,

---

[6] Quotation attributed to Albert Einstein.

[7] United Nations General Assembly, Forty-third Session, Supplement No. 53. 70th plenary meeting. 6 December 1988.

[8] IPCC Principles. www.ipcc.ch/site/assets/uploads/2018/09/ipcc-principles.pdf

black line).[9] A new scientific niche has been created that has gone from near-irrelevance to being populated by more than 25,000 scientists, accounting for 0.3% of all scientific output (fig. 1, dashed grey line).[10] And it is still growing.

The climate has always undergone natural changes, but despite claims to the contrary, we still lack a comprehensive understanding of the exact reasons and mechanisms behind these natural shifts. Various hypotheses exist, but we remain uncertain why the Little Ice Age, the coldest period in the last 10,000 years, occurred between 1300 and 1845. Similarly, we can't fully explain the pronounced warming of the early 20th century or the significant melting of the Arctic between 1915 and 1930, followed by a cooling trend until the 1980s. These are well-established facts that continue to elude a proper explanation.

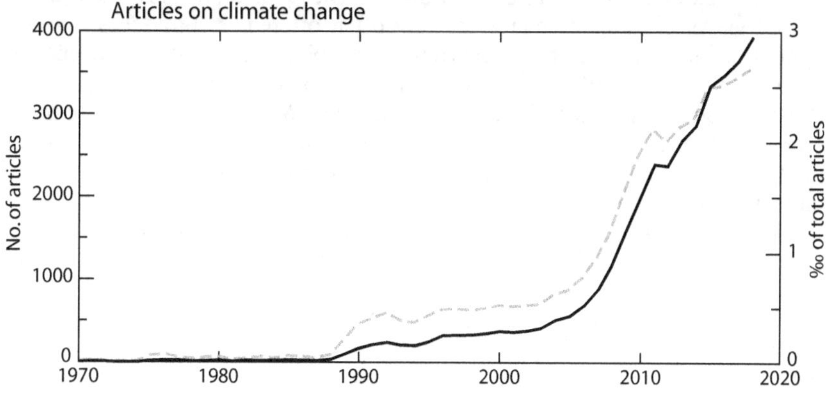

Figure 1. Number and proportion of scientific articles on climate change.

Over the past nine years, I have focused on studying natural climate change extensively. This has involved consulting thousands of scientific articles and closely examining the evidence from 800,000 years of climate history, leading me to become an expert in the field.[11] I have concluded that some essential processes involved in natural climate change still elude us. Rather than trying to fit the evidence into preconceived notions, I've followed the principles I was taught as a scientist, which involve letting the evidence accumulate and guide our understanding. This approach means being open to all available evidence and considering every possible explanation before formulating a hypothesis that fits all the evidence. This robust scientific method helps us avoid falling prey to confirmation bias, an inherent tendency of the human mind. As Sherlock Holmes explained, *"It is a capital mistake to theorize before you have all the evidence. It biases the judgment."*[12]

What I have found about how the climate changes naturally is not what I thought I would find at the beginning of the process, and it is the subject of this book. As noted above, the energetics of the climate system is paramount to

---

[9] Klingelhöfer, D., et al., 2020. Environ. Sci. Eur. 32, pp.1–21.
doi.org/10.1186/s12302-020-00419-1

[10] The number of world scientists and researchers is estimated at 8.8 million.

[11] Vinós, J., 2022. Climate of the Past, Present and Future: A scientific debate. 2nd ed. Critical Science Press.

[12] Doyle, A.C., 1887. A Study in Scarlet. Part I, Chapter 3.

climate change. Of the various processes involved in how energy flows through the climate system, one is poorly understood and almost completely neglected in relation to climate change. It is called meridional transport, and it involves the net transport of heat from the equator to both poles in the direction of the meridians. There is a wealth of evidence pointing to changes in this crucial feature of the climate as the unexpected driver of climate change that we have been missing.

A convincing sign that the hypothesis, which emerges directly from the evidence, is correct is its ability to explain the enigmatic solar influence on climate. This effect is visible in the paleoclimate record but conspicuously absent from the modern instrumental record.

This book aims to present compelling evidence that challenges the oversimplified views of climate change that are often offered. It introduces a novel hypothesis that sheds light on a previously unexplored cause of climate change: the variable amount of heat transported to the poles, which is influenced by various factors. This hypothesis, known as the "Winter Gatekeeper," emphasizes the importance of heat transport and its climatic impact, especially during the winter. Key factors such as solar activity act as gatekeepers, regulating the amount of heat transported.

The evidence-based Winter Gatekeeper hypothesis is compared with the popular model-based Enhanced CO_2 Effect hypothesis to see how well each explains known past climate changes.

No doubt, I have a personal desire for this hypothesis to be fundamentally correct. However, as a scientist, my primary goal is not to prove myself right but to uncover the scientific truth about climate change. While many scientists may believe that changes in CO_2 hold the key to explaining climate shifts, I find this answer lacking sufficient evidence. I recognize that others may disagree, as is common in scientific discourse, and I respect different interpretations of the evidence. I urge you to explore the evidence presented in this book, especially if you are open to challenging your existing beliefs. If nothing else, this book highlights a significant gap in our understanding of climate, a gap that also persists in our climate models. As a result, it should lead us to question and increase our uncertainty about how we approach the challenges posed by climate change.

Why me?

Some have argued that I lack the expertise of an earth scientist, which might seem to diminish the scientific value of my views. However, I believe the opposite is true. My background as a scientist specializing in molecular biology, neuroscience, and cancer research provides me with rigorous training in the scientific method and extensive experience in analyzing critical evidence from scientific articles. Being a non-climatologist allows me to see what other non-specialists need to understand such a complex topic. This unique perspective allows me to present climate science in a way that is accessible and understandable to a general audience.

Expertise lies in the knowledge one possesses, not in the formal education one has received. In my case, I've been studying climate for nine years, twice as long as it took me to get my Ph.D. The vast body of knowledge I've accumulated about past and present natural climate change constitutes my expertise in the field.

But what sets me apart from climate scientists in writing critically about climate change is that I am not one. Challenging the established paradigm of emissions-driven climate change can be hindered by academic training within that very paradigm. Specialized academic training can limit our ability to imagine innovative solutions to immensely complex problems like climate change. If the current climate paradigm is flawed or incomplete, it may be difficult for those trained within it to see this, while thinking outside the box becomes crucial. Moreover, daring to challenge the orthodox view carries significant career risks, from which I am entirely free.

Throughout the history of science, significant contributions have often come from outsiders to the field. Benjamin Franklin and Michael Faraday were largely self-taught. James Croll, a college janitor, proposed one of the first astronomical theories of glaciation in 1864. Milutin Milankovic, an engineer, introduced the now-accepted theory of orbital climate change in 1920. Similarly, Guy Callendar, also an engineer, was the first to link measured increases in CO_2 to a warming climate. Other examples include Alfred Wegener, a meteorologist and explorer, and Albert Einstein, a patent clerk. These individuals, among many others, demonstrate the value of diverse backgrounds in advancing scientific knowledge. Rejecting their contributions because of a lack of formal credentials would have hindered scientific progress.

My previous book on climate was written primarily for academics, making it difficult for a general audience to read. That book makes extensive use of acronyms and assumes that readers have considerable prior knowledge of climate physics. Despite these challenges, the success of the book has exceeded my expectations. Figure 2 shows a July 2023 screenshot from ResearchGate, the largest social network for scientists and researchers, and the book's page shows impressive statistics. It has a remarkable Research Interest Score, ranking in the top 6% of all research items on the platform and in the top 8% for climatology. It also secured a place in the top 1% of all items published in 2022.

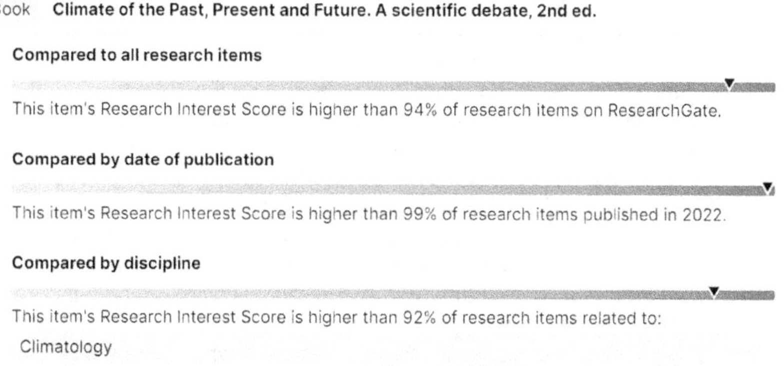

Figure 2. My previous book's Research Interest Score on ResearchGate.

The high level of interest that my previous work on climate change has received from my peers proves that the notion that I am unqualified to write about climate science for a general audience is unfounded.

How to read this book

To make climate science more accessible, I've structured this book with different reading levels. To get a quick overview, take a look at the 16 puzzle pieces at the end of some chapters and the completed puzzle at the end. This should give you a first impression in just 10 minutes. For the next level, read the chapter abstracts and summaries. They've been carefully simplified for clarity. They form an essay that can be read in just over an hour. To make the main text easier to digest, I've divided it into short chapters, each focusing on one important point and taking about 10 minutes to read. More complex or side issues have been placed in text boxes. Feel free to skip these without losing your understanding of the main points, even if they contain important evidence.

The book is divided into four parts, each with four sections. The first part deals with climate and energy, the most challenging and least engaging part. While understanding the energetics of the climate system is critical to understanding climate change, I recognize that some readers may find it daunting. I suggest that those who are not scientifically inclined skip this part and go directly to Part II.

Section 1 discusses how the climate system obtains its energy. Chapter 2 explains the nature of solar radiation and how it changes. Chapter 3 deals with the part of the incoming energy that is rejected by the planet through its reflection, or albedo. Chapter 4 explains where solar energy goes once it enters the climate system.

Section 2 discusses how the climate system gets rid of the energy it receives to maintain its thermal stability. Chapter 5 explains how energy leaves the climate system. Chapter 6 discusses the Earth's energy budget, vertical energy fluxes, and the existence of an energy imbalance. Chapter 7 explains how the greenhouse effect works, while chapter 8 analyzes the popular Enhanced CO_2 Effect hypothesis to explain climate change.

Section 3 introduces meridional heat transport, the horizontal transport of heat from the equator to the poles, responsible for the regional climates we experience. Chapter 9 introduces the temperature gradient with changes in latitude, which is the driving force of poleward heat transport. Chapter 10 explains how heat is transported through the atmosphere and ocean. Chapter 11 emphasizes the large changes in transport that occur with the seasons. Chapter 12 highlights our lack of a theory that properly accounts for heat transport.

Section 4 describes the heat transport through the different parts of the climate system, chapter 13 through the troposphere, and chapter 14 through the stratosphere. Chapter 15 deals with the important interactions between the troposphere and the stratosphere, which often strongly influence the weather, especially in winter. Chapter 16 looks in detail at the winter heat transport to the Arctic. Chapter 17 shows that although the ocean transports a large amount of heat, most of it is wind-driven.

Part II introduces the reader to natural climate change. Since this topic is rarely discussed, this part should be a discovery for many readers, and the part where the less scientifically inclined should start the book.

Section 5 discusses the natural climate changes we notice the most, caused by natural changes in ocean surface temperature. Chapter 18 explains the phenomenon of El Niño, while chapter 19 introduces the oceanic oscillations responsible for low-frequency natural climate variability.

Section 6 travels into the past, examining some periods of climate change that we cannot fully explain. Chapter 20 looks at periods millions of years ago when the poles enjoyed subtropical climates. Chapter 21 discusses the controversial relationship between distant past temperatures and their CO_2 levels. Chapter 22 reviews the frequent abrupt climate changes that occurred every few centuries during the Holocene. Chapter 23 analyzes the evidence that changes in solar activity were responsible for some of these changes.

Section 7 discusses the climatic effects of volcanic eruptions. Chapter 24 reviews the evidence for the role of volcanic eruptions in climate change. Chapter 25 reviews the evidence that some of the climate effects of volcanic eruptions are due to changes in heat transport. Chapter 26 examines the possibility that volcanoes were primarily responsible for the Little Ice Age.

Section 8 explores the limits of our knowledge of the effects of solar variations on climate. Chapter 27 reviews the effects of a solar grand minimum from various paleoclimate proxies. Chapter 28 explains the much milder effects of the solar cycle. Chapter 29 reviews the top-down mechanism that describes how the solar signal in the stratosphere is transmitted to the surface. Chapter 30 surprises us with the widely ignored effect of solar activity on planetary rotation.

Part III explains what we are ignoring about climate change, why we need a new theory, and introduces the Winter Gatekeeper hypothesis.

Section 9 explains that the climate we experience over decades is the result of climate regimes established after an abrupt climate shift. Chapter 31 illustrates how the climate changed in 1976, initiating the recent global warming trend. Chapter 32 explains how climate regimes and climate shifts were discovered and what they are. Chapter 33 presents evidence for the overlooked climate shift that occurred in 1997. Chapter 34 shows that Arctic warming is not due to global warming amplification.

Section 10 explores why we need a new theory when most scientists are satisfied with the current popular one. Chapter 35 questions the ability of the CO_2 hypothesis to explain climate changes other than the most recent one. Chapter 36 shows the failure to incorporate low-frequency internal variability into our climate theory. Chapter 37 shows how the variability of meridional transport is neglected as a climate factor despite the large amount of evidence available. Chapter 38 highlights the failure to incorporate the indirect effects of solar activity on climate, for which there is abundant evidence.

Section 11 presents the Winter Gatekeeper hypothesis, which places changes in heat transport at the center of natural climate variability. Chapter 39 illustrates the importance of the polar vortex for winter atmospheric circulation and heat transport. Chapter 40 attempts to identify the various modulators of heat transport and how they relate. Chapter 41 focuses on the Sun as the most important modulator of heat transport on centennial timescales.

Section 12 presents some evidence that strongly supports the Winter Gatekeeper hypothesis. Chapter 42 shows how observations confirm some particular tenets of the proposed role of the Sun in climate. Chapter 43 shows that the mechanism proposed by this new hypothesis is capable of altering the energy budget of the planet and causing climate change.

Part IV examines how the CO_2-based and the heat-transport-based hypotheses compare in explaining how the climate has changed in the past, is changing now, and predicts how it will change in the future.

Section 13 confronts both hypotheses with past climate change. Chapter 44 shows the superiority of the Winter Gatekeeper hypothesis in explaining the mysteries of climate change in the distant past. Chapter 45 brings the confrontation to the Holocene, where the new hypothesis again proves superior in explaining general climatic trends and specific events such as the one that occurred 2,800 years ago. Chapter 46 shows how the Enhanced CO_2 Effect hypothesis fails to explain the recent 200-year warming and the changes in warming trends observed over the past 100 years.

Section 14 examines known climate change mechanisms that are not adequately explained by the Enhanced CO_2 Effect hypothesis but are easily explained by the Winter Gatekeeper hypothesis. Chapter 47 explains a known missing cause of climate change that is capable of greatly displacing the climate equator. It also discusses the difficulty of properly assessing and accounting for the bewildering variety of low-frequency internal variability phenomena. Chapter 48 highlights the failure to properly account for climate regimes and shifts whose effects are misattributed to anthropogenic forcing.

Section 15 looks at the climate models that form the main basis of the Enhanced CO_2 Effect hypothesis. Chapter 49 reviews some of their known problems and shows how climate models do not reproduce the real climate. Chapter 50 gives examples of why we as a society are better off ignoring their predictions.

Finally, Section 16 has only one chapter, 51, which shows how the Winter Gatekeeper and Enhanced CO_2 Effect hypotheses produce very different predictions of the climate we can expect over the next 25 years, raising hopes that we should be able to falsify one of them.

What this book has to offer

The primary strength of this book is that it provides a guided tour of the astonishing evidence that scientists have amassed, illustrating the many ways in which the climate is changing due to various factors, some of which are still poorly understood. This eye-opening revelation challenges the oversimplified notion that atmospheric CO_2 is the primary regulator of Earth's temperature.[13]

A critical issue emerges from this evidence. Changes in poleward heat transport strongly influence global climate change, yet this aspect has been overlooked despite substantial evidence. While the Winter Gatekeeper hypothesis derived from this finding is relevant, its ultimate correctness is of secondary importance. What matters is the realization that we still do not understand climate change well enough to implement costly solutions that may not have the desired effects.

Whatever your views on climate change, reading this book will undoubtedly change your perspective. It unravels a stunning and extraordinarily complex set of processes that can appear stable for long periods and then change abruptly. While numerous authors have attempted to explain climate change, my goal is to bring it to life. I hope this book can convey the same sense of wonder that I have experienced in discovering the ever-changing nature of our climate.

---

[13] Lacis, A.A., et al., 2010. Science, 330 (6002), pp.356–359. doi.org/10.1126/science.1190653

PART I. CLIMATE AND ENERGY

Section 1. Climate System Incoming Energy

CHAPTER 2
SOLAR ENERGY

Solar energy powers the entire climate system. The amount of solar energy varies slightly with the solar cycle, but these changes are most relevant in the ultraviolet part of the spectrum. The differences in the amount of solar energy reaching the surface at different latitudes due to seasonal variations are responsible for the diversity of climates on Earth. These changes in solar energy are also responsible for the onset and end of glaciations.

The Sun powers the climate

The vast majority of the energy that powers the climate system[14] and supports life on Earth comes from the Sun. The amount of incoming solar radiation is staggering, estimated at 173,000 TW (terawatts, or one trillion watts). By comparison, geothermal heat flow from radiogenic decay and primordial heat is estimated at 47 TW, human heat production at 18 TW, and tidal energy from the Moon and Sun at 4 TW. Other energy sources are insignificant in comparison, such as solar wind, solar particles, starlight, lunar light, interplanetary dust, meteorites, or cosmic rays. This means that solar irradiance is responsible for more than 99.9% of the energy input to the climate system.

Nature of solar radiation

At an average distance of 150 million km (93 million miles) from the Sun, the Earth receives a radiant energy flux of 1361 W/m^2 (defined as total solar irradiance) at the top of the atmosphere, typically considered at 100 km (62 miles).[15] Nearly half of this energy arrives in the visible (400-700 nm), more than 40% in the infrared (above 700 nm), and less than 10% in the ultraviolet (UV, below 400 nm; see box 1).

The solar cycle

The Sun, like most stars, is a variable star. Its luminosity varies according to different periodicities, the best known of which are its 27-day rotation period and a more irregular 11-year period. This near-decadal periodicity is simply known as the solar cycle (fig. 3).

The cause of this cycle is a periodic shift of energy between the two solar magnetic fields generated by the solar dynamo. It manifests itself in the appearance of dark spots on the Sun's surface, known since ancient times and adequately described since the invention of the telescope. Although sunspots reduce the Sun's luminosity, they are accompanied by bright regions (faculae) that more than compensate for the loss. Therefore, more sunspots are associated with more solar radiation production.

Fortunately, the change in total solar irradiance during the solar cycle is minimal, only about 1.37 W/m^2, or 0.1%. However, this difference is unevenly

---

[14] This color indicates the first use of a glossary term.
[15] According to NASA.

    earthobservatory.nasa.gov/images/7373/the-top-of-the-atmosphere

distributed across the solar spectrum, with the UV portion changing the most and the visible and infrared portions changing little (box 1).

Radio emissions are another part of the solar spectrum that shows significant variation over the solar cycle. Daily records of solar emissions at 10.7 cm wavelength (2800 MHz) have been made to track solar activity over the past 80 years. They have the advantage that, unlike sunspot numbers, they never go to zero.

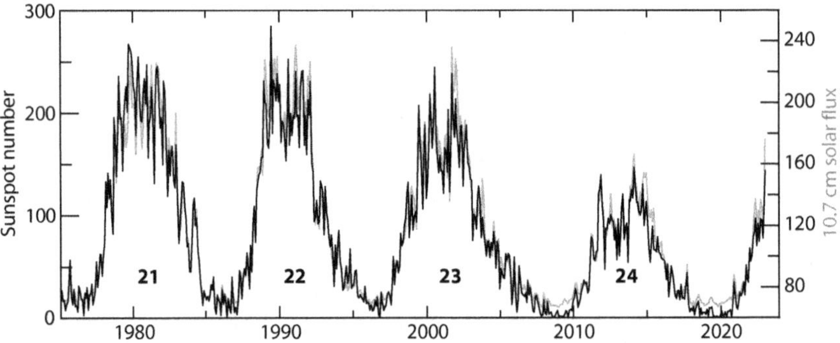

Figure 3. The 11-year solar cycle since 1975. Solar activity is measured by the monthly sunspot number (black curve, left scale) and the monthly solar flux at 10.7 cm radio frequency (gray curve, right scale).[16] Each solar cycle since 1750 has a number, and we are currently in cycle 25.

Distance to the Sun and its impact on irradiance

Total solar irradiance is calculated at the mean distance from the Sun, but the Earth's elliptical orbit causes its distance from the Sun to vary throughout the year. At perihelion (around January 4), the Earth is 5 million km (3 million miles) closer to the Sun than at aphelion (around July 4), resulting in a 6.9% difference in irradiance. This annual variation is greater than the change during the solar cycle. However, the Earth is not passive in this process and adjusts the energy it reflects and transports between hemispheres, partially compensating for this large difference. Other factors, such as the uneven distribution of continents and oceans between hemispheres, also affect how the Earth responds to solar radiation. Interestingly, the Earth is warmer when it is farther from the Sun and cooler when it is closer (ch. 5).

Box 1. Variability of the solar radiation spectrum

Although the overall variability of solar irradiance during the solar cycle is minimal, only 0.1%, this average masks important changes occurring in certain parts of the spectrum with a substantial climatic effect. The UV part of the spectrum between 200-240 nm, responsible not only for the formation of ozone but also for the existence of the stratosphere, shows a variability of 3% with the solar cycle, 30 times more than the total variability! Therefore, the main effects of solar

---

[16] Data from the Royal Observatory of Belgium, Brussels.

variability on climate should be sought in the stratosphere, not at the surface (ch. 14).

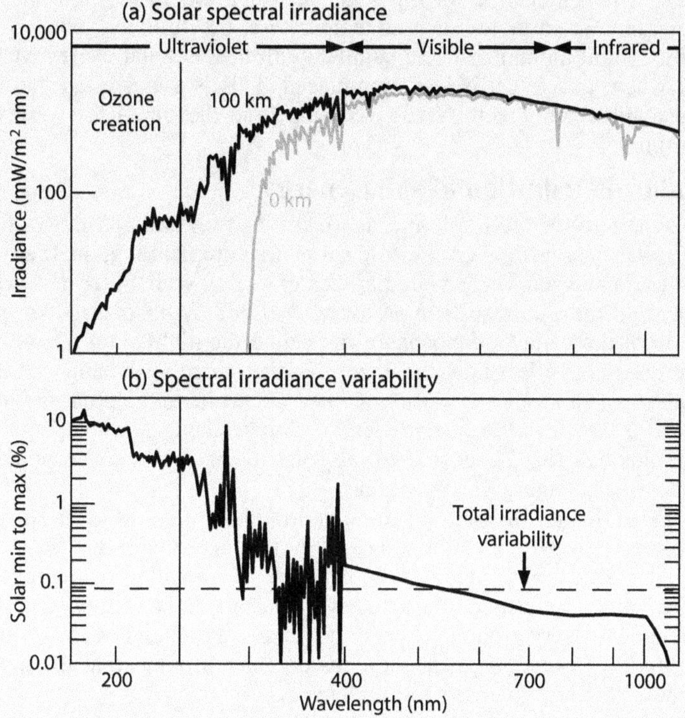

Figure B1. Solar irradiance and its variability. (a) Solar irradiance is plotted against wavelength above the Earth's atmosphere (black curve) and at the surface (gray curve) for the 200-1000 nm portion of the spectrum, which contains most of the solar energy received. (b) The fractional difference between the maximum and minimum of the solar cycle, with the dashed horizontal line indicating the total mean variability due to the solar cycle.[17] Note that there is more variation in the UV part of the spectrum than in other parts.

The effect of the seasons

The Earth's axis of rotation is tilted at an angle that varies between 24.5° and 22.1° over a period of 41,000 years, known as obliquity. The current tilt is 23.44° and will decrease over the next 12,000 years. The Earth's tilt greatly affects how the Sun's radiation is distributed over the planet's surface throughout the year, and its slow change is one of the main causes of glaciations. Over a few centuries, the obliquity can be considered almost constant, and the main effect of the tilt of the axis is that the Sun does not spend the whole year above the equator. Currently, the Sun changes its position in the sky relative to the equator (declination angle) from 23.44° (north of the equator) at the June solstice to –23.44° (south of the equator) at the December solstice. It is above the

[17] Figure after Lean, J. & Rind, D., 1998. J. Clim. 11 (12), pp.3069–3094. www.jstor.org/stable/26244250

equator (0° declination) at the equinoxes. The change in the distribution of solar radiation caused by the difference in the position of the Sun relative to the Earth's axis is responsible for the seasons, which are an essential part of the climate. Climatic variables such as temperature, precipitation, wind speed, or humidity exhibit an annual cycle coupled with the seasonal cycle, even at the equator where this effect is less pronounced. This is a clear sign that the climate responds mainly to the energy coming from the sun, a fact known since ancient times.

Irregular distribution of solar energy

The sphericity of the Earth and its axial tilt determine that most of the incoming solar energy falls on the tropical and subtropical regions (between 35 degrees north and south). The circular disk of energy with a flux of 1361 W/m^2 arriving from the Sun results in an average of 340 W/m^2 distributed over the entire top of the planet's atmosphere (including the night side). However, this average does not reflect how the energy is distributed. The annual average in the tropics and subtropics is close to 400 W/m^2, while in the polar regions, it is close to 190 W/m^2. In the current view of climate change, any variation in this average radiative flux is considered the solar radiative forcing responsible for any effect of a changing Sun on climate.

The mean annual distribution of solar irradiance says little about the profound changes in the arrival of solar energy that occur with the seasons. Near the winter solstice, high latitudes receive no solar radiation for months, while near the summer solstice, they receive constant solar radiation. Seasonal changes are also very pronounced at midlatitudes, although not as extreme as at high latitudes. In contrast, near the equator, solar irradiance at the top of the atmosphere changes little with the seasons.

Averaging solar irradiance over the entire year and the planet's entire surface greatly simplifies the calculations. Still, it obscures the profound effect of seasonal and latitudinal irradiance changes on climate.

Insolation and its latitudinal gradient

Insolation, the amount of solar energy received per unit area at the Earth's surface, is a critical determinant of surface temperature. Due to the geometry of the Earth, insolation decreases sharply with increasing latitude. Unlike solar irradiance, which measures solar energy at the top of the atmosphere, insolation is modified by various factors such as atmospheric conditions, cloud cover, and surface reflectivity.

These factors have a greater effect at higher latitudes, where there are more clouds, ice, and snow, and where the Sun's energy has a longer path and is scattered more. As a result, there is a large difference in the amount of solar energy received between the tropics and high latitudes, creating a latitudinal insolation gradient that extends from the equator to the poles. This gradient is the primary driver of temperature differences between these regions.

The temperature gradient, in turn, drives heat transport, and one of its most important forms is latent heat generated by the evaporation of water, which is then returned to the surface through condensation and precipitation. The insolation gradient, temperature gradient, and heat transport give rise to the planet's diverse climates and overall climate state, which are the central concepts explored in this book.

In summary

The amount of solar energy received varies on an annual and 11-year cycle. However, the changes due to the solar cycle are only important in the ultraviolet part of the spectrum, which is absorbed in the stratosphere.

CHAPTER 3
ALBEDO

The reflection of 29% of the incoming shortwave solar energy back to space is known as albedo. Nearly 90% of the albedo is due to the atmosphere, mainly clouds. Albedo is highest at high latitudes, contributing to their large energy deficit, which must be compensated for by heat transport. Albedo appears to be a very constrained property of the climate system, showing very little interannual variability and surprising interhemispheric symmetry. The lack of a theory to explain the value of albedo and its low variability, coupled with the poor ability of models to reproduce it, highlights our lack of understanding of one of the most fundamental properties of climate.

What is albedo?

When sunlight reaches the Earth, some is absorbed and some is reflected back into space. Albedo is the relative amount (ratio) of reflected to incoming sunlight and is expressed as a dimensionless, unitless quantity between 0 and 1. This concept is fundamental in climate because it determines how much energy the Earth absorbs. As we saw in chapter 2, the Earth receives an average of 340 W/m^2 from the Sun, absorbs 242 W/m^2 (71%), and reflects 99 W/m^2 (29%). The Earth's albedo is 0.29, which means that 29% of the incoming sunlight is reflected back into space. Scientists do not know why the Earth's albedo has this value, but they believe it must have changed very little over thousands of years; otherwise, the Earth's temperature would have been more affected. For example, an albedo of 0.32 would lead to glaciation, while an albedo of 0.27 would bring back Cretaceous conditions with palm trees at the poles.[18]

Atmospheric albedo

Clouds play a fundamental role in the Earth's albedo, with nearly 60% of the planet's surface covered by clouds. These clouds appear white from above because they reflect light of all visible wavelengths. Clouds are responsible for reflecting 45 W/m^2, or almost half of the Earth's albedo (13% reflection, fig. 4). However, it's worth noting that different types of clouds exist at different altitudes, and they also absorb infrared radiation from above and below, adding to the complexity of their role in climate.

The Earth's atmosphere is composed of gases and contains aerosols, which are liquid and solid particles in suspension (excluding clouds and precipitation). These aerosols strongly influence cloud formation by acting as condensation nuclei, a phenomenon known as indirect radiative forcing by aerosols. However, the indirect effect of aerosols is still poorly understood and is a major source of uncertainty in climate models.

Aerosols can also scatter and absorb radiation through direct radiative forcing. Scattering occurs when electromagnetic waves are deflected from their original path, which can convert direct solar radiation into diffuse radiation and

---

[18] Ramanathan, V., 2008. iLEAPS, (5), p.18.

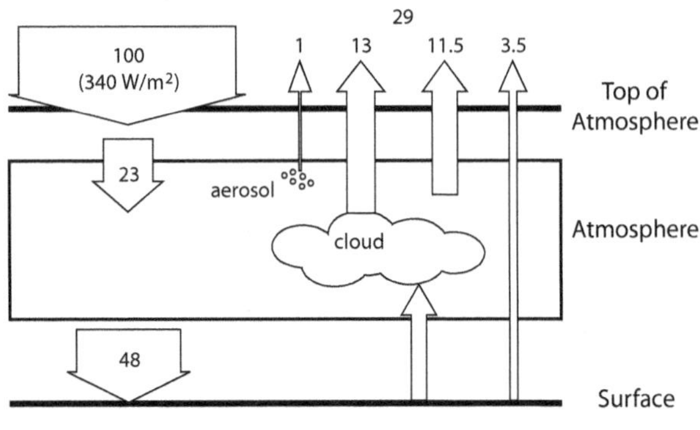

Figure 4. Schematic representation of the Earth's albedo. At the top of the Earth's atmosphere, the planet receives an average of 340 W/m² of solar radiation. Of this amount, 23% is absorbed by the atmosphere and 48% by the surface. The remaining 29% is reflected back to space by the albedo. Clouds are the largest contributor to the albedo, accounting for 13% of the reflected energy, followed by the atmospheric albedo under clear skies at 11.5%. Interestingly, the atmosphere attenuates the surface albedo, reducing it to 3.5% of the received shortwave radiation. Although aerosols contribute little to atmospheric albedo, changes in their concentration can still have a noticeable effect on Earth's albedo.

contribute to albedo. The contribution of aerosols to albedo is relatively small, accounting for only about 1% of the incoming energy reflected in this way (fig. 4). Nevertheless, albedo is so important that even a tropical volcanic eruption can cause global cooling by increasing the amount of aerosols in the stratosphere.

In addition to aerosols, any atom in the atmosphere can cause scattering. This is why the sky appears blue because scattering from oxygen and nitrogen is more likely for the more energetic blue part of the visible spectrum. As a result, more blue light is scattered from each part of the atmosphere, creating the blue sky phenomenon. This scattering also contributes to the albedo, accounting for over one-third of it (fig. 4).

Surface albedo

All surfaces have some degree of albedo, meaning they reflect some light. Oceans and vegetation have low albedo, while deserts and snow or ice have high albedo. However, surface albedo contributes little to the global mean planetary albedo because atmospheric processes attenuate the surface albedo contribution by a factor of about 3. Only 3.5% of incoming sunlight is reflected back to space by the surface (fig. 4).

The rapid decline of Arctic sea ice in the early years of this century has raised concerns about a runaway ice-albedo feedback. Loss of sea ice would reduce albedo, and additional solar energy would cause further sea ice loss. Models that reproduced the rapid loss predicted a tipping point that would lead to an ice-free Arctic by 2040, sparking public fears.[19] However, recent work suggests that up to 60% of the decline in September sea ice extent since 1979 may be due to changes in atmospheric circulation.[20] In addition, the persistence

---

[19] Holland, M.M., et al., 2006. Geophys. Res. Lett. 33 (23).
   doi.org/10.1029/2006GL028024
[20] Ding, Q., et al., 2017. Nat. Clim. Chang. 7 (4), pp.289–295.
   doi.org/10.1038/nclimate3241

of Arctic summer cloud cover significantly reduces the ice-albedo feedback.[21] The realization that internal variability is a more important factor than expected explains why the rate of decline of Arctic summer sea ice has slowed so much since 2007, contrary to all expectations.

Albedo distribution

When examined by latitude, albedo has a highly irregular distribution (fig. 5). It is highest at high latitudes and lowest in the tropics. Atmospheric albedo is highest in the Southern Hemisphere at about 60°S and in the Arctic, with a small peak in the tropics due to high cloud cover. Surface albedo, on the other hand, makes a substantial contribution only in the polar regions due to the albedo of ice and snow.

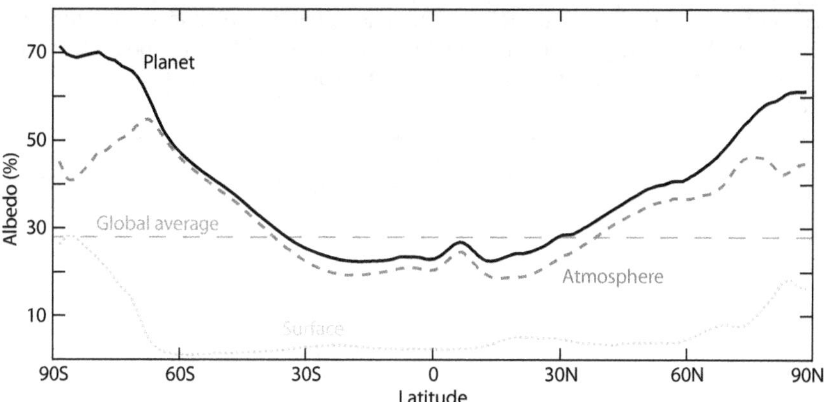

Figure 5. Albedo distribution. Latitudinal distribution of albedo divided into atmospheric and surface components. Note the low albedo value between 35°N-35°S, where most solar radiation reaches the Earth.[22]

As we learned in chapter 2, the tropics and subtropics receive most of the solar energy that reaches the top of the atmosphere. However, we now know that the mid and high latitudes, which receive less energy, reflect more of the solar energy back to space. Albedo amplifies the decrease in solar energy absorption with latitude, increasing the need to transport heat from the tropics to the poles. The latitudinal distribution of planetary albedo is intertwined with the temperature gradient between the equator and the poles and the overall heat transport in the climate system.

Albedo variability

Annual variations in albedo show a substantial semi-annual pattern with peaks in May and October. This cycle is mainly affected by changes in insolation leading up to the solstice in each hemisphere. It is due to surface albedo from snow-covered land surfaces and atmospheric albedo from changes in cloud cover. The planet's albedo rises and falls by 1% during this semi-annual

---

[21] Sledd, A. & L'Ecuyer, T.S., 2021. Front. Earth Sci. p.1067.
doi.org/10.3389/feart.2021.769844

[22] Figure after Donohoe, A. & Battisti, D.S., 2011. J. Clim. 24 (16), pp.4402–4418.
doi.org/10.1175/2011JCLI3946.1

cycle. However, models cannot realistically reproduce the observed annual cycle of albedo.[23]

What is truly impressive is the small interannual variability in global albedo. The interannual variability of the reflected flux is 0.2 W/m^2, which is only 1.4% of the annual cycle of this flux and 0.2% of the total global albedo. Cloud variability largely determines this small interannual variability, indicating that cloud changes strongly regulate albedo variability. However, models do not accurately reproduce this small interannual variability in albedo.

Because albedo is a critical aspect of the planet's climate and energy balance (ch. 6), the lack of a comprehensive theory explaining its constant value, its interhemispheric symmetry (box 2), and how clouds regulate it, combined with the inadequate representation of these properties or its annual cycle in models, represents a significant obstacle to our understanding of climate change. Albedo and heat transport are interrelated, and the inability of models to reproduce albedo correctly suggests that heat transport is also misrepresented (ch. 10).

Box 2. The Intertropical Convergence Zone and the interhemispheric albedo symmetry

A band of clouds and storms encircles the Earth near the equator, creating a small peak in albedo around 6°N (fig. 5). This region, known as the Intertropical Convergence Zone (ITCZ), is where the warm, moist trade winds from both hemispheres converge and rise by convection as part of the ascending branch of the Hadley circulation. The ITCZ is the planet's climatic equator, and its position shifts with the seasons, moving north during the boreal summer and crossing into the Southern Hemisphere during the austral summer. However, for unknown reasons, models tend to generate a false double ITCZ in the tropical Pacific.[24]

The distribution of the Earth's land surface and ice cover between the hemispheres is very asymmetric. During the austral summer (in the Southern Hemisphere), the Earth is closer to the Sun and receives 6.9% more sunlight than during the boreal summer (in the Northern Hemisphere). Despite these differences, both hemispheres reflect the same amount of sunlight within ~ 0.2 W/m^2 (0.2%). This is achieved by a change in cloud cover in the Southern Hemisphere that exactly compensates for the larger reflection caused by the larger land masses in the Northern Hemisphere. However, the models also fail to reproduce this interhemispheric albedo symmetry.

Interhemispheric albedo symmetry helps to reduce the differences in the amount of solar energy absorbed by the two hemispheres. Despite receiving more energy from the Sun, the Southern Hemisphere is about 2°C colder than the Northern Hemisphere. Several factors contribute to the Southern Hemisphere being colder despite receiving more energy: hemispheric differences in land/

---

[23] Stephens, G.L., et al., 2015. Rev. Geophys. 53 (1), pp.141–163.
   doi.org/10.1002/2014RG000449

[24] Si, W., et al., 2021. Geophys. Res. Lett. 48 (23), p.e2021GL094779.
   doi.org/10.1029/2021GL094779

ocean distribution, the coldness of Antarctica, seasonal changes in albedo, inter-hemispheric albedo symmetry, and northward heat transport across the equator (fig. 25, ch. 17). The oceans, especially the Atlantic, dominate this heat transport, carrying 0.45 PW (petawatts, one quadrillion watts) of heat northward across the equator. Meanwhile, due to the average position of the ITCZ in the Northern Hemisphere, the atmosphere transports about 0.27 PW of heat southward across the equator. The net heat transport to the Northern Hemisphere is about 0.18 PW, about 3% of the 6 PW of energy transported poleward at 35°N.[25]

In summary

Albedo refers to the reflection of about 29% of the Sun's energy by the Earth, primarily by clouds, ice, and snow. The atmosphere reflects seven times as much energy as the surface. Albedo is higher at high latitudes, which increases their energy deficit. The albedo changes very little from year to year (only 0.2%), and it appears to be controlled by changes in cloud cover. We don't understand why albedo has its specific value, its minimal changes, and why both hemispheres of the Earth have the same value despite their very different surface characteristics.

---

[25] Stephens, G.L., et al., 2016. Curr. Clim. Change Rep. 2, pp.135-147. doi.org/10.1007/s40641-016-0043-9

Chapter 4
Solar Energy Distribution

Half of the solar energy absorbed by the climate system goes to the oceans, a third to the atmosphere, and 17% to the land surface. The stratosphere receives 1.2% of the energy, almost all in the ultraviolet part of the spectrum.

Energy absorption by the stratosphere

The stratospheric ozone layer absorbs solar energy in the 200-315 nm part of the spectrum. This energy has a major effect on stratospheric temperature and circulation. Although this wavelength range accounts for slightly more than 1% of the total energy (fig. 6), it varies thirty times more with solar activity than the >320 nm range (fig. B1, ch. 2). This wavelength band is responsible for the radiative and dynamic changes taking place in the stratosphere during the solar cycle. The average UV energy absorption in the stratosphere is 3.85 W/m^2,[26] which is not insignificant. It represents 5% of the solar energy absorbed by the atmosphere. Compared to the troposphere, the stratosphere is about five times larger but contains about five times less mass. Because of its much lower density, the effect of absorbed solar energy on stratospheric temperature is very large.

Energy absorption by the troposphere

Under all-sky conditions, the shortwave atmospheric absorption is estimated to be 80 W/m^2, of which the troposphere absorbs 76 W/m^2.[27] Most of this absorption is due to water. However, due to increased albedo, clouds slightly reduce the atmospheric shortwave absorption compared to clear sky conditions.

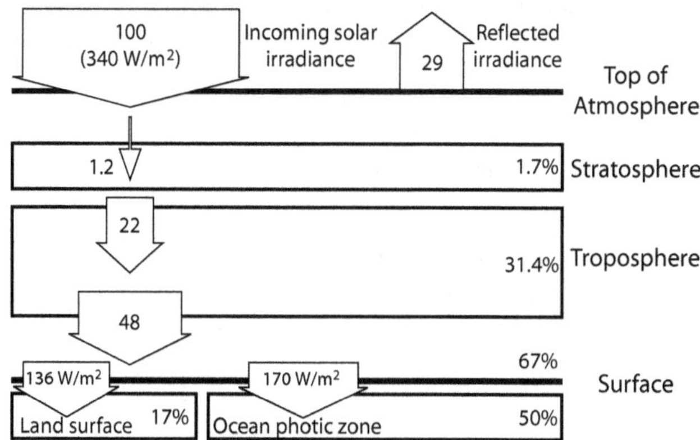

Figure 6. Schematic representation of the distribution of solar energy throughout the climate system. The top of the atmosphere receives an average solar irradiance of 340 W/m², and distributes 29% as reflected energy, 23% as energy absorbed by the

[26] Eddy, J.A., 2003. NASA LWS Sun-Climate Task Group Report of 11/3/2003
[27] Wild, M., et al., 2019. Clim. Dynam. 52, pp.4787–4812.
doi.org/10.1007/s00382-018-4413-y

atmosphere, and 48% as energy deposited at the surface. Of the solar energy absorbed by the climate system, 50% is absorbed by the ocean, 33% by the atmosphere, and 17% by the land surface.

Energy absorption by the surface

Earth's total surface area is about 510 million km^2 (197 million sq miles), of which the land surface is about 149 million km^2 (29%, 58 million sq miles). The land surface has a higher albedo than the ocean, reflecting more sunlight. As a result, the flux of shortwave solar radiation absorbed by the Earth's land surface is only 136 W/m^2, compared to the average surface absorption of 162 W/m^2.

On the other hand, the surface area of the world's oceans is nearly 361 million km^2 (140 million sq miles), or about 71% of the Earth's surface. Due to its lower albedo, the ocean receives a higher flux of shortwave radiation from the Sun at 170 W/m^2.

Distribution of solar energy

The amount of solar energy received by different parts of the climate system has important implications for the planet's climate and climate change. The cryosphere, which covers 7.4% of the Earth, receives very little solar energy due to its high reflectivity (up to 80%). However, changes in the amount of solar energy the cryosphere receives at certain latitudes can strongly influence climate change.

The ocean absorbs half of the solar energy not reflected back to space (fig. 6). This energy represents 75% of the solar energy absorbed at the surface. It enters the photic zone, which extends to about 200 m deep in the open ocean. The photic zone is the warmest layer of the ocean, and 90% of marine life lives there, supported by this energy.

About one-third of the solar energy not reflected back to space is absorbed by the atmosphere, mainly in the troposphere (31.4%). However, the stratosphere also plays a role by absorbing UV energy, which accounts for 1.7% of the non-reflected energy. As we'll see in future chapters, this has a major impact on atmospheric circulation and energy transport.

Finally, the land surface of the Earth receives the remaining 17% of the solar energy absorbed by the climate system.

In summary

Most of the sun's energy reaches the ocean, where it must be transferred to the atmosphere. In addition, the stratosphere receives a small but important amount of solar energy in the ultraviolet part of the spectrum.

SECTION 1 KEY ISSUES

The climate system receives 99.9% of its energy from solar radiation, which is remarkably consistent, varying by only 0.1% over the 11-year solar cycle. Despite being a small fraction of the total radiation, changes in UV radiation over the solar cycle have a significant impact on the stratosphere. Regional and local climates depend largely on differences in surface insolation across latitudes and seasons, while the latitudinal insolation gradient plays a critical role in shaping global climate.

The Earth reflects 29% of the solar radiation it receives, primarily through its atmosphere. This reflection, known as albedo, is most pronounced at high latitudes, contributing to their large energy deficit. Models do a poor job of reproducing albedo, which remains a complex and insufficiently understood phenomenon. It varies considerably throughout the year, but remains remarkably stable from year to year, displaying an unexpected symmetry between the hemispheres.

Half of the solar energy absorbed by the climate system goes to the oceans, one-third to the atmosphere, and one-sixth to the land surface. All of this energy must find its way back to the top of the atmosphere for the planet to maintain its temperature.

Section 2. Climate System Outgoing Energy

CHAPTER 5
OUTGOING ENERGY

The Earth's temperature is maintained by radiating the energy it receives from the Sun back to space as heat in the infrared part of the spectrum. This process occurs mainly from the atmosphere due to the presence of greenhouse gases (GHGs), which make it difficult for infrared radiation to escape. The tropics and subtropics receive more energy than they emit, while the middle and high latitudes experience an energy deficit, which becomes particularly severe during the cold season at high latitudes. The transport of heat from energy surplus regions to energy deficit regions is a fundamental feature of climate. However, the Earth's temperature and radiative fluxes throughout the year show that the notion of an energy balance at the top of the atmosphere is an oversimplification.

Thermal radiation

Thermal radiation is a type of electromagnetic radiation that results from the thermal motion of particles in matter. Any matter with a temperature above absolute zero emits thermal radiation, which consists of a wide range of frequencies. Which frequencies dominate depends on the temperature of the body. For example, because the Sun is extremely hot, its thermal radiation is predominantly in the visible part of the spectrum. In contrast, the Earth's lower temperature causes it to emit primarily in the infrared range.

If a body that doesn't produce heat receives a different amount of thermal radiation than it emits, it will adjust its temperature until it emits the same amount. In other words, matter naturally tends to balance the energy it receives with the energy it emits by changing its temperature. The critical factor in this radiative temperature change is the difference between the thermal radiation received and emitted, known as the net flux. A body that receives 1000 W/m^2 of energy and emits 900 W/m^2 will heat up at the same rate as an identical body that receives 250 W/m^2 and emits 150 W/m^2 because, in both cases, the net flux is the same at +100 W/m^2. Scientists study this property of matter to understand why the temperature of the planet changes over time.

Over the past 10,000 years, the Earth's average surface temperature has remained relatively stable, varying only within a range of about ±0.7°C (±1.25 °F), although some regions have experienced larger changes.[28] This suggests that the Earth is close to its equilibrium temperature, but it's important to note that it is never truly in equilibrium or energy balance. Factors such as the Earth's distance and orientation from the Sun, atmospheric conditions, the cryosphere, and heat transport are constantly changing, causing the Earth's temperature (box 3) and its equilibrium temperature to fluctuate. It's like saying that a person walking is in equilibrium when, in fact, they are not; instead, a series of partially compensating imbalances allows them to walk.

---

[28] Baggenstos, D., et al., 2019. Proc. Natl. Acad. Sci. U.S.A. 116 (30), pp.14881–14886. doi.org/10.1073/pnas.1905447116

From a thermodynamic perspective, the energy received from the Sun has a higher degree (shorter wavelength) than the energy the Earth emits back into space (longer wavelength). This energy degradation allows the Earth to extract work to power its climate system and support life on Earth. The entropy of the climate system, which includes the biosphere, can decrease as the entropy of the universe increases.

Distribution of outgoing longwave radiation

The Earth emits infrared radiation into space in all directions, and the amount of radiation emitted is proportional to the Earth's temperature on the Kelvin scale. Even the surface of Antarctica, which is the coldest place on Earth, emits a large amount of infrared radiation because it is still much warmer than absolute zero, which is –273.15°C (–459.67 °F) on the Kelvin scale.

The Earth's surface temperature varies with the seasons, and each hemisphere emits more radiation in the summer and less in the winter. As latitude increases, the variation in emission with the seasons becomes more pronounced. As we learned in chapter 2, the primary factor that determines surface temperature is the amount of insolation received. Therefore, the regions that receive more energy from the Sun also emit more energy into space.

The variability in solar energy received at the surface is much greater than the variability in surface thermal emission. At any given time, half of the planet is in darkness, receiving no energy from the Sun while emitting almost as much longwave energy as during the day. Because the temperature of the planet is more homogeneous than the distribution of energy from the Sun, the tropical regions of the planet receive more energy than they emit, resulting in a net energy surplus, while the mid and high latitudes receive less energy on average than they emit, resulting in a net energy deficit (fig. 7a).

Even during the long polar night when the Sun doesn't shine for months and temperatures reach –50°C (–58 °F), the pole still emits a substantial amount of longwave radiation into space while receiving none. Thus, the polar regions have the largest net energy deficit on the planet. Since the Earth only loses energy to space, we can say that the poles are the largest energy sinks on the planet during winter in terms of net energy flux. At the same time, the areas near the equator (especially the equatorial oceans) are the largest energy source for the planet (fig. 7b).

According to the principle of thermal energy equilibrium, matter tends to balance out its temperature by emitting as much energy as it receives. Therefore, the tropics should warm up to emit the same amount of energy they receive, while the mid and high latitudes should cool down to achieve the same balance. However, this balance does not occur because heat is constantly being transported within the climate system to compensate for differences in insolation. Heat transport is the most fundamental aspect of the climate system and a central theme of this book, as it forms the basis of the hypothesis presented. Without heat transport, as on the Moon, there would be no climate.

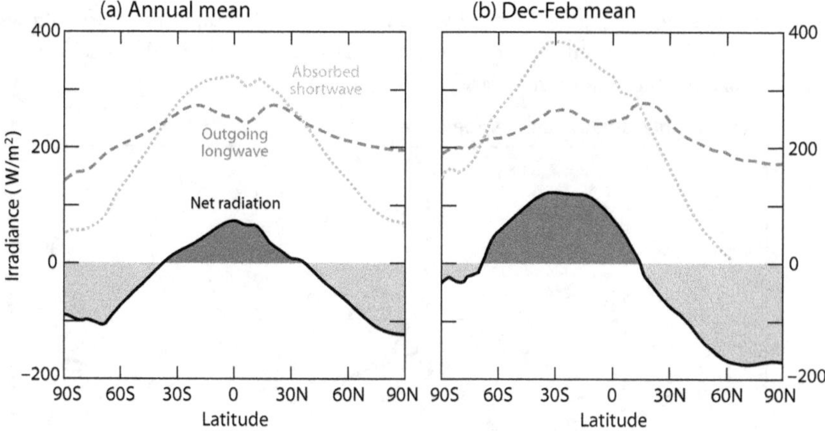

Figure 7. Latitudinal energy distribution. Plots of annual (a) and December-February (b) mean absorbed solar radiation (dotted light gray line), mean outgoing longwave radiation (dashed medium gray line), and their net difference (solid black line) around latitude circles.[29] The dark gray areas indicate a net energy gain, while the medium gray areas indicate a net loss. Due to the spherical geometry of the Earth, the areas of the graphs are not proportional.

Box 3. Seasonal changes in the Earth's temperature

The Earth's surface temperature varies greatly throughout the year, with an average temperature of about 14.5°C (50 °F). This variation occurs because most of the Northern Hemisphere is continental, while most of the Southern Hemisphere is oceanic. This causes the Northern Hemisphere to be colder in the winter and warmer in the summer. The planet's average temperature varies by 3.8°C (6.8 °F) over the course of a year, with temperatures ranging from 12.6°C in January to 16.4°C in July (54.7-61.5 °F, fig. B3).

Interestingly, the Earth is actually warmest just after the June solstice, when it is farthest from the Sun, and coldest just after the December solstice, when it receives 6.9% more energy from the Sun. While the amount of outgoing longwave radiation generally follows the temperature, the total amount of energy radiated by the planet (including the shortwave reflected by the albedo) actually increases when the Earth is cooler and decreases when it is warmer. This means that during the boreal winter, when the Earth is closest to the Sun and receives the most energy, the planet is actually at its coldest but emits the maximum amount of energy.

This observation is surprising because it contradicts the idea that a planet reaches an equilibrium temperature by balancing its incoming and outgoing radiative fluxes. If the planet has a positive radiative imbalance (as indicated by the white bars in fig. B3b), it should be warming, and if it has a negative radiative imbalance (as indicated by the gray bars in fig. B3b), it should be cooling. However, this is not what is observed. Instead, the Earth cools as the radiative imbal-

---

[29] Data from NASA's Clouds and the Earth's Radiant Energy System, CERES.

ance changes from negative to positive and vice versa. Although the Earth shows little interannual temperature variability, we don't fully understand the mechanisms that regulate its thermal homeostasis.

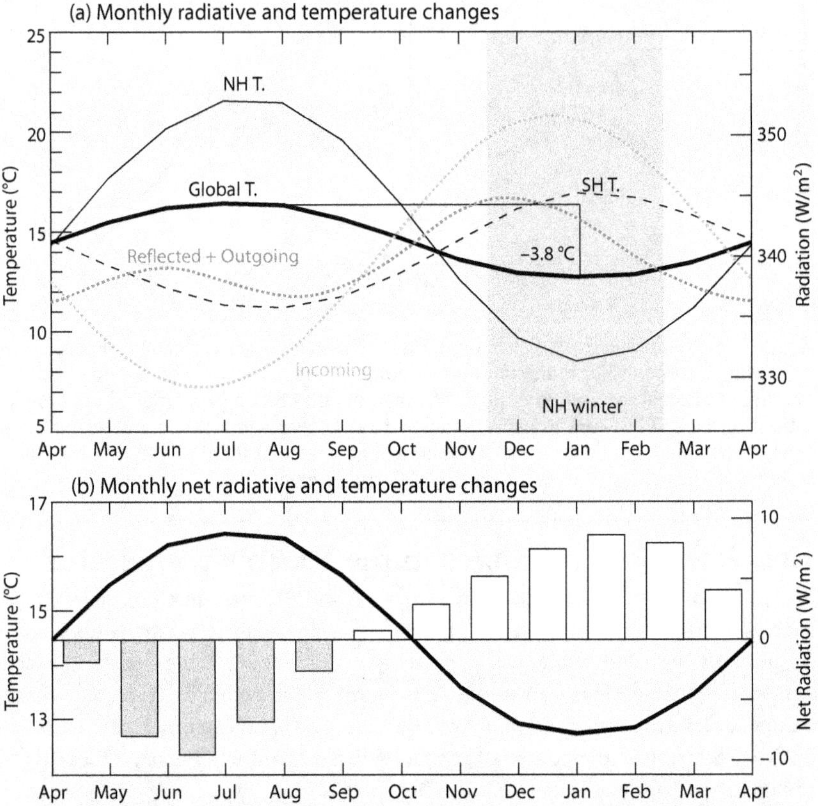

Figure B3. Annual variations in temperature and radiation. (a) The planet's global mean surface temperature (thick line) varies by 3.8°C (6.8 °F) over the year, mainly due to temperature variations in the Northern Hemisphere (NH T., thin line), which vary by 12°C (21.6 °F). Despite receiving 6.9% more total solar irradiance (dotted light gray line) in early January when it is at perihelion, the planet is cooler during January. As each hemisphere cools, the planet experiences two peaks of energy loss (reflected shortwave radiation plus outgoing longwave radiation, dotted medium gray line), with the highest during the cooling of the Northern Hemisphere. The planet emits more energy between November and January than at any other time. The Southern Hemisphere temperature (SH T.), represented by the dashed black line, experiences milder seasonal variations than the Northern Hemisphere. The gray area in the plot corresponds to winter in the Northern Hemisphere.[30] (b) Global monthly mean temperature (line) compared to monthly mean net radiative change at the top of the atmosphere (bars).

---

[30] Data from Jones, P.D., et al., 1999. Rev. Geophys. 37 (2), pp.173-199. doi.org/10.1029/1999RG900002 and from Carlson, B., et al., 2019. Geophys. Res. Lett. 46 (17-18), pp.10679-10686. doi.org/10.1029/2019GL083736

How energy leaves the climate system

A basic thermodynamic explanation of the Earth's climate system is that energy enters the system primarily at the surface, in the lower atmosphere, and in the upper ocean layers, and exits the system through the top of the atmosphere at a higher latitude than it entered. As the energy moves through the system, work is produced that manifests itself as atmospheric weather and the water cycle. Because of the variability in the time it takes for energy to leave the climate system, there is an accumulation of energy within the system, especially in the ocean (ch. 6).

Energy must be transported through the atmosphere to escape the planet as infrared radiation. This is because the Earth's atmosphere is highly opaque to infrared radiation due to greenhouse gases (GHGs). GHGs are gaseous molecules that absorb energy in the infrared part of the electromagnetic spectrum. Water vapor and carbon dioxide are the most abundant GHGs in the Earth's atmosphere, and together, they absorb most of the infrared frequencies that the Earth's surface should be emitting due to its temperature. However, an atmospheric window in the infrared frequency range of 8.5-13.5 µm allows about 17% of the longwave radiation emitted from the surface to pass through.

GHG molecules intercept the rest of the energy in the atmosphere. Molecules in the lower troposphere absorb infrared radiation and are more likely to share that energy by colliding with nitrogen or oxygen molecules than to emit it as radiation. This results in a more uniform temperature locally, and GHGs heat the lower troposphere. As altitude increases, density decreases rapidly, and in the upper troposphere and stratosphere, GHG molecules are more likely to emit the energy they receive rather than share it. This leads to an increase in outgoing radiation. When the GHG molecule finally collides, it is cooler than the other molecules and receives energy instead of giving it away. GHGs cool the upper troposphere and stratosphere.

The Earth's infrared emissions can originate at any height, from the surface to the top of the atmosphere. However, it is helpful to consider the mean emission height, also known as the effective emission height. This imaginary height has a value of about 6 km (3.7 miles) and reflects the opacity of the atmosphere to infrared emissions. Its value depends on the GHG content of the atmosphere and its vertical temperature profile (lapse rate) since the temperature of a molecule determines its ability to emit radiation. The temperature at the effective emission height is the average temperature of the Earth as seen from space. When measured from space, the Earth's emission temperature is 250 K (–23°C, –10°F), slightly lower than the 255 K calculated from theory for a blackbody.

It is important to understand how the presence of GHGs in the atmosphere affects the level and temperature of outgoing longwave radiation emissions, as this is the basis of the greenhouse effect discussed in chapter 7.

In summary

The main driver of climate change is likely to be changes in the amount of thermal energy radiated by the Earth, as solar energy and albedo remain relatively constant. The atmosphere plays an important role in this process by returning most of the energy received from the surface due to its opacity to infrared radiation. At mid and high latitudes, especially during the winter season, more energy is lost to space than is received from the Sun. This creates a con-

siderable energy deficit that must be compensated for. The Earth experiences more temperature change from month to month than in ten years. Therefore, just as a person walking is never in equilibrium, the Earth is never in radiative equilibrium.

CHAPTER 6
THE ENERGY BUDGET

For a planet to have a constant temperature, there must be a balance between the energy coming in from the Sun and the energy leaving the planet. This is called the radiative balance. However, the Earth's temperature changes every month, and the outgoing energy is never exactly equal to the incoming energy. Therefore, the energy balance is only a theoretical concept that simplifies the calculation of the Earth's energy budget. It allows us to track energy within the climate system, which is essential for understanding climate change. The warming of the Earth's ocean, atmosphere, and surface indicates an energy imbalance at the top of the atmosphere. Recent research suggests that this imbalance may be decreasing, which would have important implications for our understanding of climate change.

Earth's radiative balance

Matter naturally tends toward thermal equilibrium by adjusting its temperature, causing hot things to cool and cold things to warm until they reach the temperature of their surroundings. The same principle applies to a planet like Earth, which receives almost all of its energy from its star in the form of electromagnetic radiation. The planet will naturally adjust its temperature to ensure that the energy it emits is equal to the energy it receives, known as the energy balance. This concept is based on a theoretical balance in the energy flow coming from the Sun and leaving the Earth through the top of the atmosphere. According to climatologists, any physical process that alters this balance of fluxes and causes an imbalance is called a radiative or climatic forcing.

Framing the energy of the climate system in this way simplifies the problem, especially since the amount of energy coming from the Sun is nearly constant on an annual average (varying only 0.1% with the solar cycle; ch. 2). For the Earth to warm, the amount of energy leaving the planet (total outgoing radiation) must decrease, and for it to cool, the amount of energy leaving the planet must increase. Each change must result from a variation in one or more radiative forcings. Once a new equilibrium is reached, the planet's surface will have a different temperature.

Total outgoing radiation has two components: reflected shortwave radiation, also known as albedo, and outgoing longwave radiation. Interannual changes in albedo are minimal (ch. 3), suggesting that a reduction in outgoing longwave radiation is the initial cause of warming, as proposed by the greenhouse effect theory (ch. 7). This theory suggests that climate change occurs because of changes in the amount of energy that goes from the surface to the top of the atmosphere. In this case, the planet warms because less energy reaches the top of the atmosphere as outgoing longwave radiation.

According to this widely accepted paradigm, climate changes occur due to variations in the amount of radiatively active gases and aerosol particles in the atmosphere. However, this simplified view of climate complexity ignores that the climate system is not so simple. Looking at annual averages alone obscures the fact that the planet's temperature varies dramatically by 3.8°C (6.8 °F) over the course of a year (box 3; ch. 5). Furthermore, the Earth is warmer when it receives less energy from the Sun at aphelion, and colder when it receives more energy from the Sun, at perihelion, meaning that the supposed equilibrium does not and has never existed. In addition, changes in albedo and outgoing long-wave radiation throughout the year cause the Earth to return more energy when it is colder and less when it is warmer (fig. B3; ch. 5).

We still have much to learn about how the Earth manages to maintain such a consistent temperature over the years despite significant month-to-month variations. In addition, we do not fully understand how the Earth maintains inter-hemispheric albedo symmetry despite having such asymmetric hemispheres and experiencing a substantial reduction in snow-ice albedo in the Northern Hemisphere in recent decades. Nevertheless, the basis of the radiative model must be correct. Our measurements of surface, atmospheric, and ocean temperatures indicate that the climate system is increasing the energy it contains, which implies that the amount of energy emitted by the planet at the top of the atmosphere should be decreasing. But that is not what we are observing. Outgoing longwave radiation has actually increased over the last 40 years,[31] suggesting that if the planet has warmed, it is due to an increase in absorbed shortwave radiation, probably caused by a small decrease in albedo. This result is not what the greenhouse theory predicts for an increase in GHGs.

The energy budget

The Earth's energy budget refers to the energy flows into and out of the planet's climate system. It attempts to account for vertical energy fluxes that are not yet known with sufficient precision, so different authors provide different estimates. Our understanding of the Earth's energy budget is largely based on satellite observations of reflected sunlight and thermal infrared energy emitted by the atmosphere and surface.

To calculate the Earth's energy budget, we start with the amount of solar energy that reaches the Earth at an average distance from the Sun (one astronomical unit), evenly distributed over the entire surface of the planet. This incoming shortwave energy flux has a value of 340 W/m^2 and represents the amount of energy that the Earth must return to be in radiative equilibrium. This incoming energy is divided into several components, such as the shortwave radiation reflected by the albedo (98 W/m^2, 29%), the energy absorbed by the atmosphere (80 W/m^2, 23%), and the energy absorbed by the surface (162 W/m^2, 48%; fig. 8). These values are calculated, not measured, and may vary from one study to another. For our purposes, we rely on the values provided by NASA.[32]

All solar energy reaching the surface (48%) must be returned to achieve equilibrium. Direct radiation from the surface to space through the atmospheric window (12%) is a small fraction of the total thermal radiation emitted from the surface. Thermal radiation is exchanged between the surface and the atmosphere, but the net flux determines the temperature change. Because the atmosphere is very opaque to infrared radiation, most of it is reflected back to

---

[31] Dewitte, S. & Clerbaux, N., 2018. Remote Sens. 10 (10), p.1539.
     doi.org/10.3390/rs10101539
[32] earthobservatory.nasa.gov/features/EnergyBalance

the surface. However, because the surface is generally warmer than the atmosphere, the net flux of thermal radiation carries 5% of the solar energy from the surface to the atmosphere. Nevertheless, the surface must return the energy it receives from the Sun and is cooled mainly by evaporation (86 W/m², 25%) and by heating the air, which then rises (convection, 18 W/m², 5%).

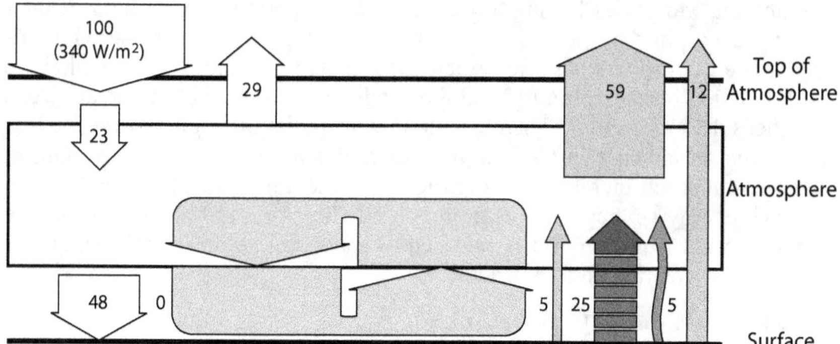

Figure 8. Schematic representation of the Earth's energy budget. The Earth receives an average solar irradiance of 340 W/m² at the top of the atmosphere. Shortwave fluxes are shown in white, net longwave fluxes in light gray, convection in medium gray, and evaporation in dark gray. Although there is a large exchange of infrared radiation between the surface and the atmosphere, the net upward transfer is relatively small. The primary mechanism by which the surface heats the atmosphere is through evaporation (latent heat).

The Earth's energy budget shows that the Sun is responsible for heating the surface, with half of this energy used to heat the atmosphere by evaporation, driving the water cycle. A quarter of the energy reaches space by direct radiation, and the remaining quarter is used to heat the atmosphere by convection and thermal radiation (fig. 8).

It is clear that the surface transfers net heat to the atmosphere, which determines the lapse rate – the decrease in temperature with height in the troposphere. However, the atmosphere does not transfer net heat to the surface. In the next chapter, we will examine how the atmosphere can alter the heat fluxes, which can lead to surface warming.

In the case of the ocean, this is even more obvious. The ocean surface is typically warmer than the atmosphere, so it loses 16 W/m² of heat to the atmosphere through conduction and convection (sensible heat). In addition, the ocean is the largest source of evaporation on the planet, losing 100 W/m² of latent heat, while the net longwave loss is 53 W/m².[33] Thus, the ocean transfers almost all of the energy it receives from the Sun to the atmosphere to maintain equilibrium. There is no net heat flux from the atmosphere to the ocean, and the atmosphere has no net warming effect on the ocean. Simply put, the Sun warms the ocean and the ocean warms the atmosphere.

To close the budget, the atmosphere must lose to space the energy it has received from the Sun, the land surface, and the ocean surface. The latent en-

---

[33] Schmitt, R.W., 2018. Oceanography, 31 (2), pp.32–40.
    doi.org/10.5670/oceanog.2018.225

ergy gained from evaporation is released through condensation when clouds and precipitation form, warming the atmosphere. As we saw in the previous chapter, the lower troposphere has a higher density, so molecules that absorb radiation are more likely to transfer that energy to other molecules, mainly nitrogen and oxygen, by collision than to emit it as radiation. This results in neighboring molecules having a more similar temperature, regardless of their infrared absorption properties. Air density and temperature decrease with altitude in the troposphere. At low air densities, GHG molecules are more likely to emit radiation before colliding, and their radiation is less likely to be absorbed by other GHG molecules. Therefore, longwave radiation begins to escape into space, and an increase in GHG molecules at this altitude results in cooling as outgoing radiation increases. At high levels in the atmosphere, molecules emit as much energy as they can get from below. Therefore, all the energy that the Earth receives from the Sun is returned to space unless the Earth changes its temperature, creating an imbalance.

The Earth's energy imbalance

When there is an annual average imbalance between the incoming and outgoing radiative fluxes at the top of the atmosphere, the Earth experiences an energy imbalance. Many scientists consider this imbalance to be the most important indicator of climate change since an excess of energy that is not returned must lead to warming, with a greater imbalance leading to greater warming. Conversely, for the Earth to cool, it must give back more energy than it receives.

The Earth's energy imbalance is estimated to be 0.75 W/m^2, and most of the additional heat generated by this imbalance, about 93%, ends up in the ocean. About 3% of the excess heat is used to melt ice, while another 4% contributes to rising land temperatures and melting permafrost. Only a fraction of this excess heat, less than 1%, remains in the atmosphere.[34] The problem, however, is that this estimated imbalance is a tiny residual of two large energy fluxes, and it is too small to measure accurately, representing only about 0.15%. In addition, the uncertainty in measuring energy fluxes at the top of the atmosphere is much larger than the imbalance itself.[35] Nevertheless, changes in ocean heat content have allowed an estimated energy imbalance of about 0.6 W/m^2.

Satellite measurements between 2000 and 2018 show a slight decrease in reflected energy and a slight increase in outgoing longwave radiation. While these two measurements should not affect the energy imbalance if they coincide, the increase in outgoing longwave radiation is greater than the decrease in shortwave reflection. As a result, there seems to be an apparent downward trend in the energy imbalance.[36] This means that the Earth should be warming at a slower rate over time, supported by a reduction in the rate of increase in ocean heat content. This possibility of a slower warming trend presents a significant challenge to our understanding of climate change, and we will explore it in future chapters.

---

[34] Trenberth, K.E. & Cheng, L., 2022. Environ. Res.: Climate, 1 (1), p.013001. doi.org/10.1088/2752-5295/ac6f74

[35] Loeb, N.G., et al., 2018. J. Clim. 31 (2), pp.895–918. doi.org/10.1175/JCLI-D-17-0208.1

[36] Dewitte, S., et al., 2019. Remote Sens. 11 (6), p.663. doi.org/10.3390/rs11060663

In summary

We know that the Earth is warming, indicating an energy imbalance at the top of the atmosphere. If the albedo remains constant, the only way the Earth can warm is by reducing its outgoing thermal radiation, pointing to an increase in greenhouse gases as the likely cause. However, observations show that the energy imbalance is mainly due to an increase in absorbed shortwave energy, suggesting instead a decrease in albedo as the likely cause. In the 21st century, the energy imbalance appears to be decreasing, suggesting that the Earth may be warming more slowly.

CHAPTER 7
THE GREENHOUSE EFFECT

The greenhouse effect warms the Earth due to the combination of greenhouse gases and a positive lapse rate in the troposphere. Water vapor is the main greenhouse gas, but its concentration depends on temperature. Carbon dioxide is a well-mixed trace gas that contributes substantially to the greenhouse effect. This effect is due to the increased opacity of the atmosphere to infrared radiation, which causes radiation into space to originate at higher altitudes. The troposphere cools with altitude, and cold molecules radiate less. The greenhouse effect increases the height of emissions and decreases them. As a result, the surface and lower troposphere must warm until the radiated energy equals the energy received from the Sun. However, the greenhouse effect is not uniform across the planet due to differences in water vapor content, so it is much weaker over the poles in winter than over the tropics.

Greenhouse gases

The Earth's temperature is regulated by the balance between the energy it receives from the Sun and the energy it radiates back into space as infrared radiation. However, a small fraction (about 1%) of the Earth's atmosphere is made up of gas molecules that absorb infrared radiation because they have two different atoms or more than two atoms. This makes the atmosphere quite opaque to infrared radiation, which causes it to warm as more energy is absorbed by these gases and shared through collisions with other molecules. These gases are known as greenhouse gases (GHGs) and are responsible for the greenhouse effect.

The most important GHG is water vapor. Its atmospheric concentration varies widely (box 4) but averages about 1%. Water vapor is more than ten times more abundant than all other GHGs combined and is responsible for about 75% of the Earth's greenhouse effect when the effect of clouds is included.[37] Water vapor is most abundant in the lower troposphere and decreases rapidly with altitude. In fact, it is 1000 times less abundant in the stratosphere (fig. 9, dotted light gray line). The decrease in water vapor abundance and atmospheric density with altitude causes a positive lapse rate (temperature decreasing with altitude) in the troposphere (fig. 9, thick black line). There are two other peculiarities of water vapor. First, its abundance depends on temperature. Second, it has the property of changing phase between solid, liquid, and gas, which requires or releases a lot of energy without changing its temperature. This energy, called latent heat, is one of the primary heat transport mechanisms in the climate system.

Carbon dioxide (CO_2) is a trace gas that makes up only about 0.04% of the atmosphere. It is well mixed in the lower and middle atmosphere and is the

---

[37] Schmidt, G.A., et al., 2010. J. Geophys. Res. Atmos. 115 (D20).
  doi.org/10.1029/2010JD014287

second most important GHG. CO_2 is responsible for about 19% of the Earth's greenhouse effect.

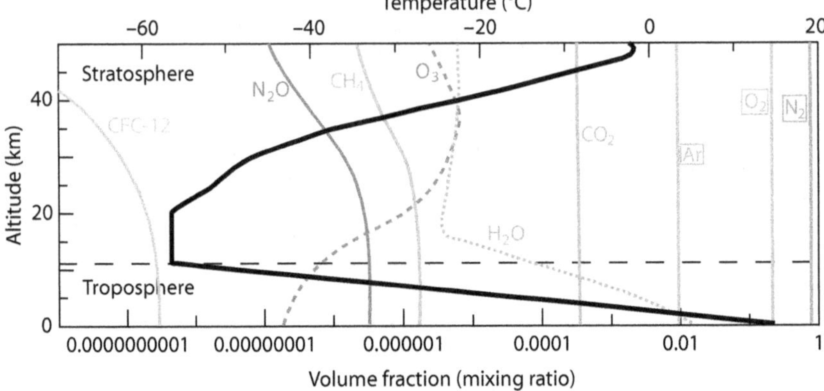

Figure 9. Atmospheric temperature and gas profiles. The thick line (upper scale) represents the vertical profile of atmospheric temperature, while the thin lines (lower scale) represent the vertical profile and abundance of atmospheric gases. Box symbols correspond to gases with insignificant infrared absorption. The dotted line represents water vapor, and the dashed line represents ozone. Both gases have highly variable vertical profiles that contribute greatly to the temperature profile.

Ozone (O_3) is the third most important GHG, and its abundance also varies greatly with altitude (fig. 9). In fact, it is 100 times more abundant in the stratosphere (the ozone layer) than in the troposphere. Although only six parts per million, it is responsible for 4% of the greenhouse effect. In addition to its role as a greenhouse gas, ozone also plays a critical role in absorbing UV radiation. It is responsible for the negative lapse rate in the stratosphere and the very existence of the stratosphere.

The remaining GHGs shown in figure 9 and table 1 are nitrous oxide (N_2O), methane (CH_4), and chlorofluorocarbons (CFCs), a group of human-made gases. Together they account for a small fraction of the greenhouse effect.

Table 1. Main greenhouse gases.[38]

Gas name	Formula	Abundance (%)	greenhouse effect attribution
Water vapor (inc. clouds)	H_2O	0–3%	75%
Carbon dioxide	CO_2	0.04%	19%
Ozone	O_3	0.00006%	4%
Nitrous oxide	N_2O	0.00005%	1%
Methane	CH_4	0.0002%	1%

---

[38] Ibid.

How the greenhouse effect works

The greenhouse effect is sometimes misunderstood as a "trapping" of heat. While it is true that the presence of GHGs results in more energy in the climate system, this additional energy is mainly stored in the ocean. In addition, the planet still returns all the energy it receives from the Sun after any necessary adjustments.

GHGs increase the opacity of the atmosphere to infrared radiation. They absorb thermal emissions from the surface, causing warming in the lower troposphere. However, they cause cooling in the upper troposphere by increasing the thermal emission to space. Because of their presence, the infrared emission to space from the surface (which occurs on the Moon) is shifted to the atmosphere. We can determine the theoretical effective emission height (Z_e in fig. 10) as the average height at which the Earth's thermal radiation is emitted. The temperature at which the Earth emits radiation is the average temperature of the atmosphere at that height. This temperature, calculated to be 255 K for a blackbody Earth, is 250 K (–23°C, –9 °F) when measured from space.[39] This corresponds to an emission altitude of about 6 km (3.5 miles).

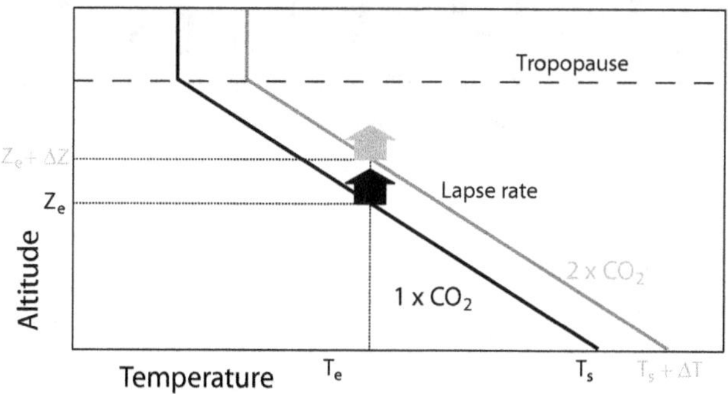

Figure 10. Schematic diagram of the greenhouse effect. The change in emission level (Z_e) due to a doubling of CO_2 (gray color) is associated with an increase in surface temperature (T_s), assuming a fixed atmospheric lapse rate. The effective emission temperature (T_e) and the outgoing longwave emissions remain unchanged.[40]

The lapse rate must be positive for the greenhouse effect to cause warming, meaning that the temperature decreases with altitude. GHGs cause the planet to emit from higher altitudes, making the atmosphere more opaque to infrared radiation, and as a result, that altitude is cooler due to the lapse rate. However, the Earth still has to return all the energy it receives from the Sun, but the cooler molecules emit less energy. Therefore, the planet goes through a period where it emits less energy than it should, causing the surface and lower troposphere to warm until the new emission altitude reaches the temperature needed to return all the energy; at this point, the planet stops warming. If GHG-induced warming is occurring, there should be a decrease in outgoing infrared radiation as the surface warms. However, this is not observed. As discussed in

[39] Peyrou-Lauga, R., 2017. 47th Int. Conf. Environ. Syst. ICES-2017-142
  hdl.handle.net/2346/72957
[40] Held, I.M. & Soden, B.J., 2000. Annu. Rev. Energy Environ. 25 (1), pp.441–475.
  doi.org/10.1146/annurev.energy.25.1.441

chapter 6 on the energy imbalance, what is observed is an increase in outgoing infrared radiation, which should cause cooling, offset by a greater increase in absorbed solar radiation, which is the cause of the observed warming.

The greenhouse effect warms the planet when an increase in GHGs in the atmosphere causes the emission height to increase. Since the emission temperature must remain constant, the temperature from the surface to the new emission height must increase, albeit by a small amount. For example, a doubling of CO_2 levels causes the emission height to increase by 150 meters. At a standard moist lapse rate of –6.5°C per km (–3.6°F per 1,000 ft), the new emission height is 1°C cooler than it should be. Therefore, if the lapse rate remains constant, the surface temperature must increase by 1°C (1.8 °F) to reach the required emission temperature at the new height.[41]

Box 4. Differences in the greenhouse effect with latitude

Water vapor is the primary GHG, but its distribution in the Earth's atmosphere is highly uneven. What matters for the greenhouse effect is the amount of water vapor per kilogram of air (specific humidity), not the humidity relative to saturation for a given temperature (relative humidity). For example, the air over deserts feels very dry because it is warm but contains much more water vapor than the air over Antarctica.

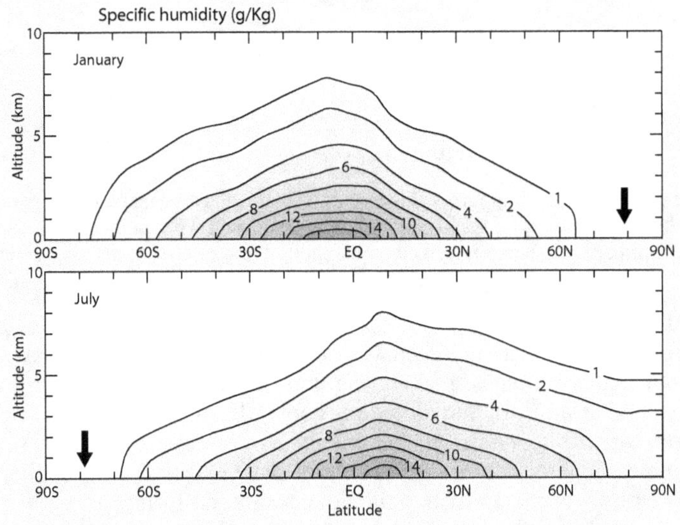

Figure B4. Specific humidity profiles as a function of latitude and altitude. Top in January and bottom in July.[42] The black arrows point to the driest areas of the planet, typically found at high latitudes during winter, where the greenhouse effect is weakest.

---

[41] Ibid.

[42] Figure after Randall, D. A., 2015. An introduction to the global circulation of the atmosphere. Princeton Univ. Press.

The driest atmosphere on the planet is found in the polar regions during winter (fig. B4). Specific humidity levels as low as 0.1 g/kg have been recorded, dropping to a thousand times less a few kilometers above the surface. Since 75% of the greenhouse effect is due to water vapor and clouds, and since clouds are also greatly reduced in polar winter conditions, the greenhouse effect is several times weaker over the high latitudes in winter than over the tropics. The large contrast in the intensity of the greenhouse effect between the polar regions in winter and the tropics is a crucial point for the climate change hypothesis explored in this book.

In summary

The greenhouse effect occurs due to the increased opacity of the atmosphere to infrared radiation when greenhouse gases are present. Water vapor and clouds are the primary drivers of the greenhouse effect, and since they are scarce in the polar atmosphere during winter, the greenhouse effect is greatly reduced in these regions during this time of year.

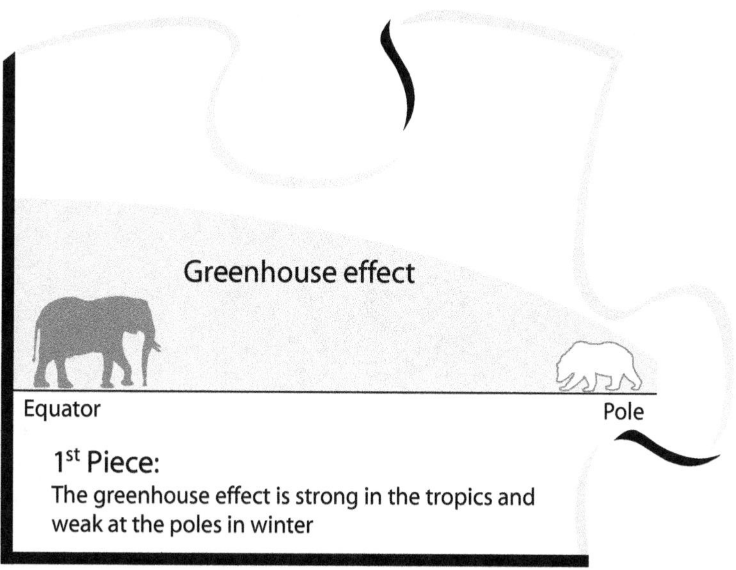

CHAPTER 8
THE ENHANCED CO$_2$ EFFECT HYPOTHESIS

The "Enhanced CO$_2$ Effect hypothesis" builds on the greenhouse effect by proposing that recent global warming is primarily due to feedback mechanisms that amplify the direct warming effect of CO$_2$. These feedbacks cannot be measured directly and are instead estimated using computer models. The warming that would occur with a doubling of CO$_2$ levels is known as climate sensitivity, which is currently estimated to be 3°C (5.4 °F) but has a large uncertainty. Despite the lack of scientific evidence, the Enhanced CO$_2$ Effect hypothesis is widely supported by climatologists and is considered the consensus hypothesis. Proponents of this hypothesis argue that natural climate change is negligible.

The Enhanced CO$_2$ Effect hypothesis

As we learned in the previous chapter, water vapor is the primary GHG in the atmosphere. However, it has an interesting property that sets it apart from other GHGs: its presence and effect are temperature dependent. When the temperature drops significantly, water vapor condenses out of the atmosphere, making temperature changes less dependent on changes in water vapor.

Svante Arrhenius proposed in the 19th century that increasing atmospheric CO$_2$ could cause substantial global warming by recruiting the effect of a temperature-induced change in water vapor. In 1939, Guy Callendar defended the hypothesis, suggesting that the warming recorded in the early 20th century was due to increased atmospheric CO$_2$ levels. Although much of the early 20th-century warming appears to have been natural, Arrhenius' hypothesis has been successful in explaining late 20th-century warming because it fits the observations better than other hypotheses.

It's important to note that the Enhanced CO$_2$ Effect hypothesis differs from the greenhouse effect theory. The latter theory states that an increase in GHGs will inevitably cause some warming. For example, doubling the amount of CO$_2$ in the atmosphere should cause a temperature increase of about 1°C (1.8 °F). The Enhanced CO$_2$ Effect hypothesis proposes that almost all of the observed warming between 1951 and the present is due to the significant rise in atmospheric CO$_2$ levels caused by human activities.[43] However, the observed warming is many times greater than the greenhouse effect theory would predict based on the increase in CO$_2$ concentrations alone. Therefore, the Enhanced CO$_2$ Effect hypothesis proposes that feedback mechanisms amplify the direct warming effect of CO$_2$.

What are these feedback mechanisms?

---

[43] IPCC, 2014: Climate Change 2014: Synthesis Report. p.5 & fig. SPM.3.

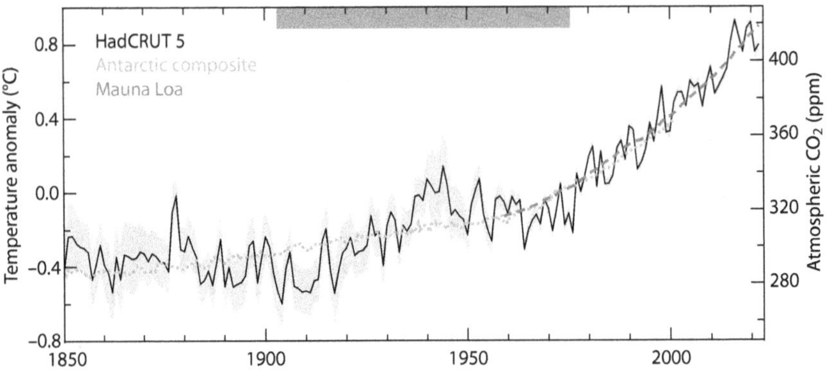

Figure 11. CO_2 and temperature increase between 1850 and 2022. Temperature anomaly is relative to the 1961-90 baseline from the HadCRUT 5 dataset (black continuous curve with gray uncertainty). 1850-2001 atmospheric CO_2 proxy from a composite Antarctic ice core (dotted light gray line).[44] 1959-2022 atmospheric CO_2 record from Mauna Loa (dashed gray line). The temperature-CO_2 correlation is generally good but weaker in the 1905-1975 period, indicated by the dark gray bar.

Large uncertainty between the hypothesis and reality

If we double the amount of CO_2 in the atmosphere and expect a temperature increase of 1°C (1.8 °F), everything in the climate system should remain constant, including the lapse rate. However, we know that this cannot happen because every change in the climate system triggers a response. For example, an increase in temperature leads to more water evaporation, increasing the amount of water vapor in the atmosphere. Water vapor is also a GHG, and its increase affects cloud formation, precipitation, and albedo. A change in the greenhouse effect causes many changes throughout the climate system, and we are unsure how much and in what ways these changes will occur. Consequently, there is considerable uncertainty about the amount of warming that could result from an increase in the greenhouse effect. This uncertainty is reflected in a wide range of predictions from different studies.

When the greenhouse effect changes, it causes a temporary change in the radiative flux at the top of the atmosphere. Scientists call this a climate "forcing". Any climate system response to this forcing, which causes further changes in the radiative flux at the top of the atmosphere, is called "feedback". Feedback is the system's output returning to its input, further modifying the output. If the feedback causes further change in the same direction as the forcing, it is positive feedback. Conversely, if it causes a change in the opposite direction, it is negative feedback. In a stable system such as the climate, negative feedback must dominate since positive feedback can cause unstable conditions that lead to runaway effects.

Climate feedbacks due to CO_2 increase

Climate feedbacks are impossible to measure because, by definition, they result from one variable affecting another. A change in the net flux at the top of

---

[44] Bereiter, B., et al., 2015. Geophys. Res. Lett. 42 (2), pp.542–549.
doi.org/10.1002/2014GL061957

the atmosphere (a forcing) caused by a change in the greenhouse effect is already within the uncertainty range of our measurements of inward and outward fluxes (ch. 6). It is impossible to distinguish feedback from the signal produced by the forcing, so feedbacks cannot be observed. Instead, they can only be inferred probabilistically or estimated with models. This model-based estimation leads to high uncertainty, sometimes even about their positive or negative sign.

Two negative feedbacks are the Planck feedback, which states that a warmer body emits more radiation, and the lapse rate feedback, which states that the atmosphere is expected to warm more than the surface, lowering the lapse rate and reducing surface warming.

The cloud feedback, on the other hand, is uncertain. Although an increase in clouds can increase the opacity of the atmosphere to infrared radiation, it can also increase the reflection of solar radiation (albedo). The overall effect is thought to be positive, but there is considerable uncertainty.

Among the positive feedbacks is the water vapor feedback, where warming increases the moisture content of the atmosphere, leading to a large increase in the greenhouse effect. However, despite its importance, water vapor measurements have not confirmed that it behaves as a positive feedback.[45] Another positive feedback is the ice-albedo feedback, which creates a loop in which melting ice causes additional surface warming, which causes more ice to melt. This feedback is thought to be important for Arctic amplification, but its overall importance is relatively small because 90% of the global albedo occurs in the atmosphere (ch. 3).

Climate sensitivity

The estimated forcing of feedbacks based on models is +1.5-2 W/m^2, but the uncertainty around this value is greater than what is usually accepted. It is possible that some feedbacks are unknown to us or that their estimated values are inaccurate. If this estimate is correct, it would suggest that most of the warming caused by changes in the greenhouse effect is due to feedbacks that cannot be measured directly.

The concept of climate sensitivity attempts to quantify how much warming would result from a doubling of atmospheric CO$_2$, but the answer is not straightforward and depends on the time scale considered. In particular, we use the term transient climate response to refer to the warming that occurs in the 20 years following an instantaneous doubling of CO$_2$. On the other hand, equilibrium climate sensitivity describes the warming that occurs after the ocean has had time to adjust and the climate system has reached a new equilibrium state, which may take centuries.

One of the earliest estimates of climate sensitivity was given in the Charney Report, prepared for the U.S. National Academy of Sciences in 1979. The report estimated a value of 1.5-4.5°C/doubling.[46] More recently, the 6th Assessment Report of the Intergovernmental Panel on Climate Change (IPCC) gave an estimate of 2.5-4°C/doubling.[47] While both reports agree on a best estimate

---

[45] Paltridge, G., et al., 2009. Theor. Appl. Climatol. 98, pp.351–359.
doi.org/10.1007/s00704-009-0117-x
[46] Charney, J.G., et al., 1979. Nat. Acad. Sci. pp.2030–2050.
[47] Forster, P., et al., 2021. Climate Change 2021: The Physical Science Basis. Cambridge Univ. Press, pp. 923–1054.

of 3°C (5.4 °F), it is important to note that after 45 years, there has been little progress in answering the question of how much warming a doubling of CO_2 should produce.

Although after 45 years of research we still do not know how much warming a doubling of CO_2 should cause, we are often presented with specific estimates of how much warming our GHG emissions will cause and how much we need to reduce them to stay within certain politically defined limits. Set initially at +2.0°C (+3.6 °F), this limit has been lowered to +1.5°C (+2.7 °F) in 2018, at which point warming is supposed to change from slightly unsafe to dangerous.[48] It is important to acknowledge the considerable uncertainty in estimating the warming caused by GHGs and not to hide this fact from the public.

A climate sensitivity of 3°C (5.4 °F) implies that two-thirds of the warming is due to poorly understood feedbacks, while only one-third is caused directly by the greenhouse effect of CO_2. However, this sensitivity value appears to be too high when we consider the pre-industrial era in climate studies, around 1750, when CO_2 levels were only 277 ppm (fig. 12). Although this was a relatively cold period, there is no evidence that it was much colder than in 1850 (fig. 11) since glaciers around the world were of similar size in both periods.[49] According to temperature records, the 1850s were one degree cooler than the 2010s. It is difficult to accept the proposition that 270 years ago the temperature was almost two degrees lower than today, as required by a 3°C sensitivity.

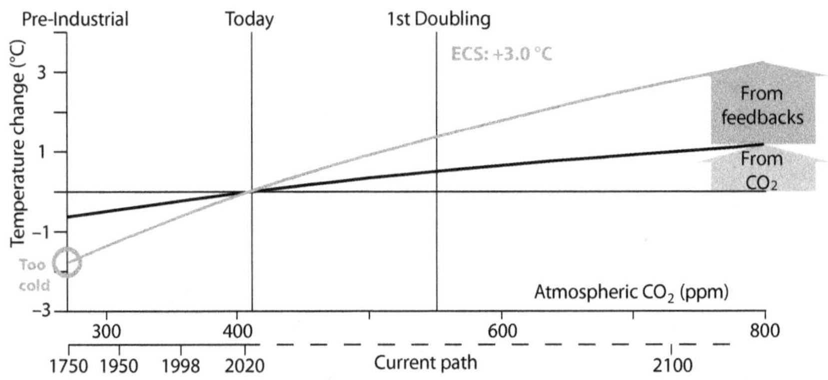

Figure 12. The effect of a 3°C sensitivity on temperatures. Relationship between atmospheric CO_2 and temperature anomaly, assuming an equilibrium climate sensitivity of 3°C (5.4 °F) per doubling of CO_2. The black curve represents the direct warming effect of CO_2, while the gray curve shows the total warming effect, including feedbacks. The lower scale shows the historical increase in CO_2, with the current rate of increase extrapolated to 2100. Backward extrapolation of this climate sensitivity projects a temperature too cold for 1750 when CO_2 levels were 277 ppm (red circle).

---

[48] Masson-Delmotte, V., et al., 2018. Global Warming of 1.5°C. Cambridge Univ. Press, pp. 3–24.

[49] Oerlemans, J., 2005. Science, 308 (5722), pp.675–677.
    doi.org/10.1126/science.1107046

The consensus

The hypothesis that the enhanced CO$_2$ effect is the primary cause of recent global warming has gained widespread acceptance, although it is based primarily on climate models and unmeasured feedbacks. However, climate models are abstractions, not scientific evidence. Surprisingly, this hypothesis, which is not supported by scientific evidence, is called the climate change consensus. It is often accompanied by a figure for the high percentage of scientists who accept it, a common marketing tactic. Scientific progress requires the confrontation of competing hypotheses, but no other hypothesis has gained sufficient acceptance among climate scientists to challenge the consensus. However, we still do not know enough about how the climate is changing to rule out any other possibility.

In my view, there are three main reasons why the Enhanced CO$_2$ Effect hypothesis is widely accepted among climate scientists. First, there is a reasonable correspondence between recent CO$_2$ increases and temperature increases (fig. 11). Scientists tend to prefer simpler explanations (Occam's razor), and the greenhouse effect is a well-established theory that identifies increased CO$_2$ as one of the causes of warming. Second, the very good correlation between CO$_2$ levels and temperatures during the Pleistocene found in ice cores supports a close relationship. Finally, the Enhanced CO$_2$ Effect hypothesis provides a plausible explanation for the necessary changes in energy fluxes at the top of the atmosphere to change the climate. Any hypothesis that does not provide this explanation is unlikely to be considered.

Box 5. The absence of natural climate change

The model-based hypothesis of an enhanced CO$_2$ effect has several problems rarely discussed publicly. One of these problems is that it does not account for natural climate change. This hypothesis relies on a high sensitivity to aerosols to explain the cooling observed in the mid-20th century, offset by a higher sensitivity to CO$_2$ to explain the warming observed in the late 20th century. As a result, these two factors account for almost all of the observed climate change since 1750. Radiative forcing calculations from then to now largely negate the role of changes in solar activity and volcanic eruptions in the climate change that has occurred (fig. B5).

Figure B5. The consensus hypothesis leaves no role for natural climate change. Change in global annual mean radiative forcing from 1750 to 2011 due to human activities, changes in total solar irradiance, and volcanic emissions.[50]

Although many climate scientists acknowledge a dominant contribution of anthropogenic CO$_2$ increases to

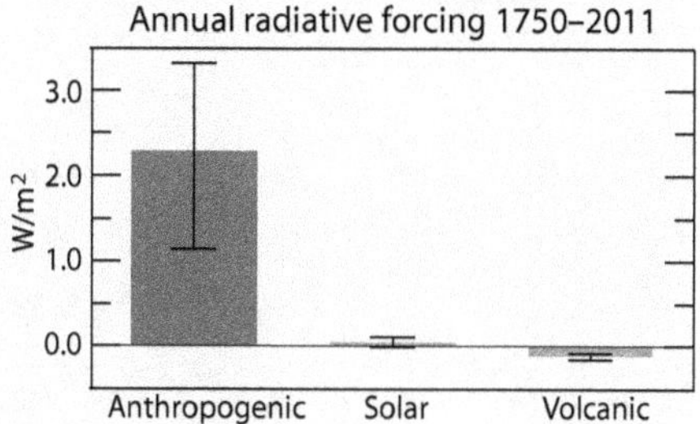

---

[50] Figure after Wuebbles, D.J., et al., 2017 Climate Science Special Report: Fourth National Climate Assessment, Vol I p.14. doi.org/10.7930/J0J964J6

recent climate change, they also recognize the importance of natural climate change over the past 270 years. However, the current formulation and modeling of the Enhanced CO_2 Effect hypothesis does not adequately explain the 14th-18th century cold period, which represents the planet's last major climate change, and the warming observed in the early 20th century. This raises questions about the ability of the hypothesis to explain natural climate change alongside anthropogenic climate change.

In summary

The Enhanced CO_2 Effect hypothesis does not rely solely on the direct warming effect of increased CO_2. It relies mainly on the secondary warming effect caused by feedback responses to the direct warming of CO_2. This secondary warming is not exclusive to CO_2 and should occur in response to any source of warming. Unfortunately, observations cannot confirm this hypothesis because the measured warming does not reveal its cause. The hypothesis is supported by computer models, which are not scientific evidence, and it explains recent climate change well but renders climate change in the past few centuries up to 1950 inexplicable.

SECTION 2 KEY ISSUES

For the planet to maintain its temperature, all the energy it receives from the Sun must be radiated back into space as infrared radiation. The Sun heats the surface, and this energy is transferred to the atmosphere mainly by evaporation and convection. Greenhouse gases, however, cause the emissions to space to originate at higher altitudes. Since the temperature in the troposphere decreases with altitude, and the emission temperature must remain constant to balance the energy, the surface must become warmer. Water vapor and clouds are the main contributors to the greenhouse effect, accounting for 75% of the total, while CO_2 is responsible for 19%. During winter in the polar regions, where water vapor and clouds are scarce, the greenhouse effect is much weaker.

The "Enhanced CO_2 Effect hypothesis" suggests that feedback mechanisms that amplify the warming from increased atmospheric CO_2 levels are the main drivers of recent global warming. These feedbacks, which cannot be measured directly, respond to the initial warming by amplifying its effects. Observations cannot confirm this hypothesis because they don't indicate the cause of the warming. Despite the lack of hard evidence and ignoring natural climate variability, the hypothesis is supported by computer models and widely accepted by scientists.

The warming of the Earth's surface indicates an energy imbalance at the top of the atmosphere. In the 21st century, this energy imbalance appears to be decreasing, suggesting that the Earth may now be warming more slowly.

SECTION 3. CLIMATE SYSTEM ENERGY TRANSPORT

Chapter 9
What We Call 'Climate' is the Transport of Heat

The difference in the absorption of solar radiation between the equator and the poles results in a latitudinal temperature gradient. This gradient determines the climatic state and average temperature of the planet and has changed significantly over time. Currently, the Earth is in an "icehouse" state. The latitudinal gradient creates a poleward (meridional) heat transport that makes high latitudes warmer than their insolation would suggest. Meridional transport redistributes heat, moisture, clouds, and momentum toward the poles. In essence, weather and climate result from variations in meridional transport.

The latitudinal temperature gradient

In the previous two sections, we talked about how energy moves vertically between the top of the atmosphere and the surface. But energy also moves horizontally in the climate system, mostly as heat transport. This process hasn't received as much attention from climatologists and is generally considered less important, as reflected in most climatology textbooks. However, heat transport is crucial to understanding the climate, and it's the central theme of this book. The climate is essentially a manifestation of heat transport. Variations in heat transport may be the key to understanding climate change.

Chapter 2 explained that the tropics absorb more shortwave radiation than the poles. This creates a temperature gradient from the equator to the poles, known as the latitudinal temperature gradient. This gradient results mainly from the latitudinal insolation gradient, with insolation being the primary determinant of surface temperature.

However, the latitudinal temperature gradient is also influenced by climatic factors. It's steeper toward the South Pole because the Antarctic Circumpolar Current and the Southern Annular Mode isolate Antarctica. These ocean currents and winds circle Antarctica, making it much colder by blocking heat from warmer areas. So the slope of the gradient isn't the only factor affecting the amount of heat transported.

The latitudinal temperature gradient is the most important factor in defining Earth's climate (fig. 13). By analyzing geological, paleontological, and isotopic climate proxies, we can estimate the changing latitudinal temperature gradient over the past 540 million years.[51] This gradient provides an estimate of the average temperature of the planet and allows us to reconstruct its climatic evolution.

The Earth's current climate in the 21st century is classified as an icehouse condition, which is among the coldest 10% of climates over the past 540 mil-

---

[51] Scotese, C.R., et al., 2021. Earth-Sci. Rev. 215, p.103503.
  doi.org/10.1016/j.earscirev.2021.103503

lion years. As a result, the latitudinal temperature gradient is very steep, driving a large amount of heat poleward.

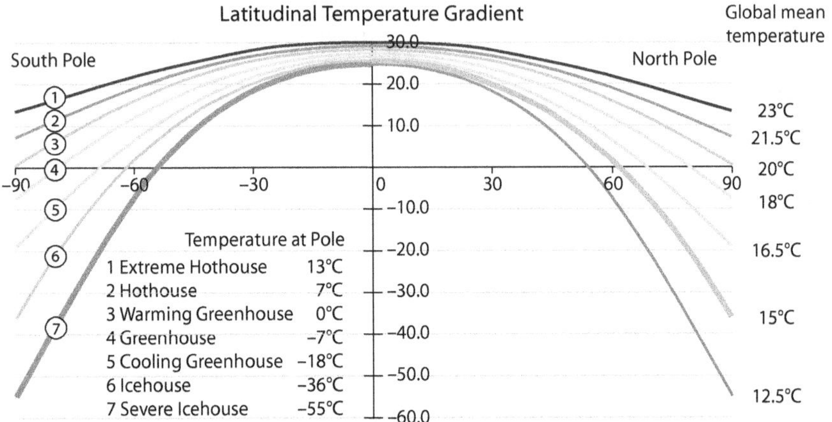

Figure 13. The latitudinal temperature gradient. Pole-to-pole curves of latitude versus temperature (°C) represent climatic conditions ranging from extreme hothouse to severe icehouse and the derived global mean temperature.[52] The current climate is described by curve 7 for the Southern Hemisphere and curve 6 for the Northern Hemisphere (thick curves).

Heat transport along the gradient

The latitudinal temperature gradient creates an accumulation of potential energy in the atmosphere, making it unstable and leading to a poleward flow of heat. This process is called meridional transport because it occurs primarily in the same direction as the meridians. The oceans also contribute to meridional heat transport, driven by varying heat input and atmospheric winds.

Heat transport reduces the temperature contrast between the poles and the equator, releasing the potential energy created by the latitudinal temperature gradient.

The tropical region between 30° north and 30° south latitude covers half of the Earth's surface but receives about 2/3 of the incoming absorbed radiation, which amounts to about 80 PW (petawatts, one quadrillion watts). In contrast, the other half of the Earth's surface, poleward at 30° latitude, receives about 40 PW. Therefore, the total heat input is about 120 PW. The tropical region radiates about 69 PW, and about 11 PW is transported to the extratropical regions. Despite the cooler temperature of Antarctica, only about 5 PW are transported to the southern extratropical region, while about 6 PW are transported to the northern extratropical region. As mentioned earlier, atmospheric circulation and ocean currents transport heat and also play a crucial role in regulating the amount of heat transported.

Meridional heat transport is responsible for keeping the poles warmer than they should be, and reduced heat transport in the Southern Hemisphere contributes to it being about 2°C (3.6 °F) cooler on average than the Northern

---

[52] Figure after Scotese, C.R., 2016. PALEOMAP Project,
        www.researchgate.net/publication/275277369

Hemisphere. Without this heat transport the poles would be, on average, 100°C (180 °F) colder than the equator, instead of the current 40°C (72 °F) difference.[53] Heat transport also makes winter conditions more bearable at high latitudes, especially near the major ocean basins where most of the transport occurs. A small net transport of about 0.2 PW across the equator to the Northern Hemisphere (box 2, ch. 3) also contributes to the uneven heat distribution between the hemispheres.

Climate is a manifestation of meridional transport.

As the old saying goes, *"climate is what we expect, weather is what we get,"* climate is defined as the average of meteorological variables over a period of time long enough to determine their variability. The meteorological variables we call weather or climate depend primarily on insolation and the meridional transport of heat and moisture.

Heat is transported in three ways: sensible, potential, and latent heat. Sensible heat is the thermal energy stored in molecules and can be measured with a thermometer. It can be transferred by radiation or by collisions between molecules. When hot desert air moves over a cooler place, it warms it through sensible heat transport. Potential heat is the heat stored as potential energy in molecules. As a parcel of air rises, it cools and expands, having a different temperature but the same potential temperature. As it moves to another location and descends, it contracts and heats up, transporting potential heat from one place to another. Latent heat is the energy required to break the weak bonds between liquid water molecules as they evaporate. This energy cannot be measured with a thermometer until the water vapor molecules condense, forming these bonds again and releasing the energy where they condense.

In addition to heat, meridional transport is responsible for the transport of water, aerosols, chemicals, clouds, and angular momentum.

The lapse rate creates an unstable troposphere as the air heated by the surface rises by convection. Moist air rises more easily than dry air due to its lower density. The latitudinal temperature gradient amplifies these processes more significantly at the equator than at the poles, creating a gradient of potential energy. This potential gradient contributes to the slope of the tropopause, which occurs at a higher altitude of 17 km (10 miles) at the equator and only 9 km (6 miles) at the poles. The rotation of the planet, combined with this potential gradient, produces turbulence in the atmosphere and drives the meridional heat transport.

The variables that make up weather and climate, including pressure, wind, clouds, temperature, and precipitation, are affected by the transport of sensible, potential, and latent heat. The distribution of solar energy at the surface, determined by the insolation pattern, is the background over which these processes operate. Therefore, changes in meridional heat transport are largely responsible for most meteorological changes. The question is whether they also play a role in global climate change.

---

[53] Lindzen, R.S., 1994. Annu. Rev. Fluid Mech. 26 (1), pp.353–378.
doi.org/10.1146/annurev.fl.26.010194.002033

Box 6. The meridional transport of momentum

Angular momentum is a property that is conserved in any rotating object and is a function of its rotational inertia and speed of rotation about its axis. For example, an ice skater who brings their arms closer to their body while spinning will reduce their rotational inertia by moving some of their mass closer to the axis and will cause an increase in rotational speed to maintain momentum. The transport of heat and moisture from the equator and tropics to mid and high latitudes is coupled to the transport of angular momentum between the solid Earth and the atmosphere. At low latitudes, surface winds flow easterly (east to west) against the Earth's rotation. This causes the atmosphere to gain momentum from friction with the solid Earth, which reduces its rotational speed. The surface winds are westerly at midlatitudes, and the atmosphere loses momentum to the solid Earth, which increases its rotational speed. Consequently, an atmospheric flow of angular momentum poleward is necessary to conserve momentum and maintain the Earth's rotational speed.

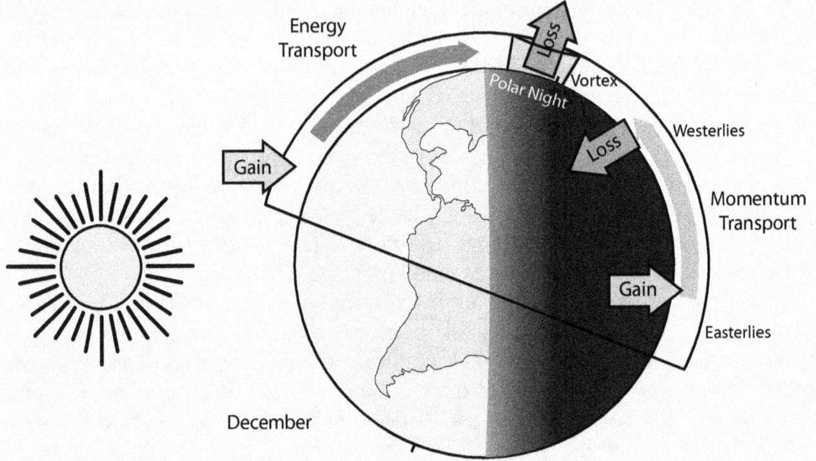

Figure B6. Meridional transport. Energy (left) and angular momentum (right) are transported due to the latitudinal temperature gradient and the rotation of the planet.[54]

To maintain momentum, any changes in atmospheric angular momentum must be balanced by corresponding changes in the solid Earth rotation rate. The seasonal variation in the zonal wind circulation is the main factor causing changes in atmospheric momentum. This circulation is strongest in winter when a deeper latitudinal temperature gradient causes more angular momentum to be present in the atmosphere. As a result, the Earth rotates slightly faster in January and July and slower in April and October, when the zonal circulation is weakest. These changes in the Earth's rotational speed are minute and result in measurable microsecond changes in the length of the day.

---

[54] Figure after Marshall, J. & Plumb, R.A., 2008. Atmosphere, Ocean and Climate Dynamics: An Introductory Text. Academic Press.

In summary

In addition to insolation, the poleward (meridional) transport of heat and moisture plays a crucial role in determining climate. This transport is mainly due to temperature differences between the equator and the pole, as well as atmospheric and oceanic circulations. Meridional transport not only moves heat and moisture but also carries angular momentum, which affects the rate at which the Earth rotates.

CHAPTER 10
HOW HEAT IS TRANSPORTED

Heat transport is critical to the redistribution of energy on the planet. Despite its importance, it remains poorly understood due to challenges in accurate measurement and inadequate theoretical models. The atmosphere is the primary driver of heat transport, and its importance increases with latitude. Oceanic transport contributes about one-third of the total heat transported, but its importance decreases with increasing latitude. In the western boundary currents, the oceans transfer much of their heat to the atmosphere, which is then transported poleward by midlatitude storms.

Atmospheric versus oceanic heat transport

The presence of two fluid masses, the ocean and the atmosphere, on Earth's solid surface transforms the planet's energetics by introducing a dynamic aspect to its radiative properties. These fluids play an essential role in the transport of energy within the climate system. Heat is primarily transported from the tropical regions, where there is a net energy surplus, to the midlatitudes and polar regions, where there is a net energy deficit. This heat transport through the ocean and atmosphere moderates the climate and produces all the atmospheric phenomena we recognize as weather.

Direct measurement of heat transport is difficult due to insufficient data on the temperature and dynamics of the atmosphere and oceans. Instead, the total heat transport is calculated from the difference between the fluxes of absorbed shortwave radiation and emitted longwave radiation at the top of the atmosphere. The oceanic heat transport is usually calculated from the net sea surface heat flux, and the atmospheric heat transport is obtained by subtracting the oceanic transport from the total. However, this approach assumes that the change in oceanic heat storage is negligible, which may not be true, and is inadequate for studying the relationship the relationship between atmospheric and oceanic transport because they are not obtained independently.

The ocean has a much greater heat capacity than the atmosphere and contains about 96% of the energy in the climate system. In contrast, the land, atmosphere, and cryosphere contain only 2%, 1%, and 1%, respectively.[55] Despite the vast difference in energy content, the atmosphere transports most of the energy in the climate system. In the Northern Hemisphere, the ocean is responsible for about 30% of the energy transport, but only 18% in the Southern Hemisphere, leaving the atmosphere responsible for 75% of global heat transport. Water vapor is the key to understanding how the atmosphere can transport so much heat despite its lower heat capacity. This is because water vapor is 100 times more efficient at transporting heat than dry air. Through latent heat transport, the atmosphere effectively doubles its heat transport capacity.

---

[55] Cuesta-Valero, F.J., et al., 2016. Geophys. Res. Lett. 43 (10), pp.5326–5335. doi.org/10.1002/2016GL068496

The latitudinal distribution of heat transport is highly asymmetric. Oceanic transport is more important at low latitudes, while atmospheric transport dominates at high latitudes (fig. 14).

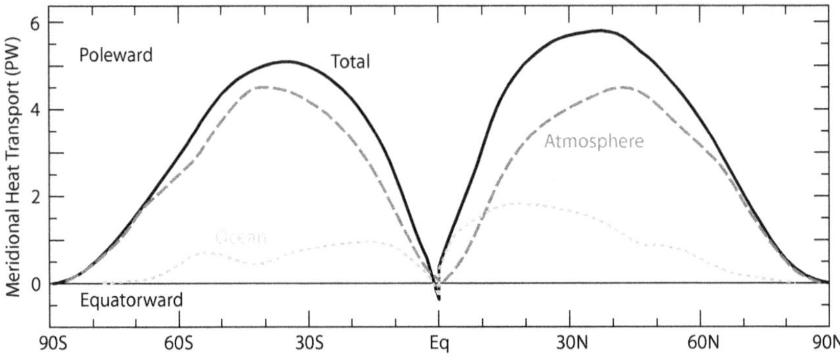

Figure 14. Meridional transport decomposition. Positive values represent poleward transport in petawatts, and equatorward transport is represented by negative values. Total transport is shown as a black solid line, atmospheric transport as a medium gray dashed line, and oceanic transport as a light gray dotted line.[56] The geometry of the Earth and the minimal cross-equatorial transport affect the shape of the curve.

Some figures in this book that decompose meridional transport, including figure 14, define poleward transport as positive and equatorward transport as negative. This is done by placing an artificial zero line at the equator rather than using the usual definition of northward transport as positive and southward transport as negative. This representation is for illustrative purposes only and allows the important hemispheric differences in transport to be better visualized.

Does the temperature difference between latitudes determine the heat transport, or does the heat transport determine the temperature difference? The answer seems to be both, but we cannot determine precisely how much each influences the other. Climate models are unreliable in representing meridional heat transport, as the results can differ by up to 20% between models. Furthermore, within each model, meridional transport remains nearly constant despite changes in ocean circulation, interannual variability, and paleoclimatic conditions.[57] Considering the large difference in the latitudinal temperature gradient between the Last Glacial Maximum and the present, we must conclude that current models do not adequately reproduce meridional heat transport.

Atmospheric transport

In this section, we will focus only on the atmospheric heat transport within the troposphere since the stratospheric transport, while small in energy terms, is important for other reasons that we will discuss in chapter 14. The atmosphere acts as a heat engine, a thermodynamic concept that refers to a system

[56] Figure after Yang, H., et al., 2015. Clim. Dynam. 44, pp.2751–2768. doi.org/10.1007/s00382-014-2380-5

[57] Donohoe, A., et al., 2020. J. Clim. 33 (10), pp.4141–4165. doi.org/10.1175/JCLI-D-19-0797.1

capable of converting thermal energy into kinetic energy. It does this by transferring heat from a hot source (the surface) to a cold sink (the upper troposphere). In the process, the atmosphere uses the energy as work to redistribute water and move air as it transports heat. It is worth noting that about one-third of the atmospheric energy budget is used to drive the water cycle.[58]

Heat transport in the troposphere involves two processes: the mean meridional circulation and turbulent flow processes known as eddies. In the extratropics, most heat is transported by transient eddies, which are storms that carry large amounts of heat, moisture, and momentum (box 6, ch. 9). These storms originate mainly over ocean basins and follow a path that brings them closer to the poles. The transient eddies of the Northern Hemisphere are organized into zonally restricted storm tracks that exist because of the zonal asymmetries created by the continents. The eastern slopes of the Himalayan-Tibetan and Rocky Mountain ranges create stationary eddies due to orographic perturbations of the atmospheric flow. These stationary eddies are crucial in shaping and intensifying storm tracks and determining where they end.[59] Stationary eddies are important in climate change because they contribute significantly to the difference in atmospheric heat transport between summer and winter. In winter, when transport toward the dark pole is greatly enhanced, these stationary eddies greatly increase the amount of heat transported.[60] Figure 15 shows the northward heat flux through the eddies during the boreal winter, following the trajectories of the storms that define the main Arctic gateways.

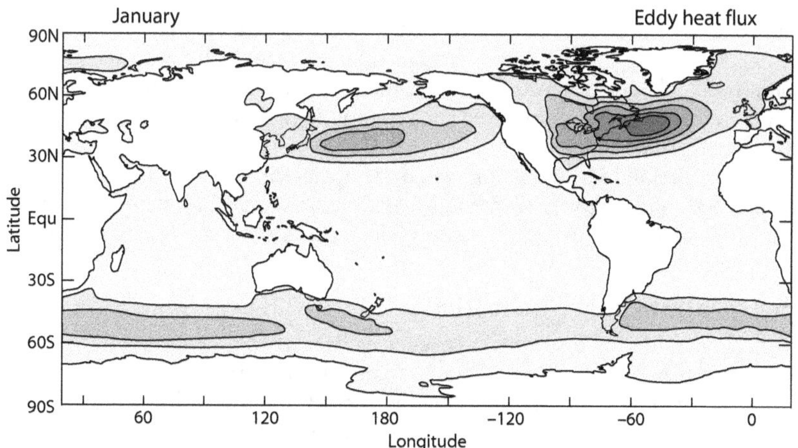

Figure 15. Eddy-driven northward heat flux in January. Each contour is 5°C m/s. Shading in the Southern Hemisphere indicates southward flow.[61] Wind speed maxima (darker shading) give rise to strong midlatitude storms.

---

[58] Laliberté, F., et al., 2015. Science, 347 (6221), pp.540–543. doi.org/10.1126/science.12571

[59] Kaspi, Y. & Schneider, T., 2013. J. Atmos. Sci. 70 (8), pp.2596–2613. doi.org/10.1175/JAS-D-12-082.1

[60] Peixoto, J.P. & Oort, A.H., 1992. Physics of climate. New York: American Institute of Physics. pp.330–336.

[61] Figure after Hartmann, D.L., 2016. Global physical climatology. 2nd ed. Elsevier.

As Leon Barry and colleagues say, *"The atmospheric heat transport on Earth from the Equator to the poles is largely carried out by the midlatitude storms. However, no satisfactory theory describes this fundamental feature of the Earth's climate."*[62] Heat transport is fundamental but poorly understood.

Oceanic transport

Unlike the atmosphere, the ocean is not considered a heat engine because it gains and loses heat through its surface, so the source and sink of heat are not separated as in the atmosphere. Instead, the ocean is driven by external mechanical energy from wind stress and tides, although it is 1000 times smaller than the heat flux.[63] Wind stress (the force per unit area that the wind exerts on the ocean) directly controls mass fluxes in the upper several hundred meters of the ocean. Surface buoyancy conditions play an important role in the transport of heat and salt, both in models and in reality, because the fluid must become dense enough to sink, but these conditions are not what actually drives the circulation.[64]

In chapter 4, we learned that 75% of the solar energy absorbed at the surface goes into the ocean. At low latitudes, the solar radiation absorbed by the ocean exceeds the heat flux from the ocean to the atmosphere, resulting in excess heat that is transported poleward and released at higher latitudes. In essence, the ocean acts as a heat storage and redistribution system. The meridional transport of heat through the ocean can be divided into two primary components: the meridional overturning circulation and the flow associated with ocean gyres. The scientific literature has debated the possibility of a slowdown in the Atlantic Meridional Overturning Circulation due to recent climate change, which could reduce heat transport. However, the evidence supports the argument that natural variability has dominated the Atlantic Meridional Overturning Circulation over the past century.[65]

Our understanding of the energetics of ocean circulation is incomplete and unbalanced, and many fundamental questions remain unanswered. Given this, it is difficult to believe that models can accurately reproduce oceanic heat transport.

Ocean-atmosphere heat flux

In chapter 6, we examined the energy balance of the ocean surface. The ocean receives 170 W/m^2 from the Sun and transfers 16 W/m^2 of sensible heat, 53 W/m^2 of longwave radiative energy, and 100 W/m^2 of latent heat to the atmosphere.[66] Considering only the latent and sensible heat fluxes, and not the radiative fluxes, it is unsurprising that the atmosphere transfers heat to the ocean only during the summer at high latitudes (fig. 16) and nowhere in the annual mean. The ocean's role is to heat the atmosphere with the solar energy it

---

[62] Barry, L., et al., 2002. Nature, 415 (6873), pp.774–777. doi.org/10.1038/415774a

[63] Huang, R.X., 2004. Ocean, energy flows in. Encycl. Energy, 4, pp.497–509.

[64] Wunsch, C., 2002. Science, 298 (5596), pp.1179–1181.
doi.org/10.1126/science.1079329

[65] Latif, M., et al., 2022. Nat. Clim. Change, 12 (5), pp.455–460.
doi.org/10.1038/s41558-022-01342-4

[66] Schmitt, R.W., 2018. Oceanography, 31 (2), pp.32–40.
doi.org/10.5670/oceanog.2018.225

receives, thus providing thermal inertia to the system.

The most important contribution to the ocean-atmosphere heat flux comes from the wind-driven gyres, particularly the warm waters of their western boundary components during the Northern Hemisphere winter: the Gulf Stream, the Kuroshio Current, and the northern North Atlantic (fig. 16). During the winter season, when heat transport is most intense, most of the heat gained by the ocean in the tropics is transferred to the atmosphere between 25° and 50° latitude for final poleward transport. Comparing figures 15 and 16 allows us to understand where the heat for the eddy flow comes from.

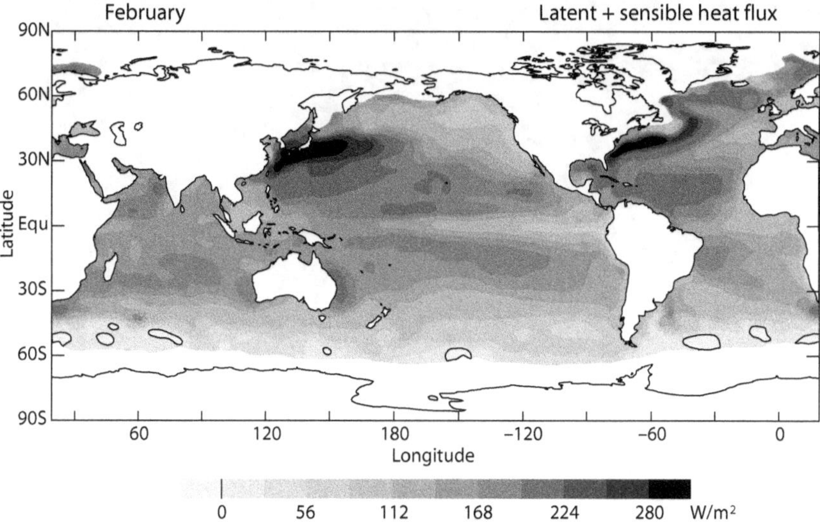

Figure 16. Ocean-atmosphere heat flux for the month of February. The 1981-2005 mean sum of the latent and sensible heat flux in the ice-free ocean is plotted. The flux is downward (negative) only in small areas of the Southern Ocean.

The interannual changes of the latent heat flux between 1981 and 2005 were analyzed, and the results show a significant increase of 10 W/m² during this period.[67] This value is ten times larger than that produced by climate models, indicating once again that meridional transport is not adequately represented in the models.

In summary

Poleward heat transport is a critical factor in climate, but our understanding of it is still poor. About one-third of the Earth's transported heat is carried by the ocean, with its importance decreasing toward the poles. Much of this heat is transferred to the atmosphere by midlatitude western boundary currents. The remaining two-thirds of the heat is transported by the atmosphere, with its importance increasing as we move toward the poles. Midlatitude storms play a critical role in transporting heat poleward. However, climate models are not yet consistent in their representation of poleward heat transport.

---

[67] Yu, L. & Weller, R.A., 2007. B. Am. Meteorol. Soc. 88 (4), pp.527–540. Source for figure 16. doi.org/10.1175/BAMS-88-4-527

CHAPTER 11
THE HEAT TRANSPORT SEESAW

At high latitudes, the amount of incoming shortwave solar radiation varies much more throughout the year than the amount of outgoing longwave radiation, resulting in large seasonal differences in regional net radiative flux. As a result, these regions experience very cold winters, and more heat must be transported to them during this time of the year. This seasonal oscillation in transport is particularly pronounced in the Northern Hemisphere, where heat transport is very low in summer and very high in winter. The Arctic has a different heat budget than the Antarctic, so it is the largest heat sink to space in winter. However, this heat loss is limited by the formation of polar vortices, walls of strong winds that surround the polar regions and restrict heat transport. During the cold season, these vortices are pummeled by atmospheric waves.

Seasonal differences in transport

In chapter 5 (fig. 7), we examined the average outgoing longwave radiation flux that occurs throughout the year from the entire surface of the planet. In contrast, during the cold season at high latitudes, absorbed shortwave radiation is very low, dropping to zero poleward of 75° during the winter polar night. This period of permanent darkness can last for months, depending on the location. As a result, the polar regions in winter are defined as the largest heat sink on the planet, where energy received from lower latitudes is efficiently lost to space through radiative cooling.

Seasonal changes in surface insolation have a significant effect on surface temperatures. As the net radiation deficit increases with latitude, temperatures become colder. The effect is most pronounced in winter when the temperature difference between the tropics and the poles is greatest. Figure 17a shows the radiation deficit at the top of the atmosphere in both hemispheres during each winter. It's worth noting that the climatological seasons differ from the astronomical seasons because they include three full months.

In the corresponding winters, the energy deficit over the North Pole (-170 W/m^2) is greater than over the South Pole (-110 W/m^2) because the North Pole is much warmer and thus produces more outgoing thermal radiation. This fact defines the Arctic as the largest heat sink to space on the planet in winter, an important climatic asymmetry that is often overlooked.

Seasonal changes in radiation and temperature strongly influence heat transport, which varies greatly throughout the year. Solar energy collected over much of the planet must be transported toward the winter pole rather than the summer pole. This means that the band of clouds and storms that marks the ascending branch of the Hadley circulation called the Intertropical Convergence Zone (box 2), follows the Sun's seasonal movement into the summer hemisphere. The location of this climatic equator bisects the atmospheric circulation and poleward heat transport. Since most oceanic transport is wind-driven, the displacement of this convergence zone largely determines the magnitude of global poleward transport. The axial tilt and precession of the planet mainly determine the mean latitudinal location of this meridional heat transport zero point.[68]

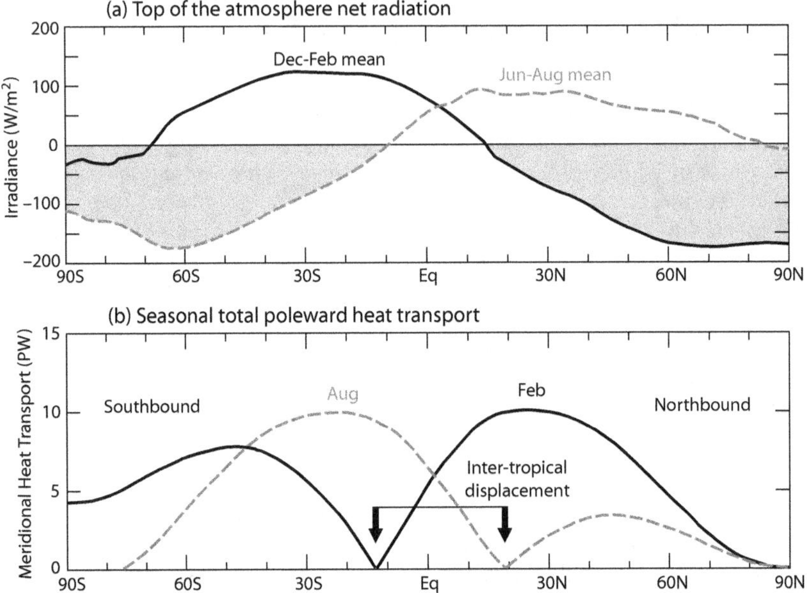

Figure 17. Seasonal energy and transport demand. (a) Differences in net top-of-the-atmosphere radiation between boreal winter (black line) and austral winter (dashed gray line).[69] Grey areas indicate energy deficits. (b) Meridional heat transport from February (black line) and August (dashed gray line) to the corresponding pole in petawatts. Arrows indicate the Intertropical Convergence Zone, where the transport is split between the hemispheres, and its seasonal change in position.

As discussed in the previous chapter, accurate measurement of the magnitude of heat transport is challenging, and the assumption that changes in heat storage are negligible is doubtful in the annual average and invalid throughout the year. The atmosphere has little capacity to store heat, and changes in the Earth's net radiation over the year show that much energy must enter the ocean at certain times and leave at others (fig. B3b, ch. 5). Although models cannot fill the knowledge gap in heat transport, one approach to estimating seasonal changes is to assume that the ocean's variable heat storage is close to zero when ocean temperatures reach their maximum or minimum near the end of August and February.[70] Figure 17b shows that the seasonal heat transport is highly asymmetric, as required by the radiation and temperature asymmetries. The maximum seasonal transport of 10 PW (petawatts) is almost twice the maximum annual mean transport, with the main difference between the hemispheres being the magnitude of the respective summer transport. This differ-

---

[68] Liu, Y., et al., 2015. Nat. Commun. 6 (1), p.10018. doi.org/10.1038/ncomms10018
[69] Figure after Randall, D. A., 2015. An introduction to the global circulation of the atmosphere. Princeton Univ. Press.
[70] Stephens, G.L. & L'Ecuyer, T., 2015. Atmos. Res. 166, pp.195–203. Source for figure 17b. doi.org/10.1016/j.atmosres.2015.06.024

ence can be explained by the much warmer high northern latitudes during summer (box 3). The Northern Hemisphere shows a larger oscillation in seasonal temperature and transport, smaller in summer and larger in winter, because the annual mean hemispheric heat transport is slightly larger in the Northern Hemisphere, as shown in the previous chapter (fig. 14).

The heat budget of the polar regions

The heat budget of the polar regions is of particular interest to us because they are very special places in terms of the greenhouse effect (box 4, ch. 7). The asymmetries observed in the seasonal hemispheric heat transport partly reflect the energetics of the regions located at 70-90° latitude. During the Arctic summer, the net radiative deficit is very small (fig. 17a and the dark gray arrows in fig. 18). Most of the heat transported across the 70° latitude band (F_{WALL}, flux across the 70° wall) is returned to the surface as downward flux at the bottom of the atmosphere (F_{BA}) and stored as sensible heat in the ocean (S_O) or as latent heat from melting ice and snow (S_{LHI}).[71] The Antarctic region behaves very differently during the austral summer. Less energy reaches the south polar region due to Antarctica's climatic isolation (ch. 9). Half of this heat is lost at the top of the atmosphere due to a larger net deficit, and the other half goes mainly to melting ice and snow since the ocean receives very little due to the large Antarctic continent.

Figure 18. Observed polar heat budget in summer and winter. Values in W/m². Numbers in parentheses are indirect estimates. See abbreviations in the main text.

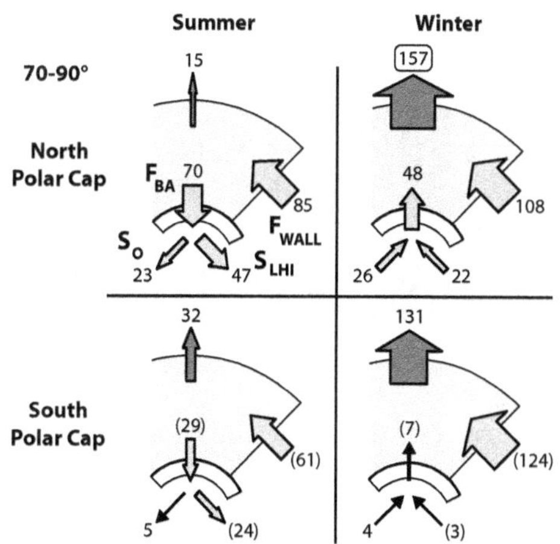

During the boreal winter, the Arctic receives 20% more heat transported from lower latitudes due to increased atmospheric eddy action (ch. 10). In addition, the surface returns 70% of the stored heat through ocean cooling and latent heat released by freezing water. The heat from the Arctic winter atmosphere is lost to space by radiative cooling, creating the largest energy deficit on the planet. On the other hand, the Antarctic region receives more heat transport from lower latitudes, although not as much as it should, given its lower temperature. However, the surface contribution from the large, permanently frozen continent is less.

Several features of the Arctic make it a region of particular interest for climate change. These are as follows:

* The Arctic has the largest net radiation deficit of any region on Earth.

---

[71] Peixoto, J.P. & Oort, A.H., 1992. Physics of climate. New York: American Institute of Physics. pp.353–364.

- The difference between summer and winter conditions in the Arctic is greater than in any other region of the planet.
- The Arctic is the most sensitive region on Earth to climate change.
- The polar vortex surrounding the Arctic is weaker and less stable than the one in Antarctica (box 7).

Because of these and other factors discussed later, our primary focus on heat transport will be on the meridional transport of heat to the Arctic during winter and the changes in this transport.

Box 7. The polar vortex

Circumpolar vortices are large-scale west-to-east (westerly) wind flows that circle the poles. High-speed winds closer to the poles create the vortex, which emerges in the fall and persists in the stratosphere until spring. The larger tropospheric vortex is present throughout the year but weakens considerably from spring to fall. The thick line in figure B7 indicates the latitude at which the westerly wind reaches its maximum in the hemisphere and forms the boundary of the polar vortex. In the case of the tropospheric vortex, the fast westerly winds that define its boundary are called the jet stream. In the stratosphere, it is the polar night jet.

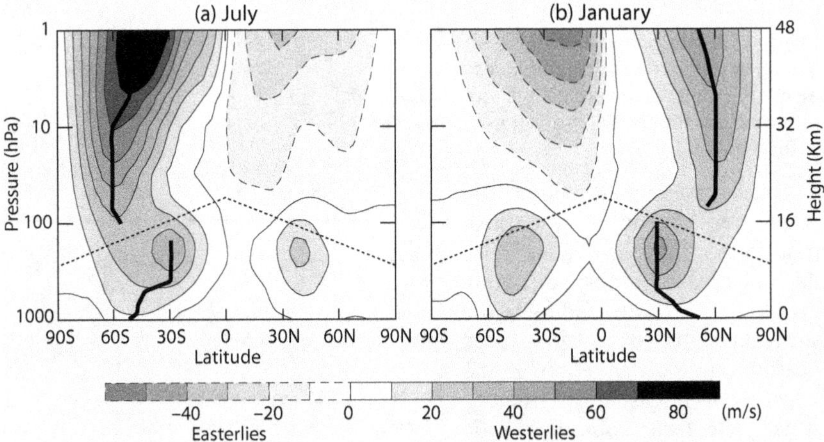

Figure B7. Mean zonal wind speed in July and January. The easterly winds (east to west) are shown with dashed lines and negative speed, blowing toward the plane of the graph. Westerly winds are shown with solid lines and positive speed, blowing away from the plane of the graph. A thick line connects the highest westerly wind speed points at each pressure level, indicating the boundary of the polar vortex. In the troposphere, this is the jet stream. The dotted line represents the tropopause, which separates the troposphere from the stratosphere.[72]

As autumn arrives, the high latitudes experience a rapid decrease in insolation. This cooling of the atmosphere leads to the formation of a low-pressure center

---

[72] Figure after Waugh, D.W., et al., 2017. Bull. Am. Meteorol. Soc. 98 (1), pp.37–44. doi.org/10.1175/BAMS-D-15-00212.1

over the pole, surrounded by strong winds. These winds hinder the transport of heat to the very cold interior of the vortex, reinforcing it and increasing the wind speed. A robust vortex acts as a shield against cold air masses in its interior, and it also limits the amount of heat lost from its interior by radiative cooling since very cold air and surfaces emit less energy.

The polar vortex can be weakened by a type of atmospheric wave known as Rossby waves, which are similar to ocean waves in the atmosphere. These waves can travel long distances quickly without air travel and provide energy and momentum against the vortex, slowing and weakening it. The upper tropospheric winds act as a barrier, deflecting smaller waves and allowing only larger planetary waves to reach the stratosphere. Any weakening or disruption of the vortex is transmitted downward, which can disrupt normal winter weather patterns. High wave activity makes the tropospheric vortex slower and more meandering, and if severe enough, can cause a winter block, resulting in extreme weather events that can last for days. The strength and frequency of cold waves during a Northern Hemisphere winter are determined by the large amount of energy that waves deliver against the vortex. In contrast, the Southern Hemisphere vortex is stronger and wave activity is weaker, resulting in a more stable vortex.

In summary

Atmospheric circulation and poleward heat transport are much stronger in the winter hemisphere, resulting in a biannual seesaw in their magnitude between the hemispheres. In the Northern Hemisphere, there is a greater oscillation in seasonal temperatures and heat transport, with the Arctic acting as the largest heat sink to space on the planet in winter. However, the polar vortex strongly limits heat transport to the polar regions during this season. The northern polar vortex is weaker and more variable due to the continuous attack of atmospheric waves originating from continental orographic reliefs.

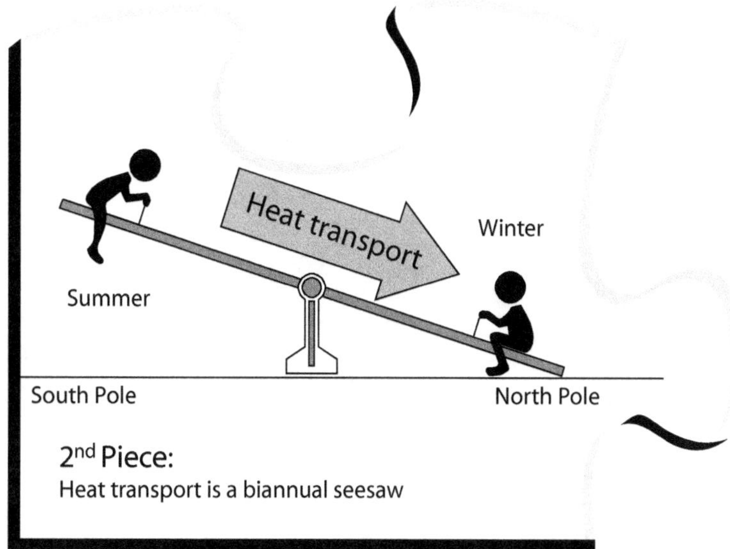

CHAPTER 12
HEAT TRANSPORT AND CLIMATE CHANGE

Meridional transport plays a crucial role in generating the Earth's many climates, but changes in heat transport are not usually considered a cause of climate change. This is because the sum of horizontal transport is zero in the global average, as heat is removed from one place (negative) and added to another (positive). However, we do not understand heat transport well, so there is no general theory to explain it. It has been proposed that the magnitude of the transport adjusts itself to the temperature difference between the equator and the pole by maximum entropy production, a passive role that would justify its neglect. According to the models, changes in the oceanic or atmospheric component of meridional transport should be compensated by opposite changes of the same magnitude in the other component. Our understanding of the role of meridional transport in climate change leaves much to be desired.

Why changes in heat transport are not considered a cause of recent climate change

The energetics of the climate system we reviewed in chapters 2-11 may be a dry subject, but its importance cannot be overstated. I hope that it has not deterred many readers. We must remember that without energy, matter is inert. Everything that happens in the climate is because energy makes it possible. Therefore, climate change is fundamentally about energy change.

We must follow the path of energy through the climate system to understand the climate and its changes. In the previous ten chapters of this book, we have reviewed the four main steps in this process. These include the arrival of energy from the Sun, the reflection of energy by albedo, the absorption and transport of energy within the climate system, and its exit as radiation at the top of the atmosphere. Of these four steps, the first two and the last are important for climate change (as forcing or feedback), while the third is not considered as such.

It is well established that variations in the way energy enters the climate system have led to climate change. The Milankovitch orbital theory is the best example of this, explaining the glacial cycle. In the 1920s, the Serbian engineer and mathematician Milutin Milanković hypothesized that changes in the Earth's orbit, which caused changes in the Earth's tilt, wobble, and distance from the Sun over thousands of years, altered the seasonal and latitudinal distribution of insolation. This, in turn, caused the Earth's climate to change between glacial and interglacial periods. The hypothesis was confirmed in 1976 when it was shown from analysis of deep ocean cores that the Earth's major climate oscillations over the last half million years have followed the Milankovitch orbital frequencies.[73]

In terms of energy reflection, albedo is one of the most critical feedback factors in the climate system. It is estimated that a 1% decrease in albedo (from 0.29 to 0.28) would have the same radiative effect as a 100% increase in CO_2 ($\approx$ 3.4 W/m^2). However, we are unable to determine changes in albedo with sufficient precision to assess their role in recent climate change. Models indicate that much of the CO_2-induced warming is due to changes in clouds, leading to increased absorption of shortwave radiation by the climate system.[74] In addition, a controversial hypothesis proposes that variable cosmic rays play an important role in cloud nucleation and albedo changes.[75]

Regarding energy output as a cause of climate change, the leading theory of climate change for the past 50 years has been the effect of changes in outgoing longwave radiation due to changes in GHGs, based on the greenhouse effect. The Enhanced CO_2 Effect hypothesis proposes that recent climate change is primarily due to feedbacks acting on the initial warming caused by changes in CO_2.

Meridional heat transport is often overlooked as a climatic energy process, to the point that most general climatology textbooks do not mention it or only briefly. While it is considered an important cause of past climate change, it is not generally considered a cause of recent climate change. Nevertheless, possible changes in meridional heat transport are often used to warn society of a fictitious shutdown of the Atlantic Meridional Overturning Circulation and the Gulf Stream, which could have dire climatic consequences for Europe.[76] Interestingly, the most abrupt and major climate changes of the past glacial period, known as the Dansgaard-Oeschger events, are thought to have resulted from changes in oceanic heat transport that caused the accumulation and sudden release of a large amount of heat stored under sea ice.[77]

Heat transport is generally not considered to be a cause of recent climate change because the energy content of the climate system can only change through changes in radiative fluxes at the top of the atmosphere. The other three stages of the energy pathway, incoming shortwave radiation, reflected shortwave radiation, and outgoing longwave radiation, are known to be capable of producing these changes. Although heat transport can have a regional effect on climate, when one place loses heat to transport, another place gains it, and the total amount of energy remains unchanged. This is often expressed as the integral (sum) of all horizontal heat transport being zero. The horizontal component of energy transport disappears in the global average, another example of an average masking the underlying complexity.

---

[73] Hays, J.D., et al., 1976. Science, 194 (4270), pp.1121–1132.
doi.org/10.1126/science.194.4270.1121

[74] Donohoe, A., et al., 2014. Proc. Nat. Acad. Sci. 111 (47), pp.16700–16705.
doi.org/10.1073/pnas.1412190111

[75] Svensmark, H., 1998. Phys. Rev. Lett. 81 (22), 5027.
doi.org/10.1103/PhysRevLett.81.5027

[76] Alley, R.B., 2007. Annu. Rev. Earth Planet. Sci., 35, pp.241–272.
doi.org/10.1146/annurev.earth.35.081006.131524

[77] Dokken, T.M., et al., 2013. Paleoceanography, 28 (3), pp.491–502.
doi.org/10.1002/palo.20042

Is there a valid theory for heat transport?

There are currently no adequate theories to explain the observed value of heat transport or how it might change, other than changes in the temperature difference between the tropics and the poles. Transport is generally considered to be a complex passive process driven by the planetary temperature difference and changing heat flux conditions.

One hypothesis with strong support is that heat transport adjusts to maximize entropy. Entropy measures the unavailability of a system's energy to do work and expresses the irreversibility of a process due to the dispersion of matter or energy. Mixing cream with coffee increases entropy due to the irreversible mixing of their matter and temperature. Similarly, the work done by the atmosphere in transporting heat leads to multiple irreversible processes that increase entropy, the amount of which depends on the temperature difference.

As heat transport reduces the temperature difference, it affects its own efficiency. When heat transport reaches an efficiency that sufficiently reduces the temperature difference, it will begin to produce less entropy. The maximum entropy production hypothesis of heat transport is illustrated in figure 19.[78]

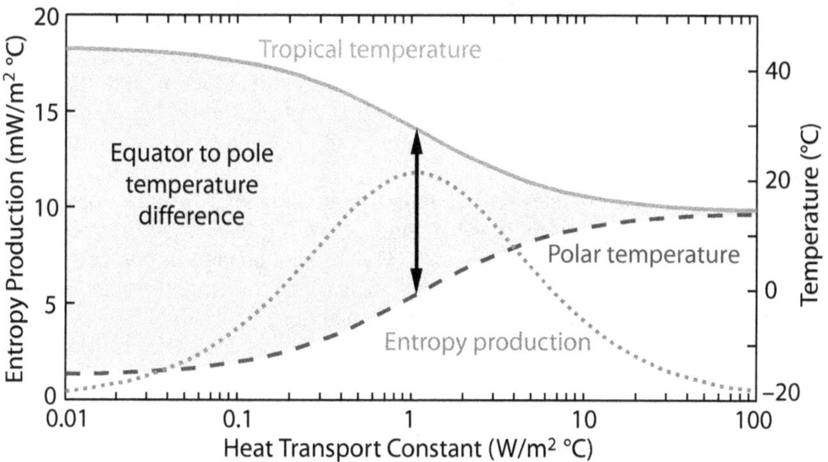

Figure 19. Maximum entropy production hypothesis of heat transport. The gray area represents the temperature difference between the tropics (solid line) and the poles (dashed line). Entropy production (dotted line) is minimum when there is no energy transport (left side of abscissa) or when transport is so efficient that there is no temperature difference (right side of abscissa), and maximum at some intermediate point. This point determines the temperature difference (arrow) and the magnitude of the heat transport.

Models support the hypothesis of maximum entropy production of heat transport. However, these models do not adequately represent heat transport, as discussed in chapter 10. Therefore, it is doubtful that model support indicates that heat transport is determined by maximum entropy production in the real world. Fitting heat transport to maximum entropy production requires a large

---

[78] Kleidon, A., & Lorenz, R., 2005. Non-equilibrium Thermodynamics and the Production of Entropy pp.1–20. Springer. Source for figure 19.

number of possible outcomes or degrees of freedom. However, it appears that meridional transport is modulated by several factors not accurately represented in computer models. This modulation substantially reduces the degrees of freedom, making the hypothesis unlikely even if entropy contributes to the determination of heat transport.

In this book, we will explore the factors that regulate heat transport and how they may affect the validity of the maximum entropy production hypothesis.

Recent climate change has not been attributed to changes in heat transport, but changes in oceanic heat transport have been used to explain some paleoclimates. The challenge is to explain warm-pole paleoclimates, which require greater heat transport (reviewed in ch. 20). However, their explanation runs into a problem within the current climate theory paradigm, where the temperature difference determines the magnitude of heat transport. The more heat that must be transported to achieve a low-temperature differential, the smaller the temperature differential to drive that transport becomes, resulting in less heat being transported, not more. As a result, known warm-pole paleoclimates are notoriously difficult to reproduce with climate models.

Meridional transport displaces heat, leaving a trace in the radiation emitted at the top of the atmosphere that is used to calculate it. However, suppose we introduce a change in one component (atmosphere or ocean) of the poleward heat transport, keeping the Sun constant, the same albedo, and the same equator-to-pole temperature difference. In that case, it should be compensated by an opposite change in the other component since the total transport should theoretically be the same.

This means that changes in net radiative heating due to solar, orbital, or compositional changes in the Earth's atmosphere can lead to different climates with different poleward heat transports. However, a change in one component of the poleward heat transport in the absence of external forcing changes should not, in theory, lead to global climate change.

A modeling study of the past 22,000 years, from the Last Glacial Maximum to the present, found that despite major changes to the Earth's climate, the total meridional heat transport remained stable, which is difficult to accept.[79] In explaining recent climate change, the accepted climate theory renders past climate changes inexplicable.

Heat transport dilemma, only a feedback or also a forcing?

After the amount of insolation, meridional transport is arguably the most important factor in determining the climate we experience anywhere in the world. Any change in heat transport can have an important effect on local and global wind patterns, cloud cover, temperature, precipitation, and the frequency of extreme weather events.

Changes in meridional transport are not considered a direct cause of climate change in accepted climate theory, although they are a critical factor in determining the climate we experience. Meridional transport may have varied along with temperature over the past 300 years, but its changes are not classified as forcing in the current paradigm of climate theory. Instead, any effects are seen as part of the feedback mechanisms that come into play in response to in-

---

[79] Yang, H., et al., 2015. Sci. Rep. 5 (1), p.16661. doi.org/10.1038/srep16661

creased CO_2-induced warming.

An unrealistic experiment was conducted using a model in which the amount of oceanic heat transport was doubled. This resulted in climate warming due to an increased water vapor greenhouse effect and decreased cloudiness.[80] Although the accepted climate hypothesis does not assign a role to changes in meridional transport as a cause of climate change, it is possible that it can modify the greenhouse effect and albedo.

Box 8. The Bjerknes compensation

In the 1960s, Jacob Bjerknes proposed that if the top of the atmosphere fluxes and oceanic heat storage remained relatively stable, the total heat transport through the climate system would also remain constant. This hypothesis is based on the understanding that in the absence of changes in the top of the atmosphere fluxes, the total energy of the climate system remains unchanged. Unless some of the energy is stored in the ocean (the atmosphere stores very little heat), the amount that must be transported also remains constant. Under these conditions, any significant change in oceanic and atmospheric heat transport should be equal and opposite. This simple hypothesis is known as Bjerknes compensation.

However, proving the hypothesis requires the capacity to calculate oceanic and atmospheric heat transport independently, which is not currently possible with observational data. Despite the lack of observational support for 60 years, the Bjerknes compensation has been extensively studied in models, where it varies widely from model to model.

A review of Bjerknes compensation in 15 Coupled Model Intercomparison Project 5 models showed its presence in all of them.[81] However, the models do not say how the mechanism works, and the authors have no choice but to say that *"the physical mechanism underlying the multidecadal variability associated with Bjerknes compensation remains unclear."*

In my opinion, the models are designed to follow a specific rule: to keep the transport proportional to changes in top-of-the-atmosphere radiative fluxes and changes in ocean storage using all available means. Thus, the models must adjust for oceanic and atmospheric transport changes in order to obtain the desired total transport. However, there is a major objection to this approach. Much of the oceanic transport is wind-driven and should vary in the same direction as the atmospheric transport. However, the models show that the non-wind-driven part, essentially the meridional overturning circulation, compensates for the increased transport from the atmosphere and the wind-driven circulation. This requires massive changes in the overturning circulation simulated by the models. This raises a critical question: how will the ocean transfer more heat to the atmosphere so that atmospheric transport can increase if oceanic transport from the tropics decreases?

---

[80] Herweijer, C., et al, 2005. Tellus A: Dyn. Meteorol. Oceanogr. 57 (4), pp.662–675.
doi.org/10.3402/tellusa.v57i4.14708
[81] Outten, S., et al., 2018. J. Clim. 31 (21), pp.8745–8760.
doi.org/10.1175/JCLI-D-18-0058.1

In summary

Poleward heat transport is often overlooked as a cause of recent climate change despite its importance and the lack of a widely accepted theory of heat transport. This is because the heat transport within the climate system does not change its total energy. Moreover, climate models assume that changes in atmospheric and oceanic heat transport balance each other out, but this assumption lacks empirical support. However, changes in poleward heat transport can contribute to climate change if they affect the radiation balance at the top of the atmosphere by changing albedo and outgoing thermal radiation. This is likely to occur if the magnitude of the heat transport changes.

SECTION 3 KEY ISSUES

Differences in insolation between the equator and the poles create a latitudinal temperature gradient. This gradient determines the climatic state of the planet, which is currently in an "icehouse" condition. It also establishes a transport of heat, moisture, clouds, and angular momentum from the equator to the poles along the meridians, or meridional transport.

Meridional heat transport is critical to energy redistribution, but it lacks a proper theory, remains poorly understood, and is inadequately reproduced by models. The atmosphere is its primary driver, while oceanic transport contributes about one-third of the total transported heat. The ocean transfers most of its transported heat to the atmosphere at the midlatitude western boundary currents. Midlatitude storms play a critical role in transporting heat poleward.

The strong seasonal variations in the high latitudes result in a strong seasonal oscillation in transport and atmospheric circulation, especially in the Northern Hemisphere. The Arctic heat budget configures it as the largest heat sink to space in winter. The polar vortex strongly limits heat loss at this time but is weakened by the action of atmospheric waves.

Meridional heat transport is ignored as a cause of climate change because it doesn't change the energy content of the climate system. Climate models assume, without evidence, that changes in atmospheric and oceanic heat transport balance each other out. However, changes in poleward heat transport can contribute to climate change if they affect the radiation balance at the top of the atmosphere. This is likely to occur if the magnitude of the heat transport changes.

SECTION 4. HEAT TRANSPORT BY THE ATMOSPHERE AND THE OCEAN

CHAPTER 13
TROPOSPHERIC TRANSPORT

Most of the heat in the atmosphere that moves toward the poles occurs in the troposphere, primarily over ocean basins. The Hadley circulation is essentially a cold season feature that mainly moves heat upward to be released as outgoing radiation. While it does transport some sensible heat toward the poles, its effectiveness is greatly reduced by the equatorward transport of latent heat. This shifts the primary responsibility for poleward heat transport in the tropics to the ocean. In the extra-tropics, atmospheric eddies such as storms and hurricanes are the primary mechanism for transporting heat in the troposphere. They carry considerable amounts of latent and sensible heat from the oceans. Wind speed plays a critical role in heat transport, not only because it determines the amount of transport but also because it is more important than temperature in determining the amount of evaporation. Global changes in wind speed over time are not well understood.

Tropospheric Meridional Circulation

In chapter 10, we discussed atmospheric transport and how heat transport is divided between the atmosphere and the ocean. We saw that about 2/3 of the total meridional heat transport is carried by the atmosphere, primarily through eddies resulting from atmospheric turbulence. Figure 15 (ch. 10) shows the eddy heat flux at its peak during the boreal winter. The troposphere makes up 85% of the atmospheric mass and is responsible for over 90% of the atmospheric heat transport. The stratosphere, on the other hand, is quite dry and therefore carries very little latent heat. Since the troposphere and stratosphere are different environments, we will focus on specific aspects of the meridional transport of tropospheric heat in this chapter.

The tropospheric circulation is typically divided into two parts: the zonal wind circulation, which moves parallel to the Earth's latitudes, and the meridional wind circulation, which moves parallel to the Earth's longitudes. The zonal circulation consists of easterly and westerly winds, while the meridional circulation consists of northerly and southerly winds. The strength of the tropospheric circulation varies throughout the year, becoming stronger during the cold season when more heat must be transported to the poles. Conversely, it becomes weaker during the warm season. However, the zonal and meridional circulations are regionally anticorrelated since an increase in wind dominance from one direction must come at the expense of other directions.

The meridional structure of the mean tropospheric circulation is usually described by three overturning cells (fig. 20). The Hadley and Polar cells are thermally driven by air rising in a warmer zone and falling in a colder zone. They are fairly strong circulations and produce prevailing winds such as the trade winds. The Ferrel cell, on the other hand, is mechanically driven by air rising in a colder zone and falling in a warmer zone. As a result, it is much weaker, and the winds it produces are more variable.

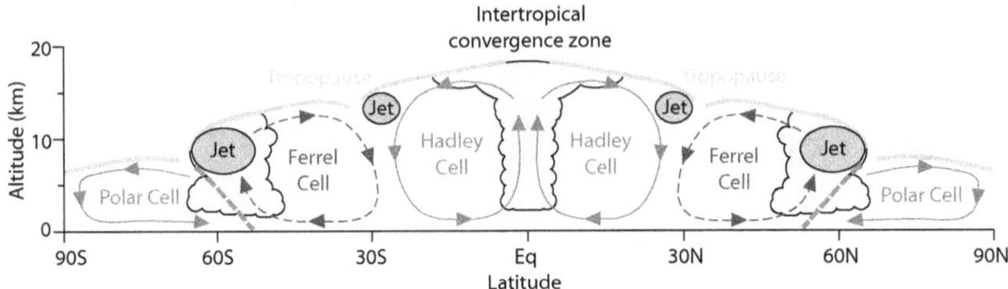

Figure 20. Schematic representation of the tropospheric mean meridional circulation near an equinox.

However, this symmetrical description of the tropospheric circulation with respect to the equator is only true for the months near the equinoxes. For the rest of the year, there is a strong winter hemisphere Hadley cell, with air rising in the summer hemisphere and descending in the winter hemisphere. The summer hemisphere Hadley cell is very weak in the Southern Hemisphere and imperceptible in the Northern Hemisphere. As a result, during the boreal winter (December-February), the Northern Hemisphere Hadley cell rises on average at 12°S and descends at 31°N. In contrast, during the boreal summer (June-August), there is no discernible Northern Hemisphere Hadley cell.

Changes in air temperature affect its humidity, causing clouds to form where the air rises and preventing clouds from forming where the air falls. Let's follow the journey of an air parcel in the Hadley cell to understand its behavior. As the parcel moves up the ascending branch, it cools and loses moisture. Once the air reaches the upper troposphere, it loses energy through thermal emission as it moves poleward through the upper branch. This happens because infrared opacity is much lower at higher altitudes in the troposphere. This mechanism, known as outgoing longwave radiation, is the primary way the tropics lose energy. It creates a poleward horizontal temperature gradient in the upper troposphere, causing surfaces of equal pressure to tilt, rising toward the equator and sinking toward higher latitudes. This, combined with the Coriolis effect of the Earth's rotation, creates a westerly (west to east) wind that increases in speed with altitude.[82] At latitudes where the temperature gradient is stronger, namely 30° and 60°, two strong westerly winds develop in the upper troposphere, known as the subtropical jet and the jet stream. However, the speed and trajectory of these winds are strongly influenced by atmospheric waves, as explained in box 7 (ch. 11).

An air parcel from the Hadley cell is likely to move eastward along the subtropical jet until, once sufficiently cooled, it begins to sink in the descending branch. During this process, the air parcel loses its moisture content and warms adiabatically (without receiving heat) through compression. This is the same process that heats up a bicycle tire when we inflate it. As the air descends and warms, it reduces its relative humidity and suppresses clouds. This very dry and warm air descending at 30° latitude is responsible for creating a belt of deserts on the planet at that latitude in both hemispheres. The air parcel eventually joins the lower branch of the Hadley cell and moves toward the equator, with the trade winds. As it moves mainly over the ocean basins, it gains a lot of

---

[82] This effect is called thermal wind balance.

moisture by evaporation and transports a significant amount of latent heat toward the equator.

The Hadley cell becomes an inefficient meridional heat transport mechanism by transporting dry static heat (potential heat + sensible heat) poleward and latent heat equatorward. Its net transport capacity is only about 10% of its potential energy transport.[83] The main effect of the Hadley cell is to transport energy upward, allowing it to be radiated as outgoing thermal energy from the upper troposphere.

Decomposition of atmospheric transport

Figure 21 shows the separation of atmospheric heat transport into two components: the energy stored in water vapor due to its changes in state, known as latent heat, and the internal energy of air molecules due to their temperature and potential energy due to their position in the gravitational field, known as dry static heat. This separation provides valuable insight into the planet's climate.

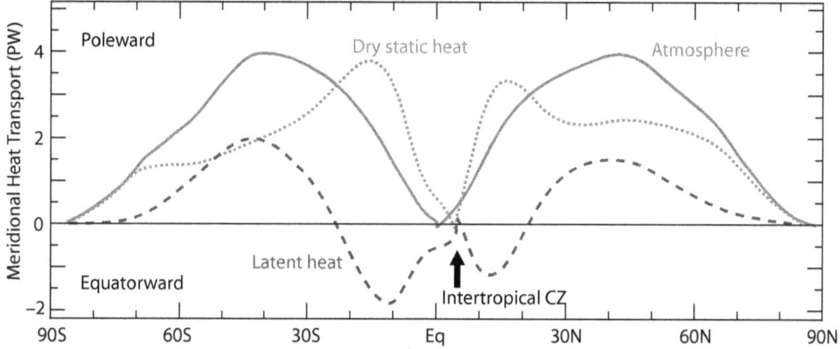

Figure 21 Atmospheric meridional heat transport decomposition. The total transport (solid curve) is decomposed into the dry static heat transport (dotted curve) and the latent heat transport (dashed curve). The poleward direction is considered from the equator for the total transport and from the Intertropical Convergence Zone for the decomposed transports, where their value is the smallest.[84]

Decomposing the atmospheric transport into dry static heat transport and latent heat transport, we can see that the Southern Hemisphere Hadley cell is much stronger than its northern counterpart, despite a similar net amount of poleward heat transport. This difference is probably due to the northern mean position of the Intertropical Convergence Zone, which splits the atmospheric transport. We can also observe that the opposing nature of the two heat transport components greatly reduces the poleward heat transport through the atmosphere in the 20°S-N tropics, where the solar energy input is greatest. This intrinsic limitation of the Hadley circulation on any planet with oceans means that the ocean must transport most of the poleward heat in the tropics. The consequences are far-reaching, and the entire El Niño-Southern Oscillation phenomenon responds to the need to transport heat poleward under low tropical

---

[83] Hartmann, D.L., 2016. Global physical climatology. 2nd ed. Elsevier. p.175.
[84] Figure after Yang, H., et al., 2015. Clim. Dynam. 44, pp.2751–2768.
   doi.org/10.1007/s00382-014-2380-5

atmospheric efficiency. When too much heat accumulates in the equatorial subsurface of the Pacific Ocean, it becomes the best predictor of El Niño, which we'll discuss in more detail in chapter 18.[85]

Latent heat transport is less in the Northern Hemisphere due to the smaller ocean surface area. However, this is offset by greater sensible heat transport from atmospheric eddies such as storms and hurricanes, which are crucial in transporting heat toward the poles. These cyclones extract large amounts of latent and sensible heat from the ocean, especially along the western boundary currents (fig. 16, ch. 10), and transport this heat toward the pole and east. In fact, these cyclones can reach the Arctic in the Northern Hemisphere and are a major cause of Arctic warming during winter.

Box 9. Evaporation and wind speed

Since latent heat transport is so important to the distribution of heat on the planet, let us look at how the atmosphere obtains this heat through evaporation. The Enhanced CO_2 Effect hypothesis of global warming focuses all the attention on the effect of temperature on evaporation, but this is only part of the story. The Clausius-Clapeyron relation specifies the temperature dependence of the water vapor pressure at the transition between two phases. It states that the water-holding capacity of the atmosphere increases by about 7% for every 1°C increase, which is important at the microscale. At the macroscale, however, evaporation saturates rapidly, and it turns out that evaporation depends more on the relative humidity of the air and even more on the wind speed that brings unsaturated air closer to the surface. Anyone who hangs clothes out to dry in the wind knows this. Despite its importance for the hypothesis, observations have not confirmed that water vapor behaves as a strong positive feedback. Wind speed plays a dual role in heat transport as a provider of mechanical energy that moves heat and as a primary cause of evaporation.

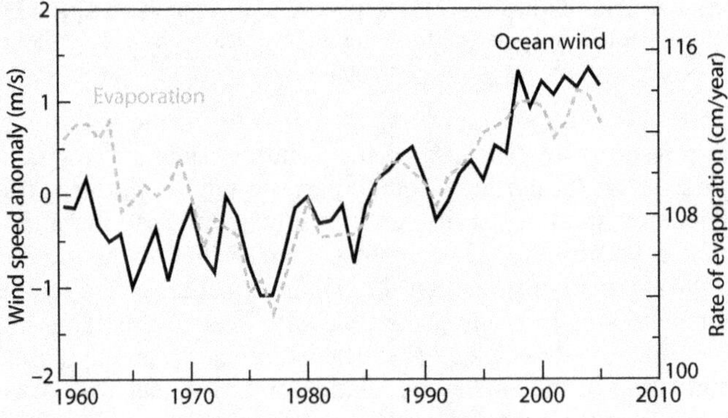

Figure B9. Evaporation and wind speed. Time series of mean annual wind speed anomaly (solid black curve) and mean evaporation rate over ice-free oceans.

---

[85] Izumo, T., et al., 2019. Clim. Dyn. 52, pp.2923–2942.
doi.org/10.1007/s00382-018-4313-1

The correlation between evaporation, the main source of latent heat, and wind speed over the oceans is consistent with the important role of wind in evaporation.[86] Figure B9 shows an increase in ocean wind speed and evaporation that is striking for several reasons. First, it implies that the magnitude or partitioning of the components of heat transport changes with time. Second, a shift in wind speed and evaporation occurred after 1976. Third, the trends in wind speed and evaporation over land are opposite to those over the ocean.[87] Unsurprisingly, models cannot explain or even reproduce the observed trends and changes in wind speed and evaporation, let alone shed light on their effect on meridional heat transport.

In summary

Most of the poleward heat transport occurs through tropospheric transport over the ocean basins. However, the Hadley circulation is less effective in poleward heat transport because it transports a significant amount of latent heat equatorward. In the Northern Hemisphere, heat transport by storms and hurricanes is critical because the smaller size of the oceans reduces their latent heat transport. Wind speed plays a dual role in heat transport; it provides the mechanical energy that moves heat and is an important cause of evaporation. Although ocean wind speed and evaporation are correlated, climate models are unable to explain the observed multidecadal trends in wind speed.

---

[86] Yu, L., 2007. J. Clim. 20 (21), pp.5376–5390. Source for figure B9.
doi.org/10.1175/2007JCLI1714.1
[87] Zeng, Z., et al., 2019. Nat. Clim. Change, 9 (12), pp.979–985.
doi.org/10.1038/s41558-019-0622-6

Chapter 14
Stratospheric Transport

The stratosphere is one of the least understood parts of the climate system. Air from the troposphere rises into the stratosphere in the tropics, where it cools and dries. In the lower stratosphere, it goes to both poles, but in the upper stratosphere, it goes from the summer to the winter pole. There, the presence of the polar vortex and strong radiative cooling in the absence of solar radiation result in temperatures of –80°C (–112 °F).

Planetary-scale atmospheric waves originate in the troposphere mostly at 30-60°N. They move vertically when the stratospheric winds move eastward, providing energy and momentum in the stratosphere. They induce a drag that drives the meridional transport of heat, air, and constituents, known as the Brewer-Dobson circulation.

Strong winds surround the Earth above the equator. They form in the middle stratosphere and slowly descend toward the tropopause. About two years later, a new belt of winds blowing in the opposite direction develops above the previous one in a Quasi-Biennial Oscillation. Although a tropical phenomenon, this Quasi-Biennial Oscillation influences the polar vortex, stratospheric circulation, and tropospheric weather. The easterly phase of the oscillation increases the effect of atmospheric waves on the polar vortex, weakening it and causing cold winters in northern Europe and the eastern United States.

The stratosphere

The stratosphere is a peculiar place that has historically surprised atmospheric scientists and is still poorly understood. Its importance for weather forecasting is a topic of interest, and its modeling is inadequate. The stratosphere exists because of ozone, which is formed from oxygen by the Sun's UV radiation. No other known planet has a stratosphere because none has free oxygen in its atmosphere.

Ozone absorbs solar energy in the UV and infrared parts of the spectrum, heating the stratosphere from above, as opposed to the troposphere, which is heated from below. As a result, the stratosphere has a negative lapse rate, and the temperature increases with altitude. This makes the stratosphere stably stratified, much like the ocean. Cooler air at the bottom does not rise and warmer air at the top does not sink. Because of the negative lapse rate, vertical transport in the stratosphere is very inefficient, and it takes months for a parcel of air to rise a few kilometers. The tropopause marks the boundary where the lapse rate becomes zero.

Air enters the stratosphere through the tropical tropopause, where it is sucked in by an extratropical pump responsible for meridional transport in the stratosphere (see below). Air leaves the stratosphere at the extratropical tropopause by sinking. As the air rises through the tropical tropopause, it expands and cools as the pressure decreases. This makes the tropical tropopause the coldest part of the tropopause (fig. 22). In the rest of the tropopause, warmer air descends into the troposphere. The tropical cold-point tropopause has two important effects. First, it causes poleward transport in the lower stratosphere against the temperature gradient until the midlatitudes are reached. Second, it cools the rising air to the point of freeze-drying so that most of the water vapor precipitates as ice, resulting in extremely dry air entering the stratosphere.

Much of the water vapor in the stratosphere comes from the oxidation of methane in the upper stratosphere, and the amount of water vapor in the stratosphere increases with altitude (ch. 7, fig. 9). Scientists do not understand well the changes in the amount of water vapor in the stratosphere and their effect on the rate of global warming. Stratospheric water vapor increased from 1980 to 2000 and decreased from 2000 to 2022 when the Hunga Tonga eruption abruptly increased stratospheric water vapor levels (ch. 24). Climate models do a poor job of simulating temperature trends in the lower stratosphere and do not consistently reproduce tropical tropopause temperatures and water vapor changes.[88]

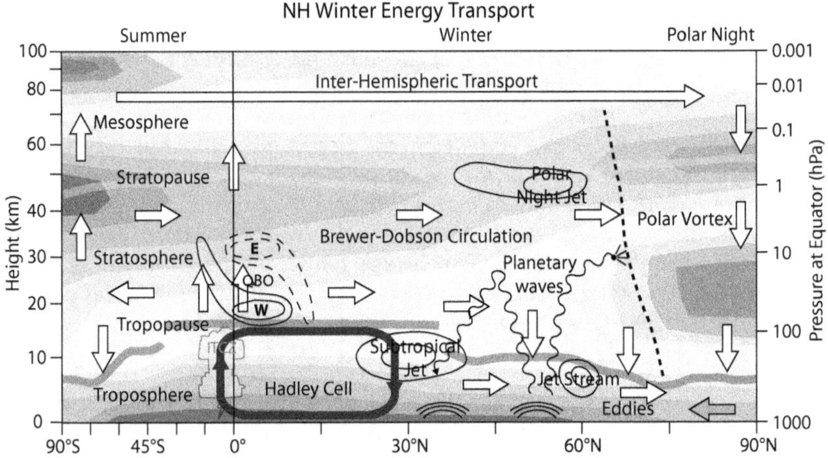

Figure 22. The atmospheric circulation around the December solstice. Temperature is gray tone-coded in 10°C increments, with a logarithmic vertical scale and a compressed latitude scale for the Southern Hemisphere. The Quasi-Biennial Oscillation (QBO) is shown with its easterly and westerly components near the equator. The Intertropical Convergence Zone (ITCZ) is shown as a high, stormy cloud. White arrows indicate heat transport by the atmospheric circulation.[89]

The meridional atmospheric circulation has a different structure with increasing altitude. Air rises from the troposphere into the lower stratosphere and moves toward both poles in a two-cell structure. Sinking occurs at mid and high latitudes. In the upper stratosphere, the circulation follows a single-cell pattern, moving from the tropics to the winter pole. Higher in the mesosphere, the air flows from the summer pole to the winter pole.

Normally, air rising at the summer pole should cool, but in the stratosphere it is exposed to UV absorption from ozone, which warms it strongly. Conversely, air sinking into the polar vortex at the winter pole should warm, but in

---

[88] Solomon, S.et al., 2010. Science, 327 (5970), pp.1219–1223.
doi.org/10.1126/science.1182488

[89] Figure after Vinós, J., 2022. Climate of the Past, Present and Future: A scientific debate. 2nd ed. Critical Science Press.

the middle stratosphere, strong radiative cooling results in temperatures as low as –80°C (–112 °F) due to the absence of solar radiation.

Horizontally, the stratosphere is divided into three zones: a tropical zone of rising air, a midlatitude zone of atmospheric wave breaking known as the surf zone, and the polar vortex.

Box 10. Atmospheric waves

The atmosphere exhibits a variety of wave motions with different spatial and temporal scales. These waves can carry large amounts of energy and momentum to distant locations in much less time than it would take for air to travel. An example in the ocean would be a tsunami, which can carry a lot of energy across the ocean in just a few hours. Atmospheric waves in the stratosphere are responsible for meridional transport, the Brewer-Dobson circulation, the Quasi-Biennial Oscillation, ozone transport, polar vortex asymmetries, polar temperatures, and sudden stratospheric warmings.

Waves are produced by the interaction of a force and a restoring mechanism, and depending on these and their size scale (wavelength), they are classified into different types. The type we are interested in for stratospheric heat transport and the new climate change hypothesis presented in this book is a Rossby wave called a planetary wave.

The potential temperature is the temperature that the air would be if it were brought to the surface, that is, if changes due to vertical motion are not considered. As we move toward the poles, there is a negative horizontal gradient in the potential temperature and a positive gradient in the spin of the air masses (vorticity) due to the increase in the Coriolis effect with distance from the equator. These two properties together form the potential vorticity, which is a conserved property. Rossby waves have the latitudinal gradient of potential vorticity as a restoring mechanism, so when air masses are forced to move, their need to conserve potential vorticity forces them to change their vorticity and enter a wave motion.

Importantly, large-scale orographic features, large eddies, and land-ocean contrasts produce wavy Rossby wave patterns around the globe in the Northern Hemisphere in the 30° to 60°N band. In contrast, the scarcity of prominent orography in the Southern Hemisphere results in a more zonal flow with fewer Rossby waves.

Planetary waves are a type of Rossby wave with very long wavelengths that can propagate vertically into the stratosphere and higher if they are large enough. These waves can only propagate upward under moderate to weak westerly zonal winds, which typically occur in winter. Easterly or strong westerly winds suppress their propagation. The wavenumber of a planetary wave expresses the number of wavelengths that fit in a complete circle around the globe at a given latitude. For example, at 60°N, a planetary wave of wavenumber 2 has two crests and two troughs over 360 degrees, with a wavelength of 10,000 km (6,200 miles). Only planetary waves of wavenumbers 1 and 2 can reach the stratosphere.

Once the planetary waves reach the stratosphere, they deposit their easterly momentum into the westerly zonal flow, slowing it down. This reduction in the stratospheric zonal flow weakens the polar night jet surrounding the stratospheric polar vortex and drives the meridional flow that transports heat and ozone poleward into the vortex. In addition to driving heat transport in the stratosphere, planetary waves also play an important role in keeping the Arctic warmer than Antarctica and preventing the formation of an ozone hole in the Northern Hemisphere.

A stronger wave flux in the Northern Hemisphere results in a stronger Brewer-Dobson circulation, which weakens the polar vortex and leads to warmer polar temperatures. In addition, the stronger circulation transports more ozone into the lower polar stratosphere in the north, resulting in higher ozone levels there. The hemispheric asymmetry in wave activity profoundly affects climate and ozone distribution.

The Brewer-Dobson circulation

The Brewer-Dobson circulation is a meridional circulation pattern that plays a key role in transporting mass and heat in the stratosphere from the equator to each pole. This circulation pattern consists of two cells in the annual mean, but during the solstices, most of the circulation in the middle and upper stratosphere is directed toward the winter pole, as shown in figures 22 and 23.

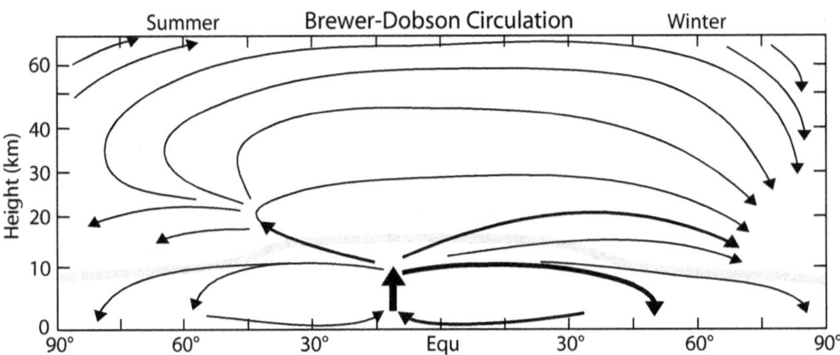

Figure 23. The Brewer-Dobson circulation. This is the meridional atmospheric circulation that takes place in the stratosphere above the tropopause (thick gray line). Most of the circulation occurs toward the winter pole.

Planetary-scale atmospheric waves drive the Brewer-Dobson circulation. These waves create a westward drag by reducing the eastward flow of wind, resulting in a poleward pumping action to conserve angular momentum.[90] In other words, each wave acts like the motion of an oar propelling a boat. Because wave activity is stronger in the Northern Hemisphere, the stratospheric circulation is stronger on average in that hemisphere (box 10). In addition to transporting heat to the winter pole, the Brewer-Dobson circulation is respon-

---

[90] Butchart, N., 2014. Rev. Geophys. 52 (2), pp.157–184. Also, the source for figure 23. doi.org/10.1002/2013RG000448

sible for exchanging stratospheric air with the troposphere and transporting and distributing constituents such as ozone, water vapor, and anthropogenic halogens.

Models predict that the Brewer-Dobson circulation will strengthen as a result of global warming. A warmer troposphere leads to a rising tropopause and wave breaking at higher altitudes, increasing drag. However, observations do not support this prediction. Until about 1995, there was no statistically significant strengthening of the lower stratospheric circulation. Then, since 1995, it has strengthened in the Northern Hemisphere, coinciding with a strong cooling of the tropical tropopause.[91] The 2018 Coupled Model Intercomparison Project 6, which includes diagnostic tools for the Brewer-Dobson circulation, confirms the inconsistency between observations and models in the middle and upper stratosphere.[92] This inconsistency is an example of a prediction from the Enhanced CO_2 Effect hypothesis that is not confirmed.

The Quasi-Biennial Oscillation

Another surprise was the discovery in the 1950s of the Quasi-Biennial Oscillation of the strong zonal winds that surround the Earth in the tropical stratosphere. These winds exhibit a cycle of alternating easterly and westerly orientation with an average period of 28 months. The wind regimes originate in the middle stratosphere and move downward at a rate of about 1 km (3,300 ft) per month until they dissipate in the tropical tropopause (fig. 22). About two years after one wind regime has formed and begun to descend, an opposing wind regime develops above it.

The Quasi-Biennial Oscillation is a tropical phenomenon that affects the global stratosphere, driven by atmospheric waves from tropical convection. It affects winds, temperature, extratropical waves, meridional wind circulation, transport of chemical constituents, and ozone distribution. During easterly phases, the jet stream weakens, resulting in cold winters in northern Europe and the eastern United States. El Niño winters also tend to coincide with the easterly phase. On the other hand, westerly phases strengthen the jet stream, leading to mild and wet winters in northern Europe and the eastern United States. The oscillation also affects the frequency of hurricanes in the Atlantic. Despite its many effects, its influence on tropospheric weather is not fully understood.

The Quasi-Biennial Oscillation modulates the Northern Hemisphere polar vortex (ch. 11, box 7), which is a persistent, large-scale, mid-tropospheric to stratospheric low-pressure zone during winter. When the polar vortex is strong, it contains a large mass of very cold, dense Arctic air; when it is weak and disorganized, it allows cold Arctic air masses to push southward, causing large temperature drops over much of the Northern Hemisphere. This modulation of the polar vortex is known as the Holton-Tan effect and is one of the most puzzling aspects of the Quasi-Biennial Oscillation.

---

[91] Young, P.J., et al., 2012. J. Clim. 25 (5), pp.1759–1772.
   doi.org/10.1175/2011JCLI4048.1
[92] Abalos, M., et al., 2021. Atmospheric Chem. Phys. 21 (17), pp.13571–13591.
   doi.org/10.5194/acp-21-13571-2021

Box 11. The Holton-Tan Effect

Another unexpected discovery in 1980 was the synchronization of the equatorial Quasi-Biennial Oscillation with the strength of the northern stratospheric winter polar vortex. During the easterly phase of the oscillation, the polar night jet winds around the polar vortex weaken, resulting in significantly warmer Arctic ice cap temperatures and higher polar stratospheric geopotential heights compared to the westerly phase. Holtan and Tan proposed that this effect is due to a 10° latitudinal shift toward the winter hemisphere of the boundary surface separating the westerly and easterly zonal winds during the easterly phase of the oscillation. This shift narrows the propagation channel of planetary waves and redirects their flow toward the winter pole since these waves can only travel through the westerly wind. In essence, the strength of the polar vortex is weaker during the easterly phase and stronger during the westerly phase of the oscillation.

In the late 1990s, researchers discovered that the Quasi-Biennial Oscillation induces a meridional circulation in the extratropics of the lower stratosphere in the winter hemisphere, driven by planetary wave drag similar to the Brewer-Dobson circulation. In the easterly phase of the oscillation, the intrusion of easterly winds from the tropics creates a barrier to planetary waves, causing them to converge more strongly on the polar vortex. The modulation of planetary wave propagation by the Quasi-Biennial Oscillation and the effect of the oscillation-induced extratropical circulation on wave convergence are two complementary mechanisms that explain the tropical-polar coupling that leads to a weakened vortex in the Northern Hemisphere winter during the easterly phase of the oscillation.[93]

Scientists have been aware of this phenomenon for over 40 years, but this long-range relationship is complex. While climate models have slowly improved in their representation of the Quasi-Biennial Oscillation, the recent Coupled Model Intercomparison Project 6 shows that models still underestimate the Holton-Tan effect, and each model represents the relationship between the oscillation and the polar vortex differently.[94] In future chapters, we'll explore why the inability of climate models to accurately reproduce the dynamic properties of the winter stratosphere hinders their ability to solve the climate puzzle or reconstruct Earth's past climates.

In summary

Most of the heat transported in the stratosphere flows toward the winter pole. This transport is facilitated by a circulation system driven by the momentum of atmospheric waves, which act as a pump. The equatorial winds in the stratosphere change direction about every two years, creating a different circulation regime that significantly affects the troposphere. During the easterly

[93] Ruzmaikin, A., et al., 2005. J. Geophys. Res. Atmos. 110, D11111.
   doi.org/10.1029/2004JD005382
[94] Elsbury, D., et al., 2021. Geophys. Res. Lett. 48 (24), p.e2021GL094083.
   doi.org/10.1029/2021GL094083

phase of this oscillation, atmospheric wave activity is directed toward the polar vortex, weakening it. This effect is more pronounced in the Northern Hemisphere, where wave activity is generally greater. As a result, the northern polar vortex is weakened and the Arctic experiences warmer temperatures during the winter. A warmer Arctic means that more heat is lost through radiative cooling.

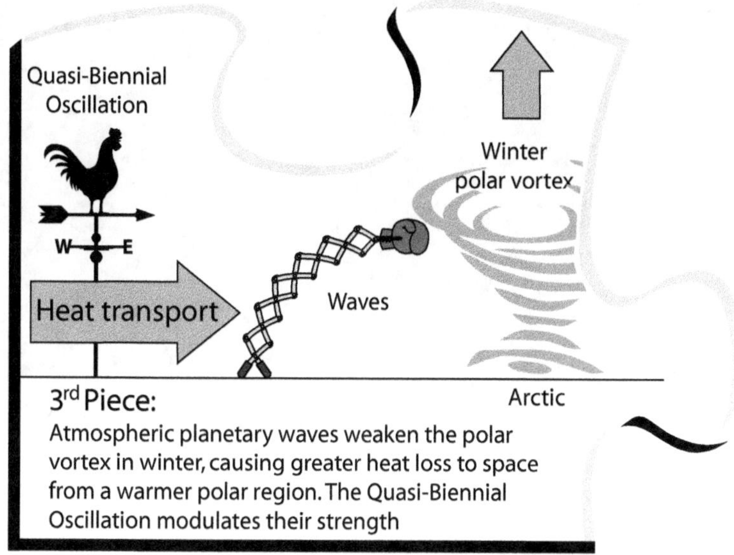

Chapter 15
Stratosphere-Troposphere Interactions

In 1999, researchers made the surprising discovery that anomalies in the stratosphere can propagate to the Earth's surface. This discovery marked the beginning of studies of what is now known as the stratosphere-troposphere coupling. These studies have shown that changes in the stratosphere significantly affect the position of tropospheric jets and storm tracks, which strongly influence sea level pressure and surface weather during the cold season. These effects are achieved through changes in the annular modes, the main patterns of climate variability in the mid- and high-latitude troposphere. However, despite all this research, the multi-decadal trends observed in the annular modes still haven't been adequately explained. Interestingly, the stratosphere-troposphere coupling responds not only to low-frequency climate variability but also to the solar cycle. Over the past two decades, researchers have made great strides in understanding how the stratosphere responds to changes in solar activity and how these responses can propagate to the Earth's surface.

Stratosphere-troposphere coupling

When the model-based Enhanced CO_2 Effect hypothesis was developed in the 1960s, and for several decades thereafter, it wasn't even considered that changes in the stratosphere could significantly affect climate at the Earth's surface. In 1998, however, scientists defined the Arctic Oscillation (also known as the Northern Annular Mode, box 12) as a circular pattern of sea level pressure anomalies centered around the North Pole. They also discovered that the Arctic Oscillation is strongly linked to the stratosphere and that anomalies originating in the stratosphere can propagate to the surface. This was a surprising finding, as previous research had shown no significant response in the lower troposphere to changes in the stratosphere.[95]

In the two decades since this discovery, researchers have found that changes in the stratosphere affect the Arctic Oscillation, causing shifts in Atlantic and Pacific jets and storm tracks and changes in sea level pressure. These effects in the lower troposphere lag behind changes in the stratosphere, indicating downward propagation. As a result, understanding the stratospheric weather has become critical for forecasting the mid-range and seasonal surface weather. There is evidence that the location of winter tropospheric jet streams, storm tracks, and pressure centers in the Northern Hemisphere depend on the stratospheric jet speed, which in turn is determined by the propagation properties of the stratosphere to planetary waves.[96]

---

[95] Baldwin, M.P. & Dunkerton, T.J., 1999. J. Geophys. Res. Atmos. 104 (D24), pp.30937–30946. doi.org/10.1029/1999JD900445

[96] Kidston, J., et al., 2015. Nature Geosci. 8 (6), pp.433–440. doi.org/10.1038/NGEO2424

Most studies of stratosphere-troposphere coupling focus on short-term changes, but a more interesting question in the context of climate change is whether this coupling responds to longer-term natural or anthropogenic variability. Recent model studies suggest that it responds to multidecadal ocean variability.[97] However, the best evidence comes from observed multidecadal changes in the Arctic Oscillation (box 12).

Over time, the Northern Hemisphere has experienced several coherent wintertime multidecadal climate trends in the stratosphere, troposphere, ocean, and cryosphere. These trends are generally attributed to anthropogenic climate change because there is little interest in attempting to disprove the Enhanced CO_2 Effect hypothesis, as the scientific method would require.[98] However, an alternative explanation has been proposed in which a coupled stratosphere/troposphere/ocean low-frequency oscillation is responsible for the observed trends.[99] In this oscillation, a positive Northern Annular Mode and associated stratospheric cooling initiates a delayed thermohaline strengthening of the Atlantic overturning circulation and extratropical Atlantic gyres, thereby increasing poleward oceanic heat transport, leading to melting of Arctic sea ice, amplification of Arctic warming, and large-scale Atlantic warming, which in turn initiates the wave-induced negative Northern Annular Mode and stratospheric warming, reversing the phase of the oscillation.

This interpretation of the low-frequency oscillatory changes in the stratosphere-troposphere coupling is consistent with the main hypothesis presented in this book and previously reported by me.[100]

Box 12. North Atlantic Oscillation or Northern Annular Mode?

To simplify the dynamic complexity of the atmosphere, scientists have identified repeating patterns of variability over time. Annular modes are the most important patterns of climate variability in the mid and high latitudes of the Northern and Southern Hemispheres. They refer to the north-south shift of a belt of strong westerly winds that accounts for 20-30% of the hemispheric variance in pressure and wind fields. These modes play a critical role in meridional heat transport, regulating the exchange of atmospheric mass between the polar regions and the midlatitudes. Changes in the annular modes significantly impact climate, particularly during the winter season in the Northern Hemisphere and the spring season in the Southern Hemisphere, when coupled with stratospheric annular variability.

The Southern Annular Mode has been recognized as an annular mode since 1999 because its three centers of action – one polar and two peripheral – act as a seesaw

---

[97] Elsbury, D., et al., 2019. J. Clim. 32 (14), pp.4193–4213.
doi.org/10.1175/JCLI-D-18-0422.1

[98] Popper, K.R., 1962. Conjectures and Refutations. The growth of scientific knowledge. Basic Books, New York.

[99] Omrani, N.E., et al., 2022. NPJ Clim. Atmos. Sci. 5 (1), p.59.
doi.org/10.1038/s41612-022-00275-1

[100] Vinós, J., 2022. Climate of the Past, Present and Future: A scientific debate. 2nd ed. Critical Science Press.

between the pole and midlatitudes. In the Northern Hemisphere, the Northern Annular Mode was first described in the 1920s as the North Atlantic Oscillation between pressure centers over Iceland and the Azores. In 1998, it was extended to include the Arctic and the North Pole pressure center and was named the Arctic Oscillation. However, there is some debate as to whether the North Atlantic Oscillation or the Northern Annular Mode better describes the pattern of pressure variability. The problem is that the Atlantic and Pacific centers lacked the necessary coordination for several decades, even though they showed it in the following or preceding decades (fig. B12). Thus, from the early 1970s onward, the pressure pattern best fit the definition of the North Atlantic Oscillation for a period of about 25 years, while in the preceding and following quarters of the century, it showed a Northern Annular pattern.

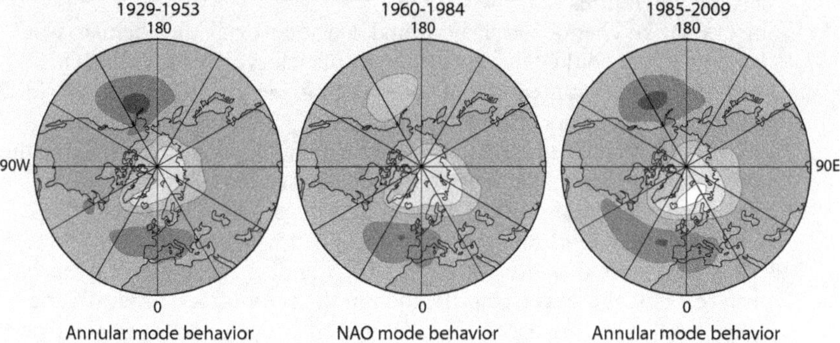

Figure B12. Winter-mean sea-level pressure anomalies for three 25-year periods. The tone scale is 1.5 hPa (positive in dark gray). NAO: North Atlantic Oscillation.[101]

During multi-decadal periods when the polar vortex is stronger than average, the Arctic, Atlantic, and Pacific sectors exhibit true Northern Annular Mode behavior. This is characterized by a seesaw relationship between the Aleutian and Icelandic Lows, which restricts heat and moisture transport to the Arctic. On the other hand, during multi-decadal periods when the polar vortex is weaker than average, the situation is better described by the North Atlantic Oscillation. This is because there is weak interannual variability of the Aleutian Low and less constrained Arctic transport. Despite the evidence supporting these findings, most scientists attempting to explain the recent trend as a response to increased anthropogenic forcing have not yet recognized the changing nature of the Northern Annular Mode/North Atlantic Oscillation.

However, some scientists recognize that the climate undergoes persistent patterns, known as climate regimes, that change from one to another through climatic shifts. This perspective, which is discussed further in Part III, section 9, is more consistent with the available evidence. However, it is not consistent with the Enhanced CO_2 Effect hypothesis.

---

[101] Figure after Shi, N. & Nakamura, H., 2014. Tellus A: 66 (1), p.22660.
  doi.org/10.3402/tellusa.v66.22660

The solar cycle and the dynamic stratosphere-troposphere coupling

For over 150 years, attempts to link the solar cycle to weather and climate changes were mostly based on surface or tropospheric observations. However, it wasn't until 1987 that Karin Labitzke discovered a relationship between the solar cycle, North Pole temperatures, and the phase of the Quasi-Biennial Oscillation in the stratosphere. This finding is discussed further in chapter 29 (box 23). Since 1987, advances in four areas have provided evidence for a solar influence on the atmosphere:[102]

* Strong statistical relationships are observed in the data record, not only in the stratosphere but also in the lower troposphere, surface, and upper ocean temperatures.
* From a radiative-chemical transport model, a solar-ozone mechanism was developed that could explain changes in planetary wave activity. This showed that the simple ideas of solar forcing energy balance, as considered by the IPCC, may be misleading.
* Model simulations have reproduced a global annual mean solar signal in the upper stratosphere that agrees well with observations, but they fail to adequately simulate seasonal patterns and the observed response in the lower stratosphere. Better results are obtained with reanalysis products.
* Substantial progress has been made in understanding the dynamical mechanism responsible for the amplification of the solar effect. Stratospheric circulation anomalies associated with the solar cycle move poleward and downward during the winter season, associated with anomalies in wave-induced momentum transport.

The solar cycle affects not only the stratosphere but also the surface through stratosphere-troposphere coupling. This phenomenon is mediated by the response of ozone to changes in UV radiation. To fully understand the solar effect on climate, it's critical to understand the meridional heat transport and planetary wave flux in the Northern Hemisphere winter stratosphere.

In summary

Recent studies have shown that the stratosphere plays an important role in determining mid to high-latitude surface temperature and pressure patterns during winter and in influencing the location of tropospheric jets and storm tracks. Stratospheric variability and polar vortex strength are associated with the annular modes, a westerly ring-like wind pattern, and north-south pressure differences around the poles that regulate poleward heat transport and mass exchange. However, these modes show multidecadal trends that models cannot fully capture. During winter, dynamic changes in the Arctic stratosphere occur that are strongly influenced by the solar cycle despite the absence of solar irradiance. This indicates that a solar effect on climate must act through changes in atmospheric circulation.

---

[102] Baldwin, M.P. & Dunkerton, T.J., 2005. J. Atmos. Sol. Terr. Phys. 67(1-2), pp.71–82. doi.org/10.1016/j.jastp.2004.07.018

CHAPTER 16
WINTER TRANSPORT TO THE ARCTIC

Atmospheric circulation and heat transport are more intense in the Northern Hemisphere during the cold season, even though the Arctic is warmer than the Antarctic. This phenomenon is due to higher atmospheric wave activity resulting from geographical differences. Wave activity increases stratospheric transport and weakens the northern polar vortex. The winter transport has several consequences, including that the Earth rotates faster in winter and the rotation period is slightly shortened. In addition, the northern polar vortex becomes weaker and more variable, and the winter heat transport is influenced by several factors, such as the Quasi-Biennial Oscillation, the El Niño-Southern Oscillation, and the solar cycle, as they affect the strength of the vortex. Typically, a few extreme events per season account for most of the heat transported to the Arctic in winter, and the frequency of such events depends on the circulation blocking caused by wave activity. While the contribution of the stratosphere to heat transport is small most of the year, it contributes 20% of the heat transported to the Arctic in winter, making it the largest heat sink on the planet as heat leaves the climate system in the form of outgoing thermal radiation.

Seasonal heat transport

It has been argued in previous chapters that the conventional view of global climate as an average of the annual radiation balance over the entire top of the atmosphere, with heat transport responding only to changes in this balance, is a gross oversimplification. This view hides significant interhemispheric asymmetries and seasonal changes. Of the four elements of climate energetics – energy input from the Sun, absorption by the climate system, transport within the system, and return to space – transport is the most variable on a seasonal timescale. However, climatologists contributing to the IPCC assessment reports do not consider it to be a driver of recent climate change. In contrast, this book argues that forced changes in heat transport are a significant – and perhaps the most important – cause of climate change. If this is the case, then the contribution of anthropogenic changes in greenhouse gases to recent climate change must be smaller than is generally believed.

In the winter hemisphere, atmospheric circulation and heat transport are stronger, and surprisingly, despite the temperature difference, more heat reaches the Arctic than the Antarctic (fig. 18, ch. 11). Recent climate change has been most intense in the mid and high latitudes of the Northern Hemisphere, where the climate is most variable. The Enhanced CO_2 Effect hypothesis attributes this to the larger proportion of land area and the Arctic amplification (Antarctica does not show polar amplification). However, changes in heat transport could be an equally valid explanation since they should be most noticeable in the mid- and high latitudes of the Northern Hemisphere during winter when transport is at its largest and most variable.

To simplify the analysis of climate change in this book, we will focus almost exclusively on heat transport during the Northern Hemisphere cold season, with the understanding that changes in heat transport during other seasons and in the Southern Hemisphere should show similar trends but of lesser extent.

We have examined why the Northern Hemisphere experiences such an increase in heat transport during winter and why the Arctic is the planet's largest heat sink to space. The geography of the Northern Hemisphere plays a significant role, with its large continents and mountain ranges resulting in higher atmospheric wave flux activity, a more variable Northern Annular Mode, and a weaker and more variable Arctic polar vortex. In addition, the Atlantic Ocean is connected to the Arctic via the wide Fram Strait, which allows warmer water to enter the Arctic, further increasing the amount of heat transported.

Seasonal changes in the Earth's rotation rate

Seasonal changes in heat transport are driven by large variations in atmospheric circulation since the atmosphere is the primary means of heat transport. The atmosphere is also responsible for transporting angular momentum poleward (box 6, ch. 9). These seasonal changes in atmospheric circulation affect the exchange of angular momentum between the atmosphere and the solid Earth, leading to changes in the Earth's rotation rate. Scientists measure these changes in Earth's rotation as small differences in the length of the day, defined as the difference between the measured duration of one revolution and 86,400 standard international seconds. Since 1962, atomic clocks with microsecond accuracy have been used to measure these changes.

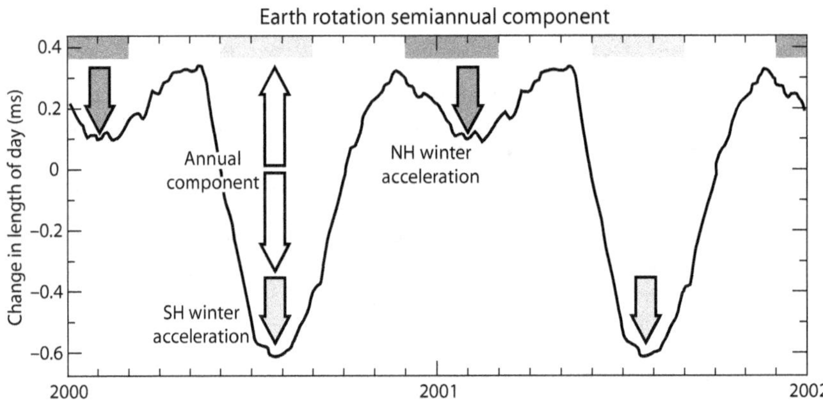

Figure 24 Seasonal changes in the Earth's rotation. The semi-annual (seasonal) component of the Earth's rotation rate is measured as changes in the length of the day in milliseconds. In medium gray is the boreal winter (NH) and in light gray is the austral winter (SH).[103]

The exchange of angular momentum between the atmosphere and the solid Earth is closely linked to changes in the atmospheric circulation and leads to a semi-annual oscillation in the length of the day. From November to January, the Earth speeds up by about 0.2 ms/day (resulting in days that are 0.2 ms shorter). In April, the Earth slows down by a similar amount before accelerating to about 1 ms/day faster than average in July. This acceleration is then fol-

---

[103] Gipson, J., 2016. IVS 2016 General Meeting Proceedings: New Horizons with VGOS. p.336.

lowed by a deceleration that brings the Earth back to its original speed in November. The semi-annual component has an average amplitude of about 0.35 ms, but the Earth rotates more slowly during the boreal winter than during the austral winter (fig. 24).

The uneven distribution of land between the hemispheres and the colder winter temperature in the Northern Hemisphere (ch. 5, fig. B3a) create an annual variation that adds to the semi-annual oscillation, resulting in more angular momentum in the atmosphere during the boreal winter. The important thing to remember is that winter heat transport leads to faster Earth rotation and slightly shorter days. In future chapters, we'll look at changes in the length of the day as an indicator of winter heat transport, specifically the boreal winter semi-annual component, as this is consistent with our focus on transport changes during this season.

The northern polar vortex and wave activity

One of the reasons why winter heat transport to the Arctic is greater than to Antarctica, despite the smaller temperature gradient, is that the northern polar vortex is weaker than the southern polar vortex. The polar region is surrounded by strong westerly winds that form the vortex and act as a barrier to meridional heat transport (fig. 59, ch. 39). As a result of the vortex, the interior temperature of the polar region is colder than it would be without the vortex or if the westerly winds were weaker. The strength of the westerly winds that form the vortex depends on the intensity of the planetary wave flux that deposits easterly momentum in the stratosphere. Vortex weakening caused by wave action propagates into the lower stratosphere.

The wave flux resulting from temperature contrasts between land and ocean and large mountain complexes is much greater in the Northern Hemisphere. Therefore, the northern polar vortex is much weaker, resulting in a higher probability of sudden stratospheric warming events. During these events, the vortex winds turn easterly and the vortex breaks up. As a result, the air is forced downward, and the temperature in the polar stratosphere can rise by 40°C (70 °F) within a few days.

These events induce a negative phase of the North Atlantic Oscillation in the troposphere, resulting in changes in storm tracks and colder temperatures in northern Eurasia and the eastern United States. Meanwhile, Greenland experiences warmer temperatures.[104] These phenomena occur every other winter in the Northern Hemisphere but only once every 20 years in the Southern Hemisphere.

The northern polar vortex is weaker and more variable than the southern. The strength of the polar vortex is a crucial factor in determining the amount of heat transported to the Arctic, and several factors, such as the Quasi-Biennial Oscillation, the El Niño-Southern Oscillation, and the solar cycle, influence its variability. Therefore, these factors can influence the amount of heat transported to the Arctic during winter.

In addition, most of the Arctic Ocean is covered with sea ice in winter, which is an excellent thermal insulator. When the ice is only 1 meter thick, it

---

[104] Baldwin, M.P., et al., 2021. Rev. Geophys. 59 (1), p.e2020RG000708.
   doi.org/10.1029/2020RG000708

reduces the heat flux from the ocean to the atmosphere by a factor of 10. In winter, heat and moisture reach the Arctic mostly through the atmosphere, with the ocean playing a secondary role.

Box 13. Atmospheric blocking and extreme intrusion events in the Arctic

During winter, a few extreme events per season associated with individual weather systems are responsible for much of the heat and moisture transport to the Arctic. Studies have shown that heat and moisture transport is closely related to large-scale atmospheric blocking that redirects cyclone tracks poleward.[105] Blocking occurs when the jet stream's capacity for the wave activity flux (a measure of meandering) is exceeded and the circulation stalls.

In winter, blocking over the Atlantic is strongly negatively correlated with the North Atlantic Oscillation. When one of these extreme intrusion events occurs, it can have a big impact on Arctic temperatures. Figure B13 shows the effect of such an event on Arctic temperatures.

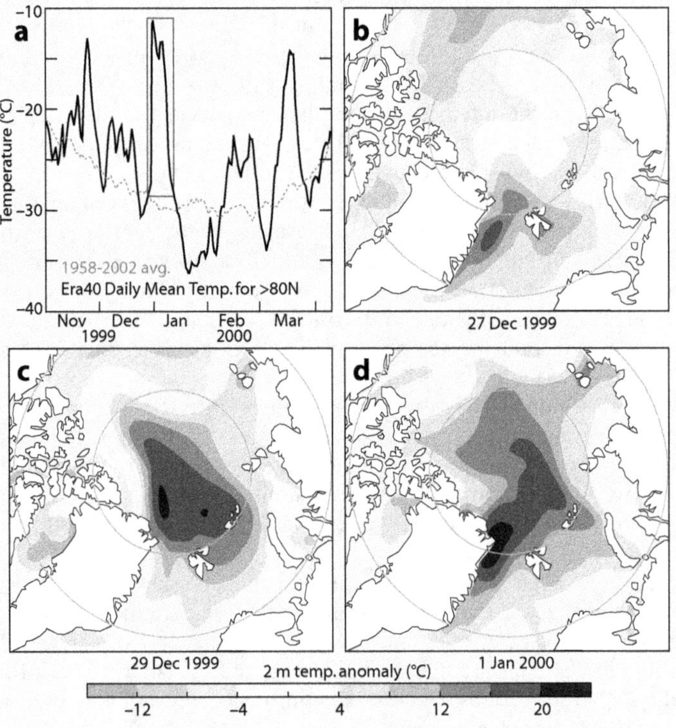

Figure B13. Intense intrusion event of moist, warm air into the Arctic in winter. a) Daily mean temperature north of 80°N for Nov 1999-Mar 2000 (black line) from the ERA40

[105] Papritz, L. & Dunn-Sigouin, E., 2020. Geophys. Res. Lett. 47 (17), p.e2020GL089769. doi.org/10.1029/2020GL089769

reanalysis, and the 1958-2002 average (gray dotted line). A rectangle marks the event. b-d) Arctic surface air temperature anomaly at different times during the intrusion event.[106]

There are three main pathways for heat and moisture transport to the Arctic: the North Atlantic (300-60°E), the North Pacific (150-230°E), and the Siberian (60-130°E). During winter, the pathways over both ocean basins are more important for meridional heat transport, with the North Atlantic pathway being the most important. These Arctic-bound transport pathways arise because large-scale blocking conditions develop to the east of each basin, redirecting midlatitude cyclones poleward.

Arctic winter energy budget and the largest heat sink

The first part of the book focuses on climate and energy, examining how heat moves within the climate system. One of the key findings is that the Arctic in winter is a unique part of the climate system. Because of its dryness and low cloud cover, the greenhouse effect is much weaker than in the tropics and mid-latitudes (fig. B4, ch. 7). In addition, the net radiative balance at the top of the atmosphere during the Arctic winter is the lowest on Earth (fig. 18, ch. 11). As a result, the Arctic is the largest heat sink on Earth.

The planet's geography leads to stronger wave activity during the Northern Hemisphere winter, which causes more stratospheric heat to be transported to the Arctic, and a weaker northern polar vortex, which allows more heat to be transported to the Arctic through the troposphere. In addition, the Atlantic Ocean efficiently transports heat to the Arctic. All of these factors contribute to the Arctic being warmer than it otherwise would be, resulting in greater energy loss from the planet in winter through outgoing thermal radiation.

The amount of heat lost from the Arctic in winter is not constant. It varies due to factors that affect the zonal atmospheric circulation, the propagation of planetary waves, and the strength of the polar vortex. These factors include the Quasi-Biennial Oscillation, the El Niño-Southern Oscillation, and solar activity. Recently, the contribution of stratospheric heat transport to the total heat transport in this region has been studied using reanalysis. This research tool combines models with an enormous amount of data on many meteorological variables from multiple sources.[107]

The results of this study are relevant to our argument that changes in the amount of heat transported to the Arctic in winter cause global climate change. The study confirms that atmospheric heat transport is the dominant factor in Arctic warming, as presented previously (fig. 14, ch. 10), while oceanic heat transport is relatively small. Although stratospheric heat transport is a small fraction of the total atmospheric heat transport due to the low mass and dryness of the stratosphere, it is responsible for 20% of the poleward heat transport at 70°N in winter, compared to only 7% in summer. Importantly, almost all of this heat is lost as outgoing longwave radiation from the top of the atmosphere, as there is little exchange of sensible and latent energy between the stratosphere

---

[106] Figure after Woods, C. and Caballero, R., 2016. J. Clim. 29 (12), pp.4473–4485. doi.org/10.1175/JCLI-D-15-0773.1 Data from the Danish Meteorological Institute.
[107] Cardinale, C.J., et al., 2021. J. Clim. 34 (11), pp.4261–4278. doi.org/10.1175/JCLI-D-20-0722.1

and the troposphere. This highlights the exceptional nature of the Arctic in winter and the importance of the Arctic stratosphere in understanding climate change.

In summary

During winter in the Northern Hemisphere, atmospheric wave activity increases, weakening the polar vortex and leading to greater heat transport to the Arctic in the stratosphere. This can also create blocking patterns in the jet stream, redirecting storms toward the Arctic. As the atmospheric circulation becomes more active, there is an increase in the transport of angular momentum, accelerating the Earth's rotation and slightly shortening the length of the day. In winter, the stratosphere contributes 20% of the heat transported to the Arctic, warming the region and leading to greater energy loss through radiative cooling.

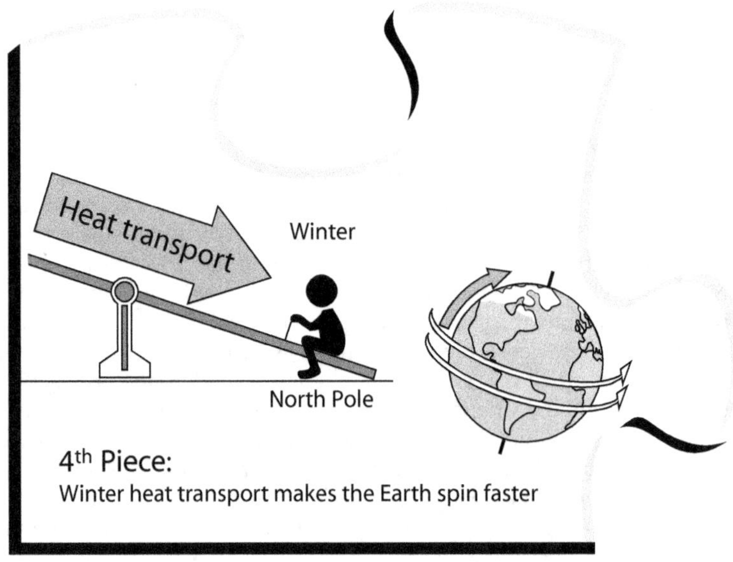

CHAPTER 17
OCEAN HEAT TRANSPORT IS LARGELY WIND-DRIVEN

The ocean is the primary source of poleward heat transport in the tropics, with the tropical Pacific being the dominant player due to its size. It exports heat to the Atlantic and Indian Oceans, which are the only ones to transport heat across the equator. However, inter-basin exchanges are relatively small, indicating that global seawater pathways play a minor role in heat transport. The Atlantic is unique in having an exclusively northward net heat transport due to its meridional overturning circulation, which accounts for about 60% of the heat transported in the North Atlantic. Oceanic heat transport from the North Atlantic to the Nordic Seas and the Arctic increased significantly between 1998 and 2002, during a period of Arctic and global climate shift.

Most of the heat transported by the global ocean is carried by water above 10°C (50 °F), located between 40°N and 40°S at depths of less than 500 m. This transport is primarily due to wind-driven circulation. Even the Atlantic Meridional Overturning Circulation is as sensitive to winds as it is to the formation of high-latitude deep water.

Analysis of the critical tropical upper-layer heat budget has revealed a remarkable 11-year variability associated with the solar cycle that is ten times larger than can be accounted for by changes in solar radiation. In addition, model studies of the Atlantic meridional circulation show that solar forcing is its most important natural determinant. These studies underscore the critical role of the Sun in modulating ocean heat transport by inducing changes in atmospheric circulation.

Ocean heat transport

The ocean plays a critical role in the Earth's climate system, providing thermal stability and storing a large fraction of the system's energy. With a total mass 265 times that of the atmosphere and a heat capacity 1000 times greater, the ocean stores 96% of the energy in the climate system and receives 75% of the energy delivered by the Sun to the planet's surface. This essential feature of the ocean has allowed the existence of complex life. However, because the Earth is currently in an ice age that began 34 million years ago (the Late Cenozoic Ice Age), the ocean has reached a cold state with an average temperature of about 4°C (39 °F), and only the upper mixed layer is substantially warmer due to solar heating and wind-induced turbulence. The sea surface temperature of the open ocean is limited to 30°C (86 °F) because deep convection occurs above 27°C (80 °F), increasing evaporation and forming clouds that effectively cool the surface. Although the upper 2.5 m of the ocean contains as much heat as the entire atmosphere, its main function in climate change is to absorb heat as the planet warms and release it as it cools, providing thermal inertia.

The ocean contributes about 25% of global poleward heat transport (ch. 10). In the tropics, the ocean is the most important heat transporter. Its contribution is even greater in the Northern Hemisphere, where it accounts for about 30% of heat transport. However, the Atlantic Ocean has a unique heat transport pattern. The South Atlantic has a net heat transport towards the equator (fig. 25).

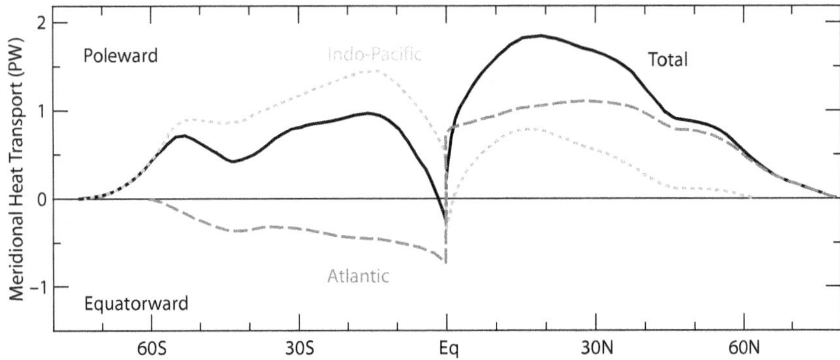

Figure 25. Ocean heat transport. Mean meridional ocean heat transport (in petawatts) for the global ocean (solid black), Atlantic (dashed medium gray), and Indo-Pacific (dotted light gray).[108]

Most of the ocean's heat is transported by water with a temperature above 10°C (50 °F), mainly in the band of the ocean between 40°S and 40°N and above a depth of 500 m. This is the main reason why meridional ocean heat transport is more important at these latitudes, where the Hadley cell is not very effective in transporting heat poleward (ch. 13).

Global ocean heat transport is dominated by heat export from the tropical Pacific, which has the largest tropical surface area and receives the most solar energy. However, it's striking how much the tropical Pacific dominates heat export to other oceans, exporting four times more heat than is imported into the Atlantic and Arctic oceans. The Atlantic and Indian Oceans transport heat north and south across the equator, respectively, but the Pacific provides this heat through the Drake Passage and the Indonesian Throughflow. While there is some exchange between the basins, it's relatively small, suggesting that global seawater pathways play a minor role in the Earth's heat budget.[109]

Poleward heat transport to the Arctic

The Atlantic Ocean has northward heat transport in both hemispheres and across the equator due to the Atlantic Meridional Overturning Circulation. This circulation is part of the thermohaline circulation, which involves the northward flow of warmer, lighter water in the upper layers of the Atlantic and the southward flow of cooler, denser water at depth. Although the two branches are mechanically driven, they are linked by the transformation of warm to cold water masses at high latitudes (ch. 10).

The uniqueness of Atlantic heat transport is highlighted in figure 25 and is related to the asymmetry of the latitudinal temperature gradient between the

---

[108] Yang, H., et al., 2015. Clim. Dyn. 44, pp.2751–2768. doi.org/10.1007/s00382-014-2380-5

[109] Forget, G. & Ferreira, D., 2019. Nat. Geosci. 12 (5), pp.351–354. doi.org/10.1038/s41561-019-0333-7

two hemispheres. Each year, the Southern Hemisphere receives more solar energy than the Northern Hemisphere. This is due to the Earth's current axial precession, which causes the Southern Hemisphere to be oriented toward the Sun when the Earth is closer to it. Albedo does not correct for this difference due to its interhemispheric symmetry (box 2, ch. 3). Despite receiving a greater annual influx of solar energy, the Southern Hemisphere is about 2°C cooler than the Northern Hemisphere, and the Earth maintains a steeper temperature gradient toward the colder Antarctic than toward the warmer Arctic (ch. 9, fig. 13). Transport theory states that more heat should flow toward the colder pole since temperature differences drive transport. However, the Atlantic transports more heat from the Southern to the Northern Hemisphere, suggesting that energy transport is not solely determined by entropy production. Rather, it is strongly influenced by geographic and climatic factors and thus may be a forcing mechanism for climate change.

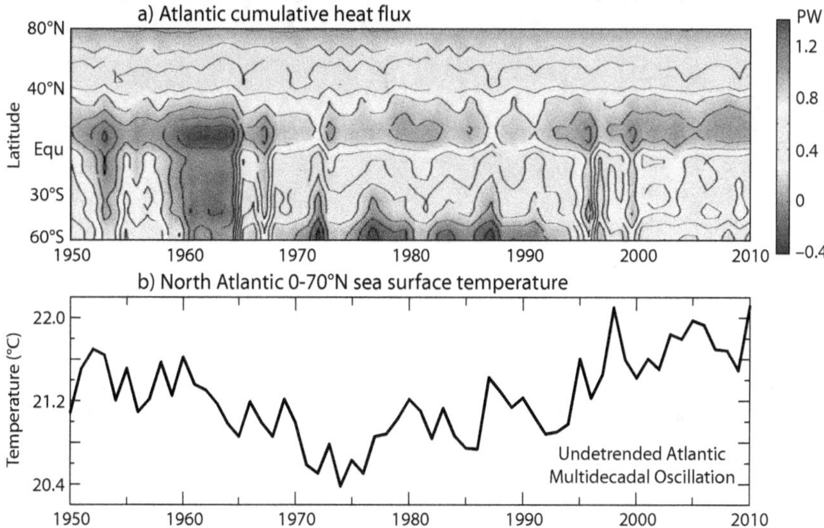

Figure 26. Atlantic heat transport and North Atlantic sea surface temperature. a) Atlantic integrated meridional heat transport over time in petawatts from reanalysis. b) North Atlantic sea surface temperature record for the same period.[110]

The exceptional nature of the Atlantic Ocean's heat transport has important implications for the climate of the surrounding regions of the North Atlantic, the Arctic, and the global climate. Sea surface temperature in the North Atlantic exhibits a multidecadal oscillation that correlates with global temperature (ch. 19).[111] Analysis of the Atlantic heat flux over time shows a clear relationship between oceanic heat transport and North Atlantic sea surface temperatures (fig. 26). This evidence supports the notion that the oscillation in North

---

[110] Top plot from Macdonald, A.M. & Baringer, M.O., 2013. Internat. Geophys. Vol. 103, pp. 759–785. doi.org/10.1016/B978-0-12-391851-2.00029-5. Bottom graph, NOAA data.

[111] Chylek, P., et al., 2014. Geophys. Res. Lett. 41 (5), pp.1689–1697. doi.org/10.1002/2014GL059274

Atlantic sea surface temperature is a result of changes in meridional heat transport. Surprisingly, despite this evidence, ocean oscillations are rarely considered in terms of heat transport.

The transport of Atlantic water to the Arctic occurs through the Nordic Seas, and the volume and temperature of the transported water strongly influence the climate of northern Europe and the Arctic. The transformation of warm to cold water masses necessary for the Atlantic Meridional Overturning Circulation occurs in the Nordic Seas and the Arctic Ocean. Although oceanic heat transport is a small part of the Arctic heat budget (ch. 11 & 16), its analysis can be very informative. A recent study of ocean heat transport in the Nordic Seas and the Arctic Ocean found a sudden increase in transport. From the 1993-98 average to the 2002-2016 average, oceanic heat transport in this important climate region, the "bellwether" for climate change, increased by 25 terawatts (9%) between 1998 and 2002 (fig. 27).[112]

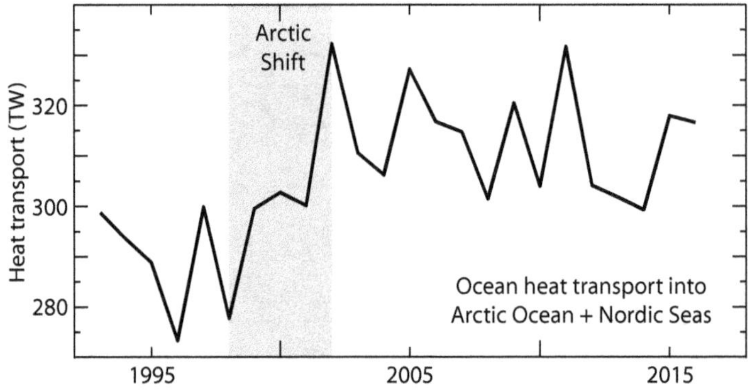

Figure 27. The Arctic Shift in ocean transport. Ocean heat transport to the Arctic and Nordic Seas during 1993-2017 shows an abrupt change during the Arctic Shift.

I refer to the period of rapid climate change in the Arctic that coincided with the change in oceanic transport as the Arctic Shift. As we will see, atmospheric heat transport to the Arctic also increased during the Arctic Shift, not showing the compensation between atmospheric and oceanic heat transport that the models predicted (box 8, ch. 12). It accelerated climate change in the Arctic, a clear demonstration of how changes in transport lead to profound climate changes that are erroneously attributed to anthropogenic forcing. The Arctic Shift was only one of the most conspicuous parts of the most significant global climate shift in 40 years. This issue is discussed in detail in chapter 33.

Wind-driven and thermohaline circulations

Ocean circulation can be divided into two types: fast circulation driven by wind stress, organized into ocean gyres, and slower circulation, related to changes in water density caused by changes in temperature and salinity (thermohaline). These two types of circulation are not independent, as the wind also affects the thermohaline circulation. It is important to note that the term thermohaline circulation, which refers to mass, heat, and salt circulation, can be

---

[112] Tsubouchi, T., et al., 2021. Nat. Clim. Change, 11 (1), pp.21–26. doi.org/10.1038/s41558-020-00941-3. Source of data for fig. 27.

misleading because heat and salt circulations are different.[113] In the Atlantic, wind-driven and thermohaline circulations contribute to poleward transport, but wind-driven gyres carry most of the heat in other oceans.

Despite its importance for understanding the climate system, our knowledge of the vertical structure of ocean heat transport is poor. This issue is fundamental to the debate over whether abyssal mixing, high-latitude deep-water formation, or winds control oceanic heat transport. This debate has led to unwarranted concerns that the Atlantic overturning circulation could be disrupted, causing significant cooling in Europe. Previous investigations of the vertical structure of oceanic heat transport, taking into account the temperature difference at the ocean-atmosphere boundary, have revealed our misunderstanding of this crucial process.[114] Such analyses show that surface circulation, which is highly sensitive to wind stress, dominates the total oceanic heat transport, while abyssal mixing has virtually no effect. High-latitude deep water formation contributes 60% of the North Atlantic heat transport, but the meridional circulation transport is also proportional to wind stress, being as sensitive to winds as to high-latitude convection.

The results of these studies challenge the common understanding of ocean heat transport as presented in books and illustrated by colorful ribbon diagrams. It is clear that winds play a critical role in ocean heat transport and that the amount of heat transported by the oceans is linearly proportional to the magnitude of wind stress. These findings lead to three controversial and far-reaching conclusions about climate change:

* Atmospheric circulation is primarily responsible for heat transport on a global scale, either directly or through its influence on oceanic transport.
* Atmospheric and oceanic heat transport cannot compensate for each other. Since they are fundamentally linked by wind action, any change in one must be accompanied by a change in the other in the same direction. Consequently, changes in the amount of heat transported poleward are not only possible but inevitable.
* Variability in global heat transport must occur on the decadal timescales typical of atmospheric and upper ocean variability, rather than the centennial or longer timescales characteristic of deep meridional overturning.

Box 14. Response of ocean heat transport to solar variability

Ocean heat transport occurs primarily in shallow tropical waters, so the heat budget of the upper layer is critical to global ocean transport. Studies of the variability of sea surface temperature and pressure have identified typical quasi-biennial and El Niño-Southern Oscillation frequencies, as well as an 11-year frequency. Although this 11-year variability is synchronous with the solar cycle, its

[113] Wunsch, C., 2002. Science, 298 (5596), pp.1179–1181. doi.org/10.1126/science.1079329

[114] Boccaletti, G., et al., 2005. Geophys. Res. Lett. 32 (10) L10603. doi.org/10.1029/2005GL022474 Ferrari, R. & Ferreira, D., 2011. Ocean Model. 38 (3–4), pp.171–186. doi.org/10.1016/j.ocemod.2011.02.013

magnitude cannot be explained by direct radiative forcing from the Sun at the surface.[115] In the global tropical ocean, the temperature in the upper layer varies by ±0.1°C in phase with the solar cycle, requiring a change of ±0.9 W/m², while the change in surface radiative forcing from the solar cycle is an order of magnitude too small, ±0.1 W/m². Therefore, the variability must be due to ocean-atmosphere mechanisms despite its synchronization with the Sun.

The effect of El Niño on ocean heat transport is characterized by the warming of the upper layer of the global tropical ocean, which then warms the overlying atmosphere. In contrast, variability associated with the solar cycle leads to the warming of the global tropical atmosphere, which then heats the underlying ocean. This process is accomplished primarily by reducing the net sensible + latent heat flux from the ocean to the atmosphere, as the increase in solar radiation in the ocean is insufficient. Evidence indicates that the effect of the solar cycle on the ocean is indirect, occurring through the atmosphere. Claims that the Sun cannot be responsible for climate change because of the small change in total solar irradiance that produces its variability ignore the abundant evidence that solar variations act indirectly by affecting atmospheric circulation.

Models agree that solar variability has a significant impact on ocean heat transport. The fully coupled atmosphere-ocean general circulation model of the UK Met Office Hadley Centre shows that solar forcing is the most important natural factor determining the multidecadal response of the Atlantic meridional circulation.[116] Solar forcing is associated with long-lasting anomalies in the atmospheric circulation over the North Atlantic caused by changes in the stratosphere due to weaker solar irradiance during the late 19th and early 20th centuries. The model does not fully capture the atmospheric response to solar variability, but it does show notable changes in the location of the Intertropical Convergence Zone, precipitation in the Amazon, and temperatures in Europe.

In summary

The ocean plays a critical role in transporting heat poleward within the tropics. Wind-driven circulation in the ocean gyres is responsible for most of the heat transport, and a global conveyor has a limited contribution. However, the Atlantic Ocean is an exception, exhibiting net northward heat transport with relevant transequatorial transport, mainly due to the Atlantic Meridional Overturning Circulation, which is sensitive to both wind stress and deep water formation at high latitudes.

The atmosphere, directly through its circulation and indirectly through the effect of wind stress on oceanic transport, is primarily responsible for most of the poleward heat transport. The multidecadal oscillation of sea surface temperature in the North Atlantic results from changes in poleward heat transport. In addition, the upper layer of the tropical ocean shows temperature changes

---

[115] White, W.B., et al., 2003. J. Geophys. Res. Oceans, 108 (C8) 3248. doi.org/10.1029/2002JC001396

[116] Menary, M.B. & Scaife, A.A., 2014. Clim. Dyn. 42, pp.1347–1362. doi.org/10.1007/s00382-013-2028-x

in phase with the solar cycle caused by changes in the atmospheric circulation that affect the heat flux from the ocean to the atmosphere.

SECTION 4 KEY ISSUES

The troposphere transports most of the heat poleward over ocean basins. The Hadley cell is ineffective because it transports latent heat toward the equator. Storms and hurricanes are the primary mechanism for transporting heat outside the tropics. Wind speed is critical because it provides mechanical energy and affects evaporation rates. Climate models have been unable to explain why wind speed shows multidecadal trends.

Poleward heat transport through the stratosphere is regulated by the strength of the polar vortex and driven by the momentum of atmospheric waves, which act as a pump. The equatorial winds in the stratosphere change direction about every two years, creating a different circulation regime that affects the stratosphere and troposphere. The easterly phase of this Quasi-Biennial Oscillation causes atmospheric waves to weaken the polar vortex. As a result, the Arctic is warmer and more heat is lost through radiative cooling.

Anomalies in the stratosphere propagate to the Earth's surface, affecting winter surface temperature and pressure patterns, the location of tropospheric jets, and storm tracks. The effect is transmitted through the polar vortex to annular modes, a polar wind pattern that regulates heat transport. These modes show multidecadal trends that models cannot explain. The solar cycle strongly influences the stratospheric dynamical changes behind these effects.

Winter atmospheric circulation and heat transport are more intense in the Northern Hemisphere due to increased atmospheric wave activity, resulting in a weaker vortex. It also creates blocking patterns in the jet stream that redirect storms toward the Arctic. The intensified winter atmospheric circulation causes the Earth to rotate faster.

The ocean transports most of the heat within the tropics, mainly through wind-driven circulation in the ocean gyres, with some contribution from the global conveyor. However, the Atlantic Ocean is an exception, with a net northward heat transport mainly due to the Atlantic Meridional Overturning Circulation. Changes in the poleward heat transport cause the multidecadal oscillation of sea surface temperature in the North Atlantic. In addition, the upper tropical ocean experiences temperature changes synchronized with the solar cycle due to atmospheric circulation changes that affect the ocean-atmosphere heat flux.

PART II. NATURAL CLIMATE CHANGE

SECTION 5. THE OCEAN

Chapter 18
El Niño

The El Niño-Southern Oscillation is an ocean-atmosphere interaction phenomenon in the Pacific Ocean that affects climate worldwide. During El Niño, a large amount of heat is extracted from the subsurface of the equatorial ocean and transported poleward through the ocean and atmosphere, ultimately resulting in surface warming but also a reduction in energy within the climate system due to increased outgoing thermal radiation. However, there is no agreed theory of the nature and causes of El Niño. In addition, the long-term changes in El Niño frequency are not fully understood. During the five-millennium warm period known as the Holocene Climatic Optimum, La Niña conditions predominated and El Niño events were rare. The relationship between El Niño and solar activity is still controversial, but there is evidence to support it. A more comprehensive view of the El Niño-Southern Oscillation as a three-mode phenomenon could lead to a better understanding of its interpretation and connection to solar activity.

El Niño-Southern Oscillation

The Earth's rotation causes the winds blowing equatorward through the lower branch of the Hadley circulation to turn westward when they meet near the equator, coming from both hemispheres. These winds are known as the trade winds, and in the Northern Hemisphere, they are northeasterly (blowing from the northeast), while in the Southern Hemisphere, they are southeasterly. Their effect is to push warm equatorial waters toward the western side of the basin. In the vast Pacific Ocean, the trade winds create the largest mass of warm water on the planet, called the Indo-Pacific Warm Pool. The accumulation of warm water in the western Pacific causes the air to rise, resulting in a pressure drop and a return circulation with air descending into the eastern Pacific, increasing surface pressure there. This equatorial wind circulation in the Pacific is called the Walker circulation.

One of the effects of this intense circulation is the strong upwelling of nutrient-rich cold water off the coasts of Peru and Ecuador, boosting fish stocks. In some years, however, the trade winds weaken and the pressure difference between the two sides of the Pacific decreases. Deprived of its support, the warm water pushes eastward, reducing the upwelling of cold water and fish populations. Peruvian fishermen called this situation El Niño (the baby boy) because it often hit them around Christmas. In other years, the opposite occurs, with stronger trade winds and an increase in the pressure difference between the two sides of the Pacific. This pushes warm water farther west, increasing upwelling and cooling the waters of the eastern equatorial Pacific. This situation is known as La Niña. The third situation is the neutral years when neither situation occurs.

The Southern Oscillation is defined in terms of pressure because it is associated with monsoon rainfall anomalies. It measures the difference in pressure between Indonesia and the eastern Pacific. In the 1960s, however, scientists realized that the Southern Oscillation was only the atmospheric aspect of the oceanic El Niño phenomenon.

The El Niño-Southern Oscillation is a strong interannual variability in the global climate system. The warm El Niño, cold La Niña, and neutral conditions that alternate during this phenomenon affect the global climate, marine and terrestrial ecosystems, fisheries, and human activities. There are two types of El Niño events, based on the location of their maximum sea surface temperature anomalies: the East Pacific type and the Central Pacific type. La Niña events, on the other hand, have no spatial variability. Moreover, the distribution of El Niño and La Niña events is irregular and can occur more or less frequently over decades, an aspect that is not yet understood.

The nature of El Niño is still a matter of debate, with several competing views on its underlying mechanism. Some see it as an unstable, self-sustaining oscillatory mode, while others see it as a stable mode triggered by stochastic forcing. Some also suggest that it may be a series of independent but coupled events.

El Niño and poleward transport

Understanding the El Niño phenomenon requires consideration of its role in poleward heat transport. The main climatic effect of El Niño is the extraction of an enormous amount of heat from the subsurface of the equatorial Pacific. The ocean then transports this heat and releases it to the atmosphere as sensible and latent heat. This process reduces the incoming shortwave radiation in the region by increasing cloudiness. The atmosphere then transports the heat out of the tropics and removes some of it from the planet by increasing outgoing longwave radiation. As a result, the equatorial region of the climate system, including the upper 500 m of the ocean, contains less heat than before, while the rest contains more. Although El Niño warms the surface, it reduces the energy within the climate system because some of the heat is lost through increased outgoing thermal radiation. Therefore, El Niño must be understood as a cooling event in the climate system, although it has the opposite effect on surface temperature. The interpretation of the El Niño system as a heat pump coupled to global poleward heat transport is not widely accepted but is the obvious conclusion from the analysis of its heat sources and sinks.[117] Thermodynamics should be the guiding principle in climate analysis.

The reinterpretation of El Niño in terms of poleward heat transport sheds new light on many open questions about this phenomenon. El Niño is tied to the boreal winter because heat transport is a winter seesaw, and the Northern Hemisphere winter atmospheric circulation is stronger due to smaller ocean basins, higher atmospheric wave activity, and greater heat transport to the Arctic than to the Antarctic (ch. 11). El Niño events are more variable in time and space than La Niña events because they respond to heat transport conditions that are also variable in time and space. This is because there are many ways to transport heat, but only one way not to transport it.

There is a controversial link between El Niño and the North Atlantic Oscillation (box 12, ch. 15), often referred to as a teleconnection, with stratospheric

---

[117] Sun, D.Z., 2000. J. Clim. 13 (20), pp.3533–3550.
    doi.org/10.1175/1520-0442(2000)013<3533:THSASO>2.0.CO;2

and tropospheric involvement.[118] The result is a weaker polar vortex and colder winters over North America and northern Europe during Niño winters. However, instead of a complex variable teleconnection that resists interpretation, what is occurring is the response of the complex global meridional transport system to the heat injected into the atmosphere from the equatorial Pacific by the Niño event. Depending on the location and amount of the heat, it affects the different components of the transport system (stratospheric Brewer-Dobson circulation, North Atlantic Oscillation, polar vortex) differently.

Box 15. El Niño during the Holocene

Interpreting El Niño in terms of poleward heat transport helps to explain why its frequency has varied so much during the Holocene. From the thermodynamic consideration that El Niño events represent cooling events in the climate system, it follows that El Niño events should increase in frequency during periods of global cooling and decrease during periods of warming if changes in oceanic heat storage are involved in the temperature change.

Analysis of sediments from an Andean lake has allowed us to reconstruct the frequency of strong El Niño events throughout the Holocene (fig. B15).[119] This reconstruction supports our interpretation. During the early Holocene warming and most of the warm Holocene Climatic Optimum, strong Niño events were very rare. However, beginning about 7,000 years ago, El Niño events became more frequent as the planet cooled and entered the Neoglacial period, characterized by the growth of glaciers over most of the world. About 1,200 years ago, as the planet warmed toward the Medieval Warm Period, there was a marked decrease in El Niño events. Conversely, about 800 years ago, as the planet began to cool toward the Little Ice Age, there was a marked increase in strong El Niño events.

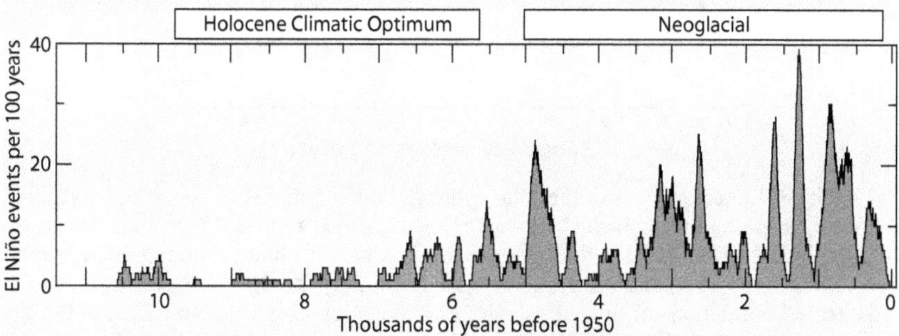

Figure B15. El Niño-Southern Oscillation activity during the Holocene. El Niño activity (number of events per century) shows a bimodal distribution, with low activity during the Holocene Climatic Optimum and high activity during the Neoglacial.

---

[118] Domeisen, D.I., et al., 2019. Rev. Geophys. 57 (1), pp.5–47.
doi.org/10.1029/2018RG000596
[119] Moy, C.M., et al., 2002. Nature, 420 (6912), pp.162–165.
doi.org/10.1038/nature01194

Today, the planet is warming, and the frequency of El Niño events is decreasing sharply. The frequency of El Niño is now very low by late Holocene standards, with a strong El Niño occurring about every 20 years.

The response of El Niño to solar activity

We have already discussed the correlation between the temperature in the upper layer of the tropical ocean and the solar cycle (box 14, ch. 17). This correlation should affect El Niño, and several studies have reported a link between the solar cycle and El Niño.[120] Surprisingly, most El Niño experts overlook this connection, and even comprehensive reviews of the El Niño-Southern Oscillation fail to mention it.[121]

To demonstrate the relationship between El Niño and the solar cycle, a statistical compositing technique called superposed epoch analysis was used to reveal a correlation between the two series. The data on solar activity (sunspots) and El Niño (Oceanic Niño Index, surface temperature anomaly) were divided into 22 bins based on fractions of the corresponding solar cycle. This method facilitates the comparison of data from the same phase of the solar cycle, even if the cycles are of different lengths. The mean and standard deviation (for the Niño index only) of six solar cycles are shown in figure 28.

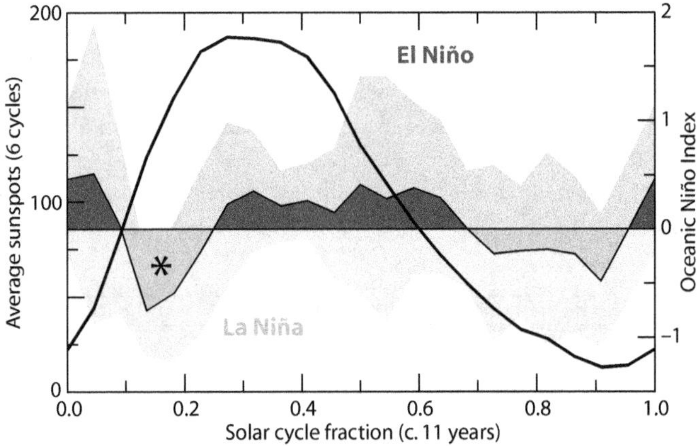

Figure 28. Epoch analysis of the solar cycle-El Niño relationship. Smoothed monthly mean sunspot number (thick black curve, left scale) for 1950-2018. Average Oceanic Niño Index for the same period (gray areas, dark positive and medium gray negative values, right scale) and standard deviation (light gray areas).[122] Relative timescale expressed in fractions of full solar cycles of varying lengths. The asterisk indicates the period for which a Monte Carlo analysis was performed.

By analyzing the data in this way, it became clear that positive Niño values tend to occur most frequently from solar maximum to when solar activity has

---

[120] See ch. 10, sect. 10.4 in Vinós, J., 2022. Climate of the Past, Present and Future: A scientific debate. 2nd ed. Critical Science Press.

[121] Timmermann, A., et al., 2018. Nature, 559 (7715), pp.535–545. doi.org/10.1038/s41586-018-0252-6

[122] Data from WDC–SILSO, Royal Observatory of Belgium and NOAA.

fallen below average, and negative values from that time to solar minimum, albeit with a delay. However, the largest anomalies occur for El Niño around the time of the solar minimum and for La Niña thereafter, when solar activity is increasing.

To assess the statistical significance of the results, I performed a Monte Carlo analysis over the 12-month period marked with an asterisk in figure 28. The mean value of the Oceanic Niño Index is –0.649, indicating La Niña conditions (i.e., a cold temperature anomaly below –0.5°C in the equatorial Pacific). I randomly extracted and averaged 100,000 groups of six 12-month periods from the Niño Index database, and in only 0.7% of them did I obtain a value equal to or less than –0.649. Therefore, the La Niña case marked with an asterisk in figure 28 has a 99.3% probability of not being due to chance, confirming the solar cycle-El Niño relationship.

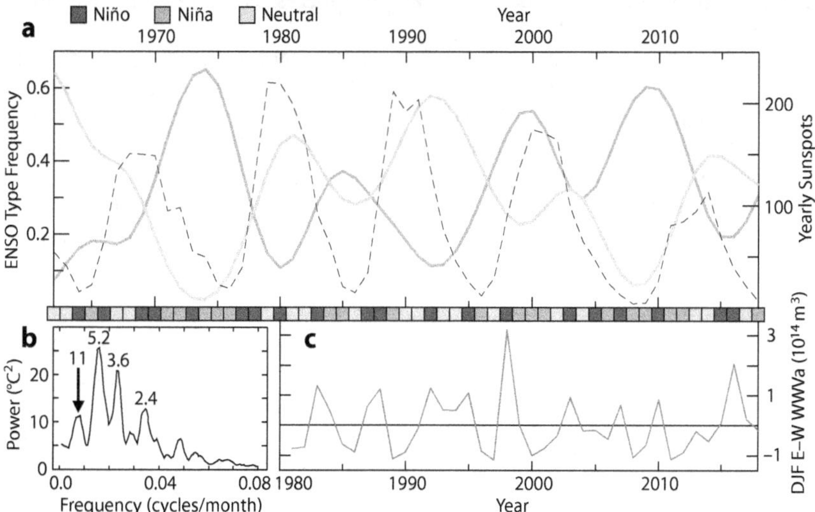

Figure 29. El Niño-Southern Oscillation modes and solar activity. a) Frequency of Niña years (thick medium gray line) and neutral years (thick light gray line), and the solar cycle (annual sunspots, thin dashed line). b) Power spectrum of the sea surface temperature anomaly time series in the Niño-3.4 region for 1900-2008.[123] An arrow marks the frequency of the 11-year solar cycle. c) Difference in warm water volume anomaly (WWVa) above the 20 °C isotherm (December-February mean) between the eastern and western equatorial Pacific.[124]

El Niño is often viewed as an oscillation between two opposing states, but this oversimplification ignores the neutral state that exists between them. This neutral state is equally important because the El Niño-Southern Oscillation functions like a heat pump associated with poleward heat transport, with three key positions: low, normal, and high. By analyzing their relative frequencies, we gain valuable insights. Interestingly, it turns out that the opposite of La

[123] Deser, C., et al., 2010. Ann. Rev. Mar. Sci. 2, pp.115–143. doi.org/10.1146/annurev-marine-120408-151453 Niño 3.4 region is 5°N-5°S, 120-170°W.
[124] Data from TAO Project Office of NOAA/PMEL

Niña in terms of frequency is not El Niño, but Neutral. While El Niño events typically occur every 2-3 years (with a range of 1-4 years), the frequency of La Niña events and neutral years is more variable, ranging from 0-4 over a 5-year period. In particular, the frequency of El Niño events (not shown in fig. 29a) does not show a significant correlation with the frequency of La Niña or neutral years, but the latter two show a clear anti-correlation, as shown in Figure 29a.

Figure 29a shows a 5-year centered moving average (Gaussian filtered) of the frequency of Niña years (blue) and neutral years (green) between 1962-2018. As can be seen, there is a good anticorrelation for the whole period. The small boxes represent the modal ranking for each year, corresponding to the official January state.[125] Niña years tend to be more frequent when solar activity is low, usually with a lag, while neutral years show the opposite trend. Figure 29b shows that this frequency pattern appears as an 11-year peak in an analysis of the entire 20th century.

This analysis does not show the frequency of Niño years because they do not follow an obvious pattern. El Niño events are more regular and frequent in the 1970s and 1980s than in the 1990s and 2000s. The primary predictor of an El Niño event is the accumulation of warm water in the eastern Pacific, suggesting that they are more likely to respond to heat transport needs than to solar signals carried by the atmosphere. Figure 29c shows the difference in the warm water anomaly between the eastern and western equatorial Pacific (detrended data), which clearly identifies El Niño events.

In summary

The results presented here point to a common misunderstanding of the El Niño phenomenon. El Niño is a heat pump system that extracts heat from the equatorial Pacific. It goes through cycles of low activity (La Niña years) and normal activity (neutral years) in response to the state of meridional heat transport, which is influenced by various factors such as solar activity. When too much warm water accumulates or the planet cools, the pump goes into overdrive, resulting in El Niño years. However, periods of low El Niño activity can last for long periods of time, from decades to millennia, because El Niño only occurs when there is a high need for heat transport. During an El Niño episode, excess heat in the atmosphere can cause a range of meteorological effects, since weather is essentially a manifestation of heat transport.

---

[125] Domeisen, D.I., et al., 2019. Rev. Geophys. 57 (1), pp.5–47. doi.org/10.1029/2018RG000596

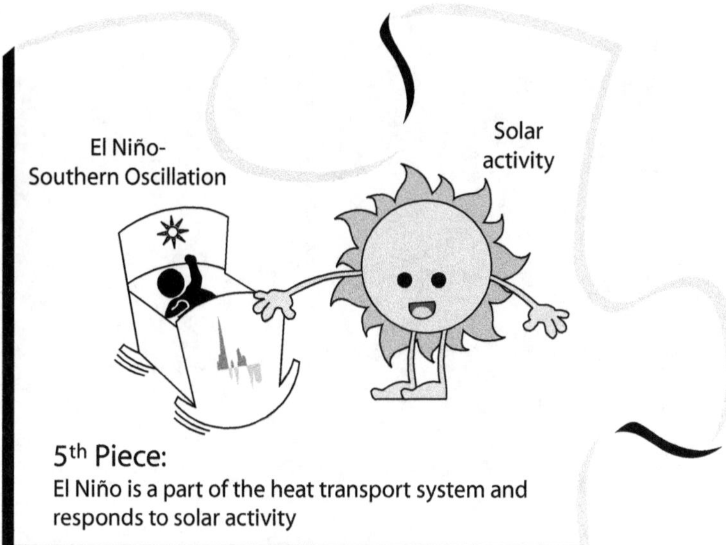

5th Piece:

El Niño is a part of the heat transport system and responds to solar activity

CHAPTER 19
OCEAN OSCILLATIONS

After the IPCC First Assessment Report, scientists discovered multidecadal ocean oscillations that revealed our incomplete understanding of the climate. While their cause is still unknown, these oscillations provide an alternative explanation for the shift from global cooling to warming in 1975, which was explained by the Enhanced CO_2 Effect hypothesis through a shift from a prevailing aerosol cooling forcing to a predominant CO_2 warming forcing.

Multidecadal ocean oscillations result from changes in the magnitude and geographic distribution of poleward heat transport, which are reflected in accompanying changes in wind speed anomalies. The Pacific Decadal Oscillation is linked to the variability of the El Niño-Southern Oscillation. The periodicity of the Atlantic Multidecadal Oscillation suggests that the 2030s may be cooler than the 2020s.

The Atlantic Multidecadal Oscillation

In the mid-1980s, scientists discovered a multidecadal oscillation in Atlantic sea surface temperature that had a significant effect on precipitation. It wasn't until 1994 that they realized it also had a strong global effect on temperature, which was unexpected because it didn't fit with the climate theory that changes in GHGs cause climate change.[126] By then, the Enhanced CO_2 Effect hypothesis had already explained the 1945-1975 cooling as due to increased cooling by industrial aerosols and the warming since 1975 as due to increased CO_2. However, the phases of the Atlantic Multidecadal Oscillation also coincide, with a cool phase from 1945-1975 followed by a warm phase since 1975, making it difficult to reconcile with existing theory.

To make matters worse, climate models don't show the observed oscillatory modes and are unable to project the expected behavior of the oscillation. While some proponents of the Enhanced CO_2 Effect hypothesis view ocean oscillations as variability caused by climate noise, this perspective prevents us from understanding their nature and causes.

The Atlantic Multidecadal Oscillation is essentially an index of sea surface temperature in the North Atlantic 0-70°N with the long-term trend removed. Figure 30 shows how this oscillation affects sea surface temperatures. It's a unitless linear regression that shows how much the global sea surface temperature would change if the index increased by 1°C (1.8 °F). However, since the index only changes by about ±0.2°C, the observed changes are about 1/3 of what the figure shows.

---

[126] Schlesinger, M.E. & Ramankutty, N., 1994. Nature, 367 (6465), pp.723–726. doi.org/10.1038/367723a0

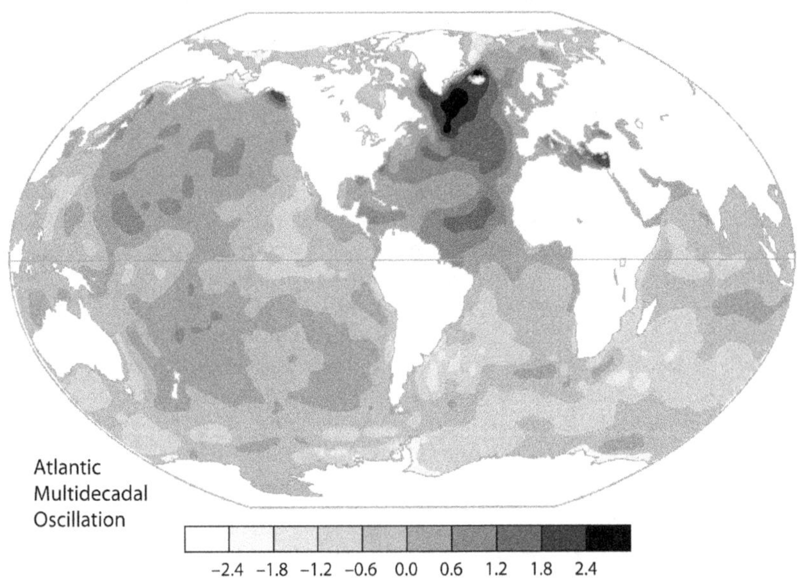

Figure 30. Spatial pattern of the Atlantic Multidecadal Oscillation. It is obtained from the 1870-2008 North Atlantic sea surface temperature anomalies after subtracting the global mean anomaly.[127]

According to the Enhanced CO_2 Effect hypothesis, human activities have been the primary cause of climate change since 1950, leaving little room for ocean oscillations to contribute. The argument is that due to the oscillatory nature of ocean oscillations, changes in opposite directions should cancel each other out, resulting in a net zero effect in the long run. However, this assumption only holds if the amplitude of the oscillations has remained the same over time, which is not the case. Ocean oscillations are not cyclic but quasi-periodic non-stationary oscillations, which means that their period and effect amplitude vary over time.

Recent evidence suggests that the Atlantic Multidecadal Oscillation had a smaller amplitude and shorter period for at least five centuries until about 1850, when modern global warming began.[128] This was about a century before our emissions began to accelerate. Some scientists believe that the positive phase of the Atlantic Multidecadal Oscillation is responsible for up to one-third of the global warming since 1975.[129] If this is true, it challenges the explanation that nearly all climate change is due to human emissions of aerosols and CO_2, especially since the phases of the oscillation coincide with the observed cooling before 1975 and warming after 1975.

---

[127] Figure after Deser, C., et al., 2010. Ann. Rev. Mar. Sci. 2, pp.115–143. doi.org/10.1146/annurev-marine-120408-151453

[128] Moore, G.W.K., et al., 2017. Sci. Rep. 7 (1), p.40861. doi.org/10.1038/srep40861

[129] Chylek, P., et al., 2014. Geophys. Res. Lett. 41 (5), 1689–1697, doi.org/10.1002/2014GL059274

The Pacific Decadal Oscillation

In 1997, just three years after the discovery of the Atlantic Multidecadal Oscillation, another oscillation was discovered in the Pacific.[130] This oscillation has a shorter period of 20-30 years and is associated with coordinated changes in the climate and ecology of the Pacific Ocean that occurred in 1976 and twice earlier in the 20th century. However, it also has a longer periodicity of 50-70 years, similar to but not coincident with the Atlantic Multidecadal Oscillation. The Pacific Decadal Oscillation can be defined by sea surface temperatures or sea level pressure and is similar in effect to the El Niño-Southern Oscillation but with a much longer period. It is sometimes referred to as a long-lived El Niño climate variability. The phases of the Pacific oscillation are related to the multi-decadal changes in El Niño frequency discussed in the previous chapter. Warm phases are characterized by a higher frequency of El Niño events, while the opposite is true for cold phases.

The causes of ocean oscillations are still unknown, and their potential for climate forecasting remains uncertain. Climate models do not reproduce them accurately, and although some models produce similar ocean oscillations, they often do so for different reasons.

Ocean oscillations and poleward heat transport

The energy supplied by the Sun in the tropics remains relatively constant on an annual basis, leading to the logical hypothesis that multidecadal changes in the surface energy of large ocean basin regions must be due to changes in poleward heat transport. However, this perspective is rarely considered in the scientific literature on multidecadal variability.

In the previous chapter, we discussed how the El Niño-Southern Oscillation acts like a heat pump, drawing subsurface heat from the equator and transporting it toward the poles. This leads us to believe that the Pacific oscillation is related to the intensity of poleward transport in the Pacific over multidecadal periods. Figure 30 shows the oscillating warming and cooling in the North Atlantic, but it also shows coordinated warming and cooling in the other oceans outside the equatorial zone, suggesting that global variations in poleward heat transport also play a role in this oscillation.

Further support for the idea that multidecadal oscillations are related to poleward heat transport comes from comparing the Atlantic oscillation with oceanic wind anomalies (fig. 31b). Since wind is the primary heat transporter on the planet, both directly and through the wind-driven ocean circulation, the correlation between global oceanic wind changes and changes in the Atlantic Oscillation supports the argument that oceanic oscillations represent a transport phenomenon. However, the interpretation of wind changes in terms of transport is complicated because zonal (east-west) winds impede poleward heat transport, while meridional (north-south) winds facilitate it. Determining how the observed wind changes affect poleward heat transport is impossible without knowing whether zonal or meridional winds are increasing or decreasing. However, we will return to this topic in future chapters.

---

[130] Mantua, N.J., et al., 1997. Bull. Am. Meteorol. Soc. 78 (6), pp.1069–1080. doi.org/10.1175/1520-0477(1997)078<1069:APICOW>2.0.CO;2 Minobe, S., 1997. Geophys. Res. Lett. 24 (6), pp.683–686. doi.org/10.1029/97GL00504

Figure 31 shows the correspondence between global temperature oscillations (fig. 31a) and the Atlantic oscillation (fig. 31b) after removing their long-term warming trend. This suggests an alternative natural explanation for the 1945-1976 cooling trend, which has been attributed to industrial aerosols by the Enhanced CO_2 Effect hypothesis and to a 1963 volcanic eruption in climate models.

Although not synchronous with the Atlantic oscillation, the Pacific oscillation also plays a role in this correspondence. Its cumulative value can be used to determine when the sign of its dominant state changes (fig. 31c), which occurs in anticipation of the change in the sign of the oscillation. By analyzing the maximums and minimums of the three graphs, a "W" pattern emerges in the 20th century. This pattern shows minimums in the 1920s and 1970s and maximums around 1900, the 1940s, and around 2000. The significance of these changes will be discussed in section 9.

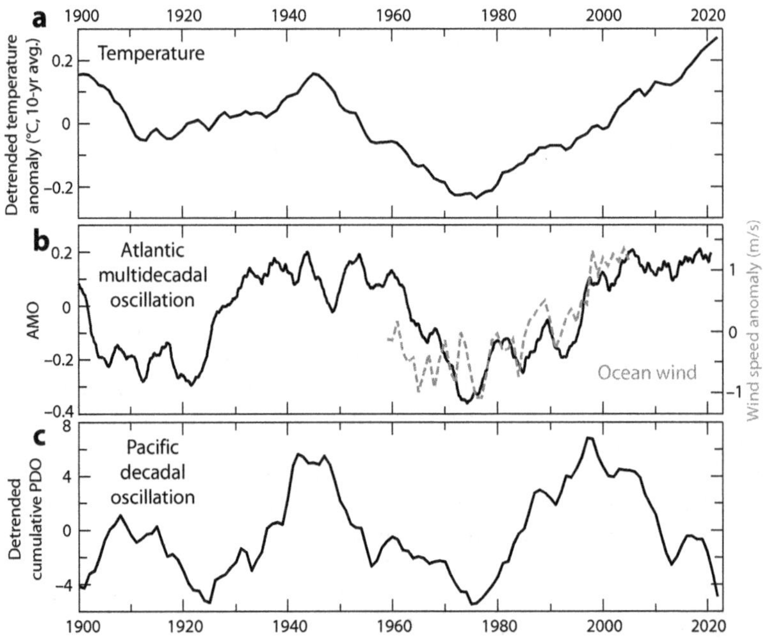

Figure 31. Multidecadal climate variability. a) 10-year average of the detrended annual global mean surface temperature anomaly. b) The black line is the 4.5-year average of the Atlantic Multidecadal Oscillation index, and the dashed gray line is the global ocean wind speed anomaly. c) Detrended annual mean cumulative Pacific Decadal Oscillation index.[131]

Box 16. The Stadium Wave hypothesis

The correspondence observed in figure 31 between the Atlantic and Pacific oscillations suggests they are somehow related. Meridional heat transport is a

---

[131] Figure 31 data: a) from Met Office. b) from NOAA and Yu, L., 2007. J. Clim. 20 (21), pp.5376–5390. doi.org/10.1175/2007JCLI1714.1 c) from NOAA (ERSST).

global phenomenon characterized by high geographic variability. In the Northern Hemisphere, we have seen that it occurs mainly through two routes: the Atlantic and Pacific basins, with a less significant route over the Eurasian continent (box 13, ch. 16). The distribution of poleward heat transport among these routes appears to vary over time. The variability in the position of El Niño between the central and eastern Pacific is an example of longitudinal variability in meridional heat transport over time.

We have discussed the three most important oscillations: the El Niño-Southern Oscillation and the Pacific and Atlantic multidecadal oscillations. However, other long-term atmospheric and oceanic oscillations have also been described. I have proposed that they are all manifestations of an oscillation in the distribution and magnitude of global poleward heat transport.[132] But the idea that they are all part of an underlying signal transmitted through the climate system was already presented in 2012 as the Stadium Wave hypothesis.[133] This hypothesis proposes that a multi-decadal hemispheric climate signal propagates through the climate system, generating a sequence of multi-year delayed atmospheric and oceanic teleconnections that give rise to the observed climate oscillations. Figure B16 shows the behavior of seven oscillations transformed into indices, highlighting their coordination.

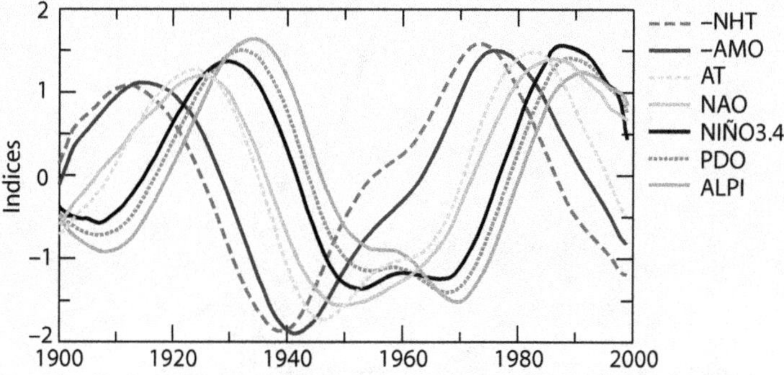

Figure B16. The stadium wave hypothesis. Normalized behavior of a network of seven indices: Northern Hemisphere temperature (NHT, inverted), Atlantic Multidecadal Oscillation (AMO, inverted), atmospheric mass transfer anomaly (AT), North Atlantic Oscillation (NAO), El Niño-Southern Oscillation (NIÑO3.4), Pacific Decadal Oscillation (PDO), and Aleutian-Low Pressure Index (ALPI).

The stadium wave hypothesis is notable for being the first to propose that multidecadal oscillations are a globally coordinated phenomenon, a concept that has yet to gain widespread acceptance. It implies the existence of an unknown global climate phenomenon occurring in the background, separate from changes in GHGs. Evidence suggests that the underlying cause is variability in meridional

---

[132] Vinós, J., 2022. Climate of the Past, Present and Future: A scientific debate. 2nd ed. Critical Science Press.

[133] Wyatt, M.G., et al., 2012. Clim. Dyn. 38, pp.929–949. doi.org/10.1007/s00382-011-1071-8 Source for figure B16.

heat transport. The lack of interest in this hypothesis can be attributed to the fact that the climate community generally does not seek alternative explanations for observations that have already been attributed to increases in aerosols and CO_2.

The coming change in the Atlantic oscillation.

Given the widespread concern about future climate, it is worth discussing what might happen next with the Atlantic oscillation, which has an important influence on global temperatures. However, ocean oscillations are not stationary, so there is a great deal of uncertainty about the next phase of the oscillation. If the period of the last oscillation is maintained, the Atlantic Multidecadal Oscillation should begin to decline around 2025 ± 1 year. As a result, sea surface temperatures over the North Atlantic may decrease by about 0.4°C (0.7 °F) over 1-2 decades, with a smaller effect on the global mean surface temperature. It is possible that a global cooling effect of 0.1-0.3°C could occur. If this projection holds, the 2030s could be slightly cooler than the 2020s.

In summary

Multidecadal ocean-atmosphere oscillations reveal that internal climate variability is essentially variability in the magnitude and geographic distribution of poleward heat transport.

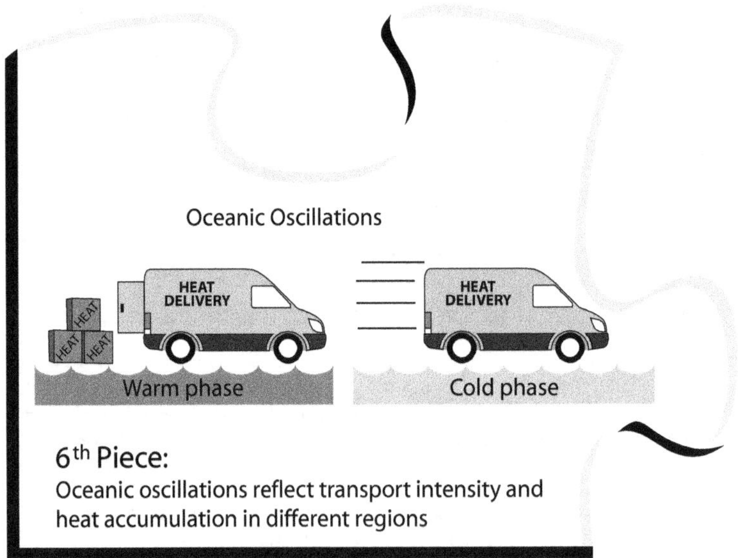

SECTION 5 KEY ISSUES

The El Niño-Southern Oscillation is an ocean-atmosphere interaction that acts as a heat pump system in the equatorial Pacific Ocean. During El Niño, heat is extracted from the ocean and transported poleward, resulting in surface warming but also in energy loss to the climate system due to increased outgoing thermal radiation. The frequency of La Niña and neutral conditions is strongly related to the solar cycle. In the short term, El Niño responds to equatorial warm water accumulation, while in the long term, it responds to the long-term temperature trend of the planet.

Multidecadal ocean oscillations are a recent discovery highlighting our incomplete understanding of climate. Their cause is unknown, but they strongly influence global temperature trends. They result from changes in the magnitude and geographic distribution of poleward heat transport, which is reflected in changes in wind speed. The Pacific Decadal Oscillation is linked to the variability of the El Niño-Southern Oscillation.

SECTION 6. PAST NATURAL CLIMATE CHANGE

Chapter 20
The Equable Climate Problem

The planet is currently in an ice age, and scientists have developed climate models and theories to explain the current climate. However, at certain times in the past, the planet was extremely warm and humid, with palm trees and crocodiles even in the polar regions. This type of climate, characterized by little temperature variation across the globe and seasons, is known as an equable climate. It remains a mystery how equable climates were possible since the low temperature gradient between the equator and the poles made it impossible to transport the heat necessary to keep the poles warm. The answer is not simply high CO_2 levels but rather fundamental changes in atmospheric heat retention and transport. Understanding that the polar regions act as cooling radiators is key to solving the equable climate paradox. Interestingly, the more heat that is transported to the polar regions, the more the planet cools. Over the past 50 million years, a series of tectonic changes have transformed a warm, heat-retaining planet into a cold, heat-radiating planet.

Icehouse and hothouse climates

Throughout Earth's history, its climate has changed constantly. Although this book is primarily concerned with modern climate change, it is necessary to examine past climate changes to understand why and how they occur. In this sixth section, the book briefly discusses specific climates and changes that have occurred in the past, with this chapter focusing on the distant past.

It may come as a surprise, but we are living in an ice age – one of the coldest in the last 500 million years. Ice ages are defined as periods of extensive ice sheets over continental areas, and we currently have not one, but two: one over Greenland and one over Antarctica. So our climate qualifies as an ice age twice over. Despite our species' tropical origins, we have adapted to this reality, and change always brings new challenges.

Our planet's global climate is currently in an icehouse period, characterized by an average surface temperature of about 14.5°C (58 °F). However, for most of the past 500 million years, the Earth has been in a greenhouse climate, with average temperatures ranging from 17-21°C (62.5-70 °F). At the other extreme, there have been periods when the planet experienced a hothouse climate with average temperatures of 22-26°C (71-79 °F). During these periods, the poles were completely ice-free and had average annual temperatures as high as 14°C (57 °F). In contrast, the tropics maintained temperatures similar to today's in a hothouse climate because the climate became very humid, and temperatures did not rise much above 30°C (86 °F) due to abundant rainfall. We know that mammals cannot survive wet-bulb temperatures above 35°C (95 °F), so we can infer that temperatures did not reach that threshold in the hothouse climate.

The climate of the world's warmest periods is defined as equable, meaning that temperatures did not vary significantly around the globe or among the seasons. In figure B17, we can observe three distinct periods of equable climate, represented by vertical gray bands: the early Triassic (252-238 million years ago), the late Cretaceous (95-80 million years ago), and the early Eocene (60-50 million years ago).

Box 17. Earth's temperature in the deep past

Our knowledge of past planetary temperatures is limited, but several lines of evidence suggest that warm and cool periods have alternated roughly every 75 million years. This evidence comes from the study of rocks (geology), fossils (paleontology), and isotopes (physical chemistry). Three cool periods during the Phanerozoic Eon have resulted in ice ages, including the present (late Cenozoic) and the Karoo Ice Age, which occurred 300 million years ago.

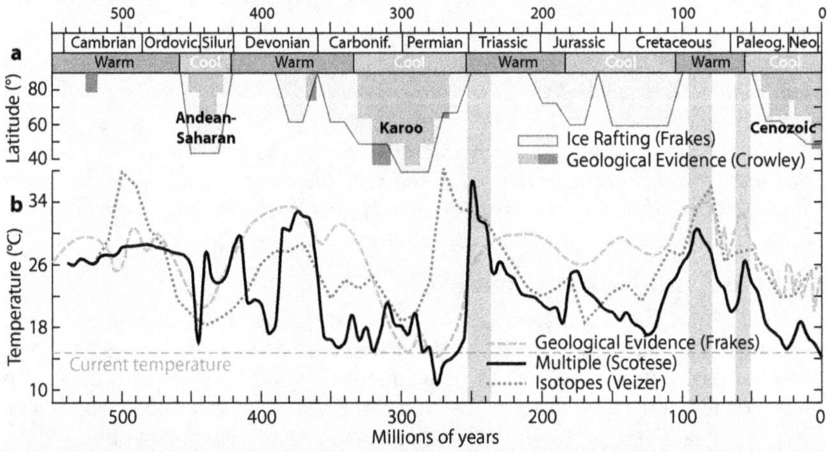

Figure B17. Temperatures of the Phanerozoic Eon. a) Evidence of ice rafting and ice sheets reaching lower latitudes from the poles helps define four cool periods, including three ice ages. b) There is general agreement among several temperature reconstructions of alternating warm and cool periods. The dashed gray line corresponds to the present temperature and the vertical gray bands mark periods of equable climate.[134]

Understanding the reasons for the Earth's temperature changes in the deep past is difficult, and several hypotheses have been proposed to explain them. These hypotheses can be divided into two groups based on whether they attribute the cause to external or internal factors. For example, one theory in the first category proposes that the Sun's orbit around the galactic center causes the Earth to pass through different regions of density that affect cosmic radiation levels, leading to climate changes.[135] Another popular theory, which falls into the second

---

[134] Data for the figure from Crowley, T.J. & Burke, K. 1998. Tectonic boundary conditions for climate reconstructions. Oxford University Press. Frakes, L.A. & Francis, J.E., 1988. Nature, 333 (6173), pp.547–549. doi.org/10.1038/333547a0 Frakes, L.A. et al. 1992. Climate modes of the Phanerozoic. Cambridge University Press. Scotese, C.R. 2018. Phanerozoic Temperatures. Smithsonian Workshop. Veizer, J., et al., 2000. Nature, 408 (6813), pp.698–701. doi.org/10.1038/35047044

[135] Shaviv, N.J. & Veizer, J., 2003. GSA today, 13 (7), pp.4–10. doi.org/10.1130/1052-5173(2003)013<0004:CDOPC>2.0.CO;2

category, suggests that fluctuations in CO_2 levels have primarily driven temperature changes in the past.[136] Meanwhile, the idea that tectonic movements and continental configuration played a significant role in shaping past climates has also received much attention.[137] The climate change mechanism presented in this book emphasizes the importance of the Earth's continental configuration in determining climate through its influence on heat transport to the poles over millions of years.

The early Eocene equable climate paradox

The early Eocene, which occurred between 56 and 47.8 million years ago, is the most recent and best-studied period of equable climate. During this time, forests stretched as far north as Ellesmere Island, Canada, at 80°N. The Arctic was home to palm trees and crocodiles, and Antarctica was covered in forests. Recent discoveries, including the first Antarctic frog fossil, indicate that even the coldest month on the Antarctic Peninsula had an average temperature of over 3°C (37 °F) at that time (fig. 32).[138] The flora and fauna in these regions were specifically adapted to the polar night, which lasted for several months and during which temperatures remained above freezing.

Figure 32. The early Eocene in Antarctica. Reconstruction of an Eocene pond in a forest on the Antarctic Peninsula, with fossil frog.[139]

The equable climates of the past pose a conceptual challenge to our understanding of climate, known as *the low-gradient paradox*. As explained in previous chapters, the temperature difference between the equator and the poles drives poleward heat transport, keeping the poles warmer than they would otherwise be. However, during the early Eocene, the poles were about 50°C (90 °F) warmer than today (14°C in the Arctic versus –35°C

[136] Royer, D.L., et al., 2004. GSA today, 14 (3), pp.4–10.
doi.org/10.1130/1052-5173(2004)014<4:CAAPDO>2.0.CO;2
[137] Eyles, N., 2008. Palaeogeogr. Palaeoclimatol. Palaeoecol. 258 (1-2), pp.89–129.
doi.org/10.1016/j.palaeo.2007.09.021
[138] Mörs, T., et al., 2020. Sci. Rep. 10 (1), p.5051. doi.org/10.1038/s41598-020-61973-5
[139] Credit: Pollyanna von Knorring, Simon Pierre Barrette, José Grau, and Mats Wedin (CC BY-SA 3.0).

today, or 57 °F versus –31 °F). Meanwhile, the tropics experienced little temperature change, with a temperature range of about 30-35°C (86-95 °F). According to theory, heat transport should have been greatly reduced during this period. So how could the poles be much warmer with much-reduced heat transport?

Climate models have been developed to simulate present-day conditions, and they tend to perform poorly when applied to very different climates, such as those of the early Eocene or the Last Glacial Maximum. This suggests that our understanding of climate may not be as complete as we think. Climate models have failed to reproduce the early Eocene, leading researchers to conclude that some important climate processes are either missing or not properly represented. Attempts have been made to tweak models to fit the early Eocene, but even if successful, we cannot be sure that the resulting simulations reflect actual climate processes.[140]

There are two obstacles to the idea that high CO_2 levels were responsible for the equable climate of the early Eocene. First, high CO_2 levels should have increased temperatures globally, not just at high latitudes. Second, there is evidence that CO_2 levels during the early Eocene were moderate, in the range of 450-600 ppm, and not nearly enough to explain the warm climate of that period.[141]

To resolve the equable climate paradox, it appears that differences in the way heat was transported from the tropics to the poles, or in the ability of the atmosphere to absorb heat, or both, must be considered. However, transport hypotheses require different mechanisms that pose theoretical problems and lack evidence for their feasibility. Radiative hypotheses based on changes in clouds, on the other hand, are more promising because they specifically target high latitudes and do not require significant changes in heat transport.[142]

A new solution to the equable climate paradox

The equable climate paradox was a turning point in my understanding of climate change. Paradoxes often arise from false assumptions about the limits of the physical world. Once we abandon those assumptions, we can resolve the contradiction. In this case, the paradox arises from the assumption that the poles require heat transport to be warm when in fact the opposite is true. The less heat is transported to the poles, the warmer the planet and the poles become. Conversely, the more heat that is transported, the colder the planet and the poles become. This happens because the poles are highly efficient cooling radiators in winter, similar to the radiator of a car engine. Heat must be transported by a fluid from the engine to the radiator, and the more heat that is transported, the colder the engine stays. Thus, the amount of heat radiated at the poles in winter determines the temperature of the planet.

During the early Eocene, the poles were warm because the entire planet was warm. When winter came to the poles, radiative cooling from the surface

---

[140] Valdes, P. J., 2000. Warm Climates in Earth History. Cambridge University Press, pp.3–20. doi.org/10.1017/CBO9780511564512

[141] Steinthorsdottir, M., et al., 2019. Geology, 47 (10), pp.914–918. doi.org/10.1130/G46274.1

[142] Abbot, D.S. & Tziperman, E., 2008. Q. J. R. Meteorol. Soc. 134 (630), pp.165–185. doi.org/10.1002/qj.211

caused a thermal inversion that cooled moist air near the ground to its dew point. Moisture from a warm ocean and a moist surface evaporated into the air, raising the dew point of the stable layer and accelerating the formation of radiative fog. A permanent thick fog and a dense cloud layer of warm, moist air from lower latitudes kept the polar regions warm in winter, severely limiting radiative cooling to the top of the cloud layer. The polar atmosphere became very opaque to infrared radiation, and the greenhouse effect became stronger due to a larger contribution from water vapor and the cloud effect. The fog only dissipated with the arrival of the Sun in the spring. The warm polar atmosphere prevented the formation of a strong tropospheric polar vortex, making heat transport more efficient despite its lower intensity.

This low-transport/warm-world, high-transport/cool-world paradigm has great explanatory power and resolves many long-standing questions about climate change.

The making of an ice age

We can now explain the planet's transition from the equable climate of the early Eocene to the Quaternary (Pleistocene) glaciation in terms of changes in meridional heat transport. We no longer need to rely on coincidental CO_2 changes or galactic events for an explanation. In the early Eocene, there was very little meridional transport to the Arctic, and global zonal transport dominated due to open circulation through the Panama and Indonesian passageways and the Tethys Sea. This kept the planet warm, warming both poles. However, a series of tectonic changes (fig. 33) completely altered the poleward transport of heat through the atmosphere over the ocean basins:[143]

1. The Arctic Gateway began to open about 55 million years ago, initiating global cooling as more heat was transported into the Arctic.[144]
2. The Tasman Gateway opened between 36 and 30 million years ago, beginning the partial isolation of Antarctica and causing it to freeze.
3. The Drake Passage opened between 30 and 20 million years ago, completing the isolation of Antarctica. This was bad for Antarctica but good for the planet because less heat was directed there. The planet warmed until it reached the Mid-Miocene Climate Optimum.
4. The Himalayas reached their present height about 15 million years ago, which favored poleward heat transport by perturbed eddy flow and increased planetary wave flux. Cooling resumed.
5. The Indonesian Throughflow is still open, but major restrictions have occurred from about 11 million years ago, favoring meridional circulation.
6. The Bering Strait opened about 5.3 million years ago, providing another gateway to the Arctic.
7. The Panama Gateway closed about 3 million years ago, eliminating the last open tropical zonal passage between basins.

---

[143] Lyle, M., et al., 2008. Rev. Geophys. 46, RG2002, doi.org/10.1029/2005RG000190
[144] Vahlenkamp, M., et al., 2018. Earth Planet. Sci. Lett. 498, pp.185–195.
   http://doi.org/10.1016/j.epsl.2018.06.031

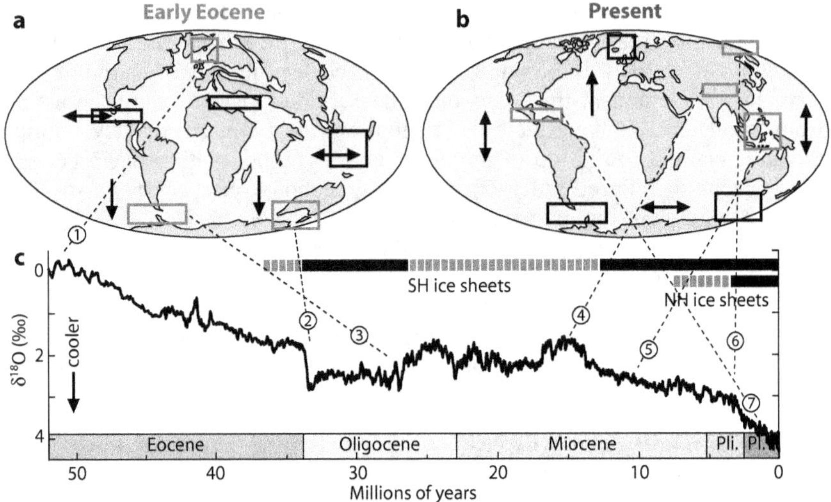

Figure 33. Meridional transport is the main determinant of climate evolution. a) & b) Early Eocene and modern circulations. Horizontal arrows indicate zonal circulation while vertical arrows indicate meridional circulation. Boxes indicate mountain ranges and oceanic gateways that influence heat transport, with gray boxes representing those mentioned by number (see main text). c) Global benthic $\delta^{18}O$ data as a proxy for temperature and continental ice. Solid bars above represent ice volume >50% of the present, and dashed bars ≤50%.[145]

The planet's circulation shifted from predominantly zonal to predominantly meridional, leading to one of the coldest glaciations in the last 540 million years. The heat from the tropics is now being dissipated at the poles, meaning that the next glaciation in a few thousand years is inevitable due to the current tectonic setting, regardless of CO_2 levels.

In summary

The late Cretaceous and early Eocene were characterized by a warm, equable climate with minimal poleward heat transport due to a low temperature gradient between the equator and the poles. During the months-long polar night, the polar atmosphere retained heat due to a strong water vapor-dependent greenhouse effect. This led to deep cloud cover and fog while the rest of the warm planet provided sufficient heat. Since the early Eocene, however, several changes in the planet's geography have resulted in more heat being directed to the Arctic during winter. These changes include the opening of the Arctic Gateway, the closing of the Panama Gateway, and the rise of the Himalayas. As a result, the planet has been losing more heat, leading to progressive cooling. Regardless of CO_2 levels, the planet's current tectonic setting will determine the occurrence of the next glaciation, which is inevitable in a few thousand years.

[145] Zachos, J., et al., 2001. Science, 292 (5517), pp.686–693.
    doi.org/10.1126/science.1059412

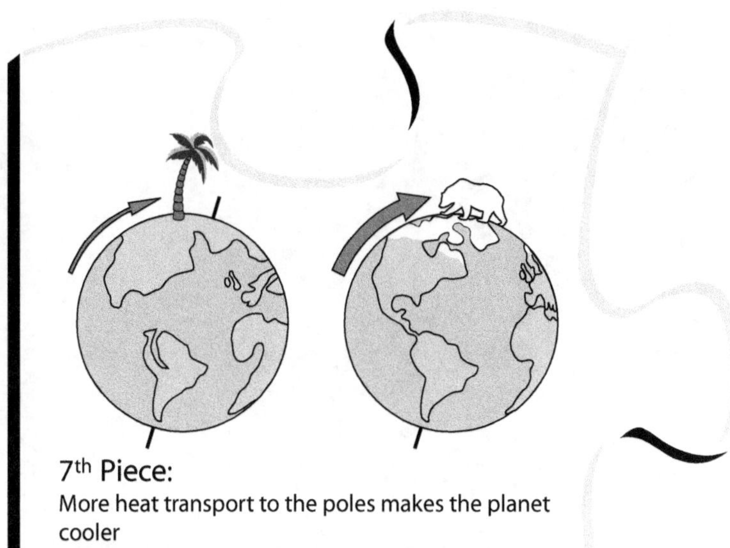

7th Piece:
More heat transport to the poles makes the planet cooler

Chapter 21
Past Climate Change and CO_2 Levels

It is often claimed that CO_2 is the main driver of the planet's climate. However, this claim is difficult to support based on past climate change. The reason is that the evidence we have from the past is limited and uncertain, making it difficult to determine CO_2 levels accurately. We only have reliable CO_2 records from ice cores for the last 800,000 years. Although there is generally a strong correlation between CO_2 levels and temperature during the Pleistocene, there are discrepancies. During some interglacial periods, temperatures dropped sharply for thousands of years while CO_2 levels remained high, which is considered impossible under current climate conditions. In addition, there has been a growing discrepancy over the past two centuries between the rise in CO_2 levels and the lack of warming in Antarctica, where the longest Pleistocene records exist. This discrepancy suggests that past correlations between CO_2 and temperature may not have been due to CO_2 changes driving temperature changes. An alternative evidence-based explanation for the glacial cycle is based on changes in the summertime poleward transport of heat and moisture driven by changes in the Earth's axial tilt.

Throughout the Holocene, there has been a stark contrast between changes in CO_2 levels and changes in temperature. This means that it's clear that CO_2 has not been the primary cause of climate change over the past 10,000 years. This raises the possibility that the current scenario of rising CO_2 levels leading to global warming is an anomaly. Alternatively, it could indicate that we don't fully understand all the factors contributing to the observed warming.

The lack of evidence in the Phanerozoic Eon

The idea that atmospheric CO_2 is the primary factor controlling Earth's temperature has been proposed and widely accepted.[146] If this hypothesis is true, we would expect the evidence from the past to support it or at least not contradict it. In the previous chapter (box 17), we reviewed the evidence for past temperatures over the past 540 million years (the Phanerozoic Eon). We found a general pattern of alternating warm and cold periods, with three major glaciations. Global temperatures probably ranged from 32°C to 9°C (89 to 48 °F). While some suggest that temperatures may have been even higher during the early Phanerozoic, such high temperatures are biologically implausible. In a very wet environment, these temperatures would have decimated terrestrial animal life by overheating. We already worry about corals surviving modest warming on a planet of about 14.5°C (58 °F), so how could they have survived on a planet with temperatures well above 32°C (89 °F)? In any case, past tem-

---

[146] Lacis, A.A., et al., 2010. Science, 330 (6002), pp.356–359.
  doi.org/10.1126/science.1190653

peratures are constrained by latitudinal temperature gradients reconstructed from biogeographic and geological information.

Determining CO_2 levels in the distant past is a major challenge. There is little evidence in soils, fossils, isotopes, and organic remains to help us understand the composition of the atmosphere in the past. Since the 1980s, scientists have relied on geochemical models of the carbon cycle to estimate past CO_2 concentrations. The problem with these models, however, is that we can't be sure that they accurately assess all the important processes that have determined past CO_2 concentrations. In a 159-page review published in 2001, researchers found that these models were inadequate and that there was *"considerable disparity...between all the models and the geological climate evidence indicating changing climate gradients throughout the Phanerozoic."*[147] Even in their latest versions, the two major models still disagree substantially in their estimated CO_2 levels over the past 350 million years.

Looking at the proxy evidence for CO_2 levels throughout the Phanerozoic period, Mark Twain's quote about science rings true: *"There is something fascinating about science. One gets such wholesale returns of conjecture out of such a trifling investment of fact."*[148] A recent study analyzed CO_2 levels from proxies for the last 430 million years, and for 77% of that time, there was an average of only one record per million years. For almost half of the 430 million-year periods, there were no records at all.[149] This means that we have very little information about CO_2 levels in the distant past.

To make matters worse, only 3% of the 430 million years contain 42% of the data. You'd think we'd have a better understanding of CO_2 levels during these epochs, but we don't. For example, during the period 220-201 million years ago there were an average of 15 records per million years, but the central CO_2 values ranged from 600 to 3,700 ppm, and the low-to-high estimates ranged from 300 to 7,000 ppm. The authors of the study tried to average out this mess, but it's still impossible to know with any degree of certainty what atmospheric CO_2 levels were in the distant past.

Amazingly, one of the authors of this study claimed, based on such inadequate data, that CO_2 was the main driver of the Phanerozoic climate.[150] If the climate science community accepts this, it's probably because it fits the current climate paradigm. However, we should remember that the unquestioning acceptance of poorly supported scientific postulates is a major source of scientific error.

When it comes to estimating Phanerozoic CO_2 levels, the most we can say about the model or proxy data is that it neither supports nor contradicts the primary role of CO_2 in determining climate. This is because the available data are too sparse and uncertain, and the models don't even agree. It is, therefore, unjustified to make definitive claims based on the available evidence.

---

[147] Boucot, A.J. & Gray, J., 2001. Earth Sci. Rev. 56 (1–4), pp.1–159.
doi.org/10.1016/S0012-8252(01)00066-6
[148] Mark Twain, 1883. Life on the Mississippi.
[149] Foster, G.L., et al., 2017. Nat. Commun. 8 (1), p.14845.
doi.org/10.1038/ncomms14845
[150] Royer, D.L., et al., 2004. GSA today, 14 (3), pp.4–10.
doi.org/10.1130/1052-5173(2004)014<4:CAAPDO>2.0.CO;2

Box 18. The Cenozoic disagreement

The Phanerozoic CO_2 proxy data suffer from significant quality problems, a fact that is not openly acknowledged when studying the relationship between past climate change and CO_2 variability. However, as we approach the present, the data quality improves. In particular, our understanding of CO_2 changes during the Cenozoic Era, also known as the Age of Mammals, is more reliable.

The last 50 million years of the Cenozoic are of particular importance because of the remarkable transition from the hothouse equable climate of the early Eocene to the severe icehouse climate of the Pleistocene. During this period, there was a gradual and sometimes abrupt decrease in global temperatures, accompanied by a corresponding decrease in CO_2 levels. As a result, both global temperature and CO_2 levels are now well below their Phanerozoic averages.

The "simultaneous" decline observed between CO_2 and temperature is often interpreted as evidence that the decline in CO_2 levels led to a decline in temperature, supporting the hypothesis that CO_2 is the primary driver of climate change. However, a careful examination of the evidence tells a different story. The problem lies in the crucial disagreement within the data about the timing of temperature and CO_2 declines over the past 50 million years, a point that is rarely acknowledged.

Figure B18 shows Cenozoic temperature and CO_2 data from a recent publication.[151] In figure B18a, the authors' temperature reconstruction (shown by the thick black line) is based on oxygen isotope variations in deep-sea benthic foraminifera and agrees well with other climate-related evidence. The graph illustrates a cooling trend from the early Eocene climatic optimum to the Eocene-Oligocene transition, which marked the freezing of Antarctica and resulted in a sudden global cooling event. This was followed by a long period of warming from the mid-Oligocene to the Mid-Miocene Climate Optimum, characterized by fluctuations in the size of the Antarctic ice sheet. This warm period was followed by another cooling period that lasted until the Pleistocene, during which Greenland underwent extensive glaciation. This helped establish the cold conditions that persist today.

Figure B18a also shows the authors' reconstruction of CO_2 levels (thick gray line) based on data from several studies using different proxies. Notably, this reconstruction shows that the decline in temperature did not consistently follow the decline in CO_2 levels. Surprisingly, most of the CO_2 decrease occurred during the Oligocene.

The clearest illustration of this discrepancy can be found in figure B18b, which corresponds to figure 2D of the scientific paper. This scatterplot shows the relationship between temperature and CO_2 changes. According to the hypothesis that CO_2 and temperature should move in sync, they should align along a diagonal line on the graph. However, this diagonal pattern only appears since the mid-Pliocene.

---

[151] Westerhold, T., et al., 2020. Science, 369 (6509), pp.1383–1387.
  doi.org/10.1126/science.aba6853

Throughout the rest of the period, either temperature changes (resulting in a horizontal shift) or CO_2 changes (resulting in a vertical shift). The plot shows a weak correlation between these two variables. It also shows that CO_2 changes precede temperature changes by tens of millions of years, while no existing hypothesis can explain such long lags.

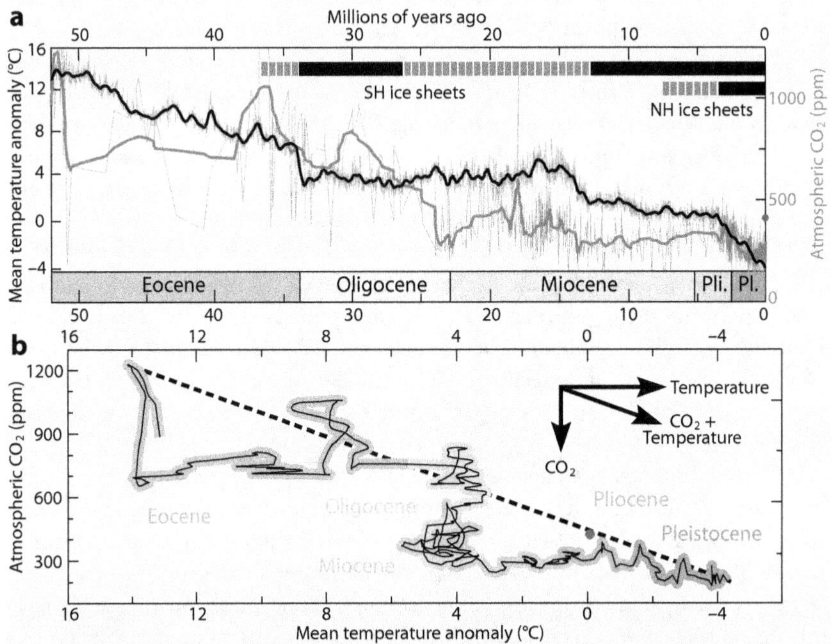

Figure B18. Relationship between temperature and CO_2 over the past 52 million years. a) Reconstructions of temperature (black line) and CO_2 (gray line), with thick lines showing smoothed data. Black bars indicate periods when the Northern or Southern Hemisphere ice sheets were more than 50% of their present extent, and dashed gray bars indicate less than 50% extent. The grey dot marks the current CO_2 level. b) Scatterplot of the thick lines in a) with the epochs indicated. Synchronous changes in CO_2 and temperature should follow a diagonal path. The gray dot indicates the current CO_2 level and temperature.

While it is worth noting that the CO_2 proxy data suffer from quality problems and future reconstructions may bring changes, the existing data provide a clear picture. They show that CO_2 levels at the end of the Oligocene and at the end of the Pliocene, which occurred 20 million years later, were not significantly different. This observation holds despite the large cooling that occurred between these two periods.

The current situation is represented by a gray dot in the figure, aligned with the diagonal line on the scatterplot. This indicates that the current temperature and CO_2 levels are what we would expect from a Cenozoic perspective. It is curious that the authors of the study claim that *"if CO_2 emissions continue unmitigated until 2100, Earth's climate system will be moved abruptly from the Icehouse into the Warmhouse or even Hothouse climate state."* Their own evidence does not support such a claim.

The prevailing belief, supported by most scientists and the IPCC, is that declining CO$_2$ levels during the Cenozoic Era was the primary cause of the cooling that led to the Ice Age. This belief also supports the Enhanced CO$_2$ effect hypothesis. However, it is important to recognize that the available data do not support this widely accepted view.

Concordance between CO$_2$ and temperature in the Pleistocene

We only have reliable CO$_2$ records from Antarctic ice cores for the last 800,000 years. These records show a consistent correlation between temperature and CO$_2$ changes (fig. 34a). While some argue that the lag between temperature and CO$_2$ at the end of glaciations is significant, the differences are too small and variable to draw definitive conclusions. The cause of glaciations has been debated since their discovery in the 1830s, with some attributing the cause to external factors such as orbital theories and others to internal factors such as the greenhouse effect. In 1976, the former group prevailed when it was discovered that glaciations follow orbital frequencies.

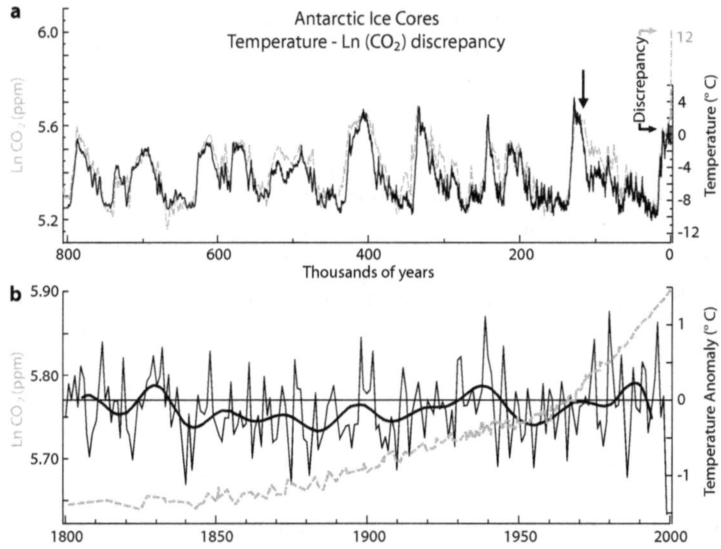

Figure 34. Temperature-CO$_2$ discrepancy in Antarctica. a) Temperature (black curve) and natural logarithm of CO$_2$ (dashed gray curve) for the last 800,000 years from Antarctic ice cores, updated with atmospheric CO$_2$ data after 2001. The vertical arrow points to the discrepancy 123.5-115 thousand years ago, and the horizontal arrows to the current discrepancy. b) As in (a) for 19th and 20th-century data. No temperature change is observed in response to the massive increase in CO$_2$.[152]

---

[152] Data from Jouzel, J., et al., 2007. Science, 317 (5839), pp.793–796. doi.org/10.1126/science.1141038 Bereiter, B., et al., 2015. Geophys. Res. Lett. 42 (2), pp.542–549. doi.org/10.1002/2014GL061957 NOAA annual mean CO$_2$ data. Schneider, D.P., et al., 2006. Geophys. Res. Lett. 33 (16), L16707. doi.org/10.1029/2006GL027057

Since the exchange of CO_2 between the ocean and the atmosphere is affected by ocean temperature, any change in ocean temperature will result in a corresponding change in CO_2 levels. Changes in CO_2 levels, in turn, affect temperature through the greenhouse effect. However, it's important to note that the reduction of ice cover at glacial terminations also leads to increased volcanic activity as the crust adjusts to the weight loss. This process could contribute up to half of the observed CO_2 increase.[153] As a result, the relationship between CO_2 and temperature goes both ways, and the correlation between the CO_2 and temperature records during the Pleistocene is insufficient evidence to support or refute the Enhanced CO_2 Effect hypothesis.

However, the record also shows that 120,000 years ago, at the end of the previous interglacial period, the temperature fell by 4°C (7 °F) over 8,000 years, while CO_2 levels remained elevated (fig. 34a arrow; see fig. 56, ch. 35, for a closer view). This supports the idea that temperature during the glacial cycle is under orbital control rather than CO_2 control. At the very least, it shows that temperatures can decrease substantially over thousands of years despite elevated CO_2 levels. This is an important consideration for the future.

The Enhanced CO_2 Effect hypothesis faces a major challenge in Antarctica. According to the Pleistocene CO_2-temperature relationship (fig. 34a), current CO_2 levels should correspond to Antarctic temperatures 12°C (22 °F) higher than they are. However, as figure 34b shows, Antarctica has not warmed over the past 200 years despite rising CO_2 levels. This discrepancy poses a serious problem for the hypothesis. While explanations have been proposed to account for this anomaly, the fact remains that if the CO_2-temperature relationship does not hold now, it cannot be used to defend past causality or to predict future climate outcomes. Therefore, claims that our climate could become like that of the Miocene, the last time CO_2 levels were this high, cannot be substantiated.

Box 19. Gradients and transport in a glacial world

The tilt of the planet's axis (known as obliquity) varies slightly, from 22.1° to 24.3°, with a cycle of about 40,000 years. This change has an important effect on the amount of solar radiation received at high latitudes throughout the year and the seasons while having a minimal effect at low latitudes. These changes in solar radiation lead to a change in the amount of energy delivered to each latitude over thousands of years, resulting in a substantial impact on climate. When the obliquity of the planet becomes greater than 23° every 40, 80, or 120,000 years, there is an opportunity to move out of the typical glacial state and into an interglacial state, which is not possible when the obliquity is less than 23°. During periods of high obliquity, summers at high latitudes are warmer and ice melts faster. However, the difference in solar radiation energy due to changes in obliquity decreases rapidly with latitude, and most of the planet is minimally affected. As a result, many climatologists believe that changes in axis wobble (precession) are more important in producing an interglacial, despite evidence supporting obliquity as

---

[153] Huybers, P. & Langmuir, C., 2009. Earth Planet. Sci. Lett. 286 (3-4), pp.479–491. doi.org/10.1016/j.epsl.2009.07.014

the main factor.[154]

Although it has little effect on the amount of solar radiation received in the tropics and midlatitudes, obliquity surprisingly has a significant effect on many tropical and subtropical paleoclimate records. The migration of the Intertropical Convergence Zone, the Earth's climatic equator (box 2, ch. 3), is also strongly and unexpectedly influenced by obliquity.[155] In addition, analysis of the source of moisture in the Greenland and Antarctic ice sheets during glacial periods indicates that obliquity has a strong effect associated with changes in the latitudinal temperature gradient.[156]

This unexpected evidence shows that obliquity has a surprising influence on atmospheric circulation, the water cycle, and moisture transport. This observation can be explained by two critical themes of this book: the control of the latitudinal temperature gradient and the resulting poleward transport of heat and moisture. The changes in tropical climate caused by obliquity are due to the changes it causes in the summer insolation gradient.[157] This explanation addresses one of the primary mysteries of glaciation. Because the seasons are reversed between the hemispheres, changes in summer insolation have opposite signs at the two poles (fig. B19a). Nevertheless, glaciations and interglacials are symmetrical phenomena, with the entire planet going into or out of glaciation. The amount of seasonal insolation at a given latitude depends primarily on precession. However, the summer pole faces the Sun, and the amount of solar radiation at high latitudes during the summer varies significantly with obliquity. When obliquity is high, the summer latitudinal insolation gradient flattens; when it is low, it steepens (fig. B19b). Changes in the summer insolation gradient follow changes in obliquity almost exactly in both hemispheres, and it is summer conditions that are critical for the onset and end of glaciation.

High obliquity results in a flatter summer insolation gradient, which favors energy conservation by the planet. In contrast, low obliquity causes the gradient to become steeper, as indicated by the more negative values in figure B19b. A steeper gradient causes more energy and moisture to flow toward the poles, leading to planetary cooling, ice growth and, ultimately, the transition from interglacial to glacial periods. As explained in the previous chapter, greater poleward heat transport causes the planet to cool.

The evidence supports the view that the glacial-interglacial cycle results from changes in the summer latitudinal insolation gradient, which subsequently cause shifts in the temperature gradient. These shifts, in turn, cause changes in the poleward transport of heat and moisture necessary for the formation and melting of ice sheets. This interpretation of the glacial cycle emphasizes the importance

---

[154] Tzedakis, P.C., et al, 2017. Nature, 542 (7642), pp.427–432.
  doi.org/10.1038/nature21364

[155] Liu, Y., et al., 2015. Nat. Commun. 6 (1), p.10018. doi.org/10.1038/ncomms10018

[156] Masson-Delmotte, V., et al., 2005. Science, 309 (5731), pp.118–121.
  doi.org/10.1126/science.1108575

[157] Bosmans, J.H.C., et al., 2015. Clim. Past, 11 (10), pp.1335–1346.
  doi.org/10.5194/cp-11-1335-2015

of summer heat and moisture transport from the tropics to the poles, a factor influenced by obliquity. According to this hypothesis, the tropics, which have a large heat and moisture capacity, play a primary role in the growth and decay of ice sheets, orchestrated by changes in obliquity. Other factors, including CO_2, play a secondary role. Thus, changes in poleward heat and moisture transport in response to changes in temperature gradients due to orbital variations are strong candidates as the ultimate cause of the glacial cycle.

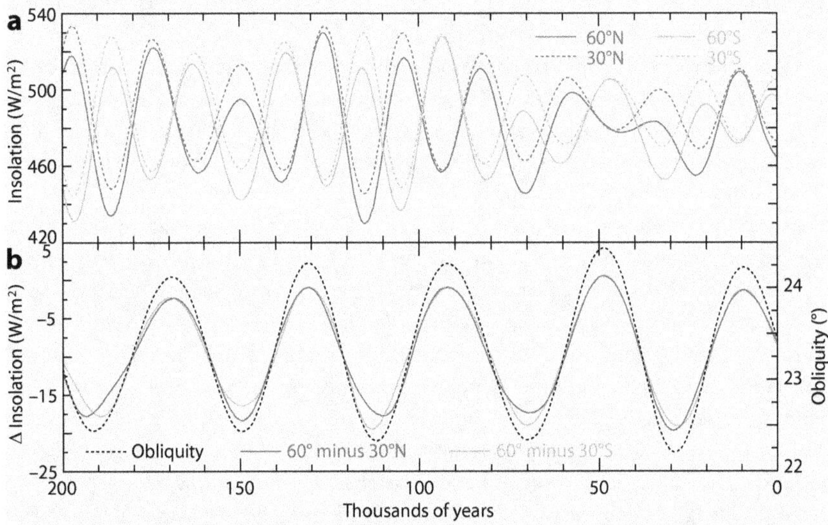

Figure B19. Changes in the latitudinal summer insolation gradient as a function of obliquity. a) Mean summer insolation depends mainly on precession. July insolation in the Northern (dark gray) and Southern (light gray) Hemispheres at 60° (solid curves) and 30° (dotted curves). b) The summer insolation gradient, shown as the difference in insolation between 60° and 30°, depends on obliquity (black dotted curve).[158]

The Holocene temperature-CO_2 conundrum

The Pleistocene temperature and CO_2 records are in fairly good agreement, but they are in complete disagreement during the Holocene. This has led to much debate among climatologists about different reconstructions of Holocene temperature. Biological and glaciological evidence shows large temperature changes over the past 10,000 years, while CO_2 changes have been relatively small and in the opposite direction.

Figure 35a shows a well-known reconstruction of Holocene temperature from 73 proxies.[159] Unlike the published version, I have not changed the original dating of the proxies, and they are expressed as an anomaly to their mean before averaging. The reconstruction is expressed in Z-score – the distance of the data from the mean in standard deviations – which helps to avoid uncertain

---

[158] Data from Laskar, J., et al., 2004. Astron. Astrophys. 428 (1), pp.261–285.
doi.org/10.1051/0004-6361:20041335
[159] Marcott, S.A., et al., 2013. Science, 339 (6124), pp.1198–1201.
doi.org/10.1126/science.1228026

temperature attributions. The reconstruction presented here ends in 1920 because the number of proxies is limited after that. This reconstruction is consistent with an independently obtained record of glacier advances over centuries,[160] indicating that the Holocene was divided into a warm period of about five millennia (known as the Holocene Climatic Optimum), followed by a cooling period of about five millennia (known as the Neoglaciation). This overall pattern is punctuated by several well-known cooling episodes that have left a strong imprint on both records, such as the Pre-Boreal Oscillation, the 8.2 and 5.2 kiloyear events, and the Little Ice Age. The close agreement between the two independent global records gives us confidence that the general features of Holocene temperature evolution are reflected in the reconstruction.

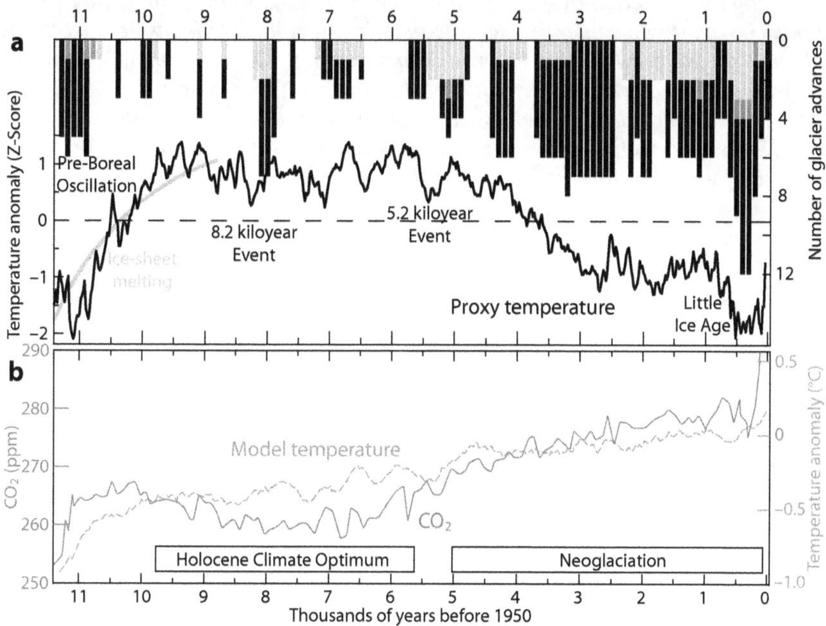

Figure 35. Holocene temperature and CO₂ evolution. a) Top inverted, global glacier advance in 17 areas of the Northern Hemisphere (black bars), Southern Hemisphere (light gray bars), and low latitudes (medium gray bars) for each century of the Holocene. The Holocene global temperature reconstruction from 73 proxies, expressed as Z-score or distance from the mean in units of standard deviation, is shown below. b) CO₂ levels (medium gray curve) measured in an Antarctic ice core and global temperature simulated from an ensemble of three models (dashed light gray curve).

During most of the Holocene, CO₂ levels were typically between 260 and 280 ppm. Interestingly, they decreased during the Holocene Climatic Optimum but increased during the Neoglaciation, the opposite of what would be expected if CO₂ were the driving force behind temperature changes.[161] This mis-

[160] Solomina, O.N., et al., 2015. Quat. Sci. Rev. 111, pp.9–34.
doi.org/10.1016/j.quascirev.2014.11.018
[161] Monnin, E., et al., 2004. Earth Planet. Sci. Lett. 224 (1–2), pp.45–54.
doi.org/10.1016/j.epsl.2004.05.007

match between CO_2 levels and temperature is known as the *"Holocene temperature conundrum"* because many climate models are so sensitive to CO_2 changes that they reproduce a warming trend throughout the Holocene that is contradicted by the evidence.[162]

In summary

The evidence from the past does not support the claim that changes in CO_2 are the main driver of climate change. Although a correlation is expected and would not prove causality, such a correlation is difficult to find outside the Pleistocene due to data quality issues. Within the Pleistocene, a correlation is present during the pronounced changes in the glacial cycle, but the increasing disparity in Antarctica over the past two centuries casts doubt on its interpretation. The available evidence suggests that the glacial cycle is instead driven by changes in the Earth's axial tilt, which drive changes in the amount of heat and moisture transported poleward during the summer. Over the past 10,000 years, CO_2 and temperature have changed in opposite directions. The situation since 1975, in which CO_2 increases may have become the primary driver of global warming, would be the exception, not the rule.

---

[162] Liu, Z., et al., 2014. PNAS, 111 (34), pp.E3501–E3505.
doi.org/10.1073/pnas.1407229111

CHAPTER 22
ABRUPT CLIMATE EVENTS IN THE HOLOCENE

The climate of the Holocene was highly unstable. Scientists have identified nearly two dozen abrupt climate events, occurring at a rate of two per millennium, by analyzing climate proxies. This is challenging because changes in CO_2 levels during the Holocene have been minimal until recently, so the cause of most of these events remains a mystery. Some experts suggest that changes in solar or volcanic activity may have triggered them. However, existing climate models and the IPCC consider these natural factors to be too weak to explain the abrupt events.

Four of the major abrupt climatic events of the Holocene occurred with an irregular quasi-periodicity of 2,500 years. These events marked the boundaries of distinct paleoecological periods. The climatic changes observed during these four events indicate that there was a shift in atmospheric circulation that affected the Northern Hemisphere more intensely. This led to a contraction of the tropics and an expansion of the polar regions, increasing the temperature gradient between the equator and the poles. The resulting atmospheric reorganization caused more heat to flow toward the poles, leading to a cooling of the planet and changes in precipitation patterns.

The unstable Holocene climate

In the previous chapter, we learned that for most of the Holocene, which spanned more than 10,000 years, CO_2 levels fluctuated within a narrow range of 20 ppm. Compared to today, where it can rise by 20 ppm in as little as eight years, this tiny change could not have had a significant impact on the climate. However, data collected over the past century show that the climate changed substantially over the same period.

By the early 20th century, researchers had identified a general pattern of Holocene climate change characterized by three distinct phases: a warming phase, a warm period (known as the Holocene Climatic Optimum), and a cooling phase (known as the Neoglaciation). Scandinavian researchers later divided the Holocene into five phases based on paleoecological studies of vegetation changes driven by temperature and precipitation shifts. These periods include the Pre-Boreal warming phase, which was shorter than the following four periods: Boreal, Atlantic, Sub-Boreal, and Sub-Atlantic, each of which lasted about 2,500 years (fig. 36). Notably, the changes between these periods were relatively abrupt, as evidenced by the transition from the Sub-Boreal to the Sub-Atlantic about 2,800 years ago, described in chapter 45.

For the Holocene climate, we define a change in climate parameters as abrupt if it occurs much faster than the long-term trend but still takes several decades or even a few centuries. By this definition, the current climate change that has occurred over the past two centuries qualifies as abrupt.

Holocene abrupt climate events have been identified using a variety of proxies in different locations around the world, such as iceberg activity in the North Atlantic, methane trapped in Greenland ice cores, precipitation records from the Middle East, and changes in the Asian monsoon. As research has progressed, the number of identified events has increased, and recent estimates suggest that at least 23 abrupt climate events took place at a rate of about two events per millennium.[163] The Holocene climate has been consistently unstable, with the current abrupt climate event being the latest in a long series of fluctuations.

The problem is that we are unable to determine the cause of almost all but one of the Holocene's abrupt climate events. Having ruled out CO_2 as a significant factor, we are left with only one credible hypothesis to explain a single event. The sudden discharge of a huge volume of icy meltwater from a massive glacial lake system in North America into the North Atlantic 8,300 years ago is thought to have contributed to one of the most significant abrupt events of the Holocene, known as the 8.2 kiloyear event.

There is no consensus among scientists as to the cause of the numerous abrupt climate events during the Holocene, and climate models have not been useful in explaining them because they cannot replicate them. Some paleoclimatologists argue that they were caused by regular variations in solar activity and sporadic changes in volcanic activity, but models attribute only a small effect to these natural forcings.

Quasi-periodic abrupt cooling events

To simplify the complexity of Holocene climate change, let's focus on four well-known and studied events with important climatic effects according to proxies. These events are the Boreal Oscillation (10,300 years ago), the 5.2 kiloyear event (5,200 years ago), the 2.8 kiloyear event, and the Little Ice Age (1300-1845 AD). They are separated by paleoecological periods identified in mid-20th-century pollen studies, as shown in figure 36. Three of these events are separated by 2500 ± 300 years, and the fourth by twice as long, forming a quasi-cycle with a missing beat. This quasi-cycle can be extended with other identifiable abrupt events to 20,500 years ago, during the Last Glacial Maximum, without missing another beat.

Many climatologists view climate as long-term weather, a chaotic, internally generated phenomenon that changes as conditions affecting the energy flux at the top of the atmosphere change. They reject the possibility that climate can be cyclical and externally determined on shorter scales than the Milankovitch orbital cycles, which last tens of thousands of years. Their position is understandable, given the historical failure to link climate changes to changes in the Sun or Moon. The evidence is also less convincing because changes in the Sun and climate are not strictly cyclical but quasi-periodic. For example, the length of the 11-year solar cycle ranges from 9 to 14 years, and its amplitude can vary widely. Despite this irregularity, the solar cycle is well established because it has been repeated many times over the past 200 years. But an irregular cycle that takes 2,500 years to complete and is sometimes undetectable is harder to

---

[163] Vinós, J., 2022. Climate of the Past, Present and Future: A scientific debate. p.61. Critical Science Press.

accept, even if that is how the Sun behaves. As a result, climatologists may remain skeptical of evidence linking climate change to quasi-cyclical variations in external factors such as solar activity.

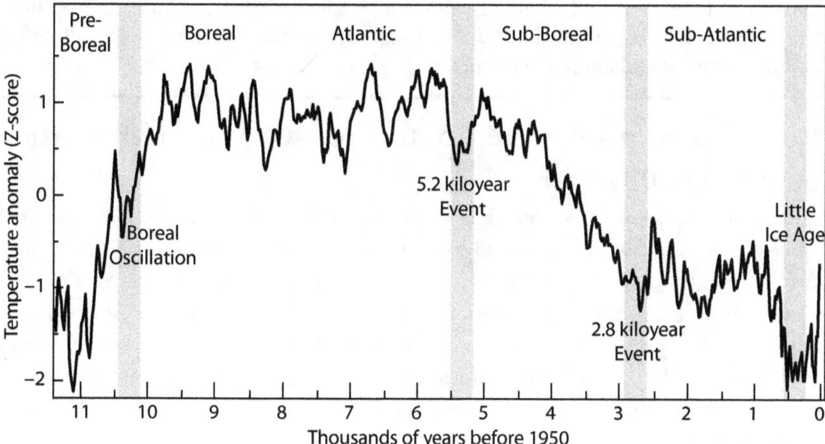

Figure 36. Four major abrupt climate events. They are indicated by vertical bars in a Holocene temperature reconstruction (ch. 21). They separate some previously recognized paleoecological periods indicated by their names at the top.

The four abrupt events shown in figure 36 had different impacts on the evolving Holocene climate because the background climate state was different during each event. However, there are some changes that all four events had in common, which are supported by various scientific studies.[164] These changes include:

* A specific increase in precipitation at mid and high latitudes
* Decreased precipitation in tropical and subtropical regions
* Weaker tropical monsoons
* Increased wind strength at mid and high latitudes
* Enhanced polar atmospheric circulation
* Winter cooling
* Cooling of the ocean surface
* Increase in the temperature difference between the equator and the poles
* Advance of glaciers
* Increased iceberg activity

All of these factors point to an atmospheric effect consistent with an expansion of the polar cells leading to a southward shift of the jet stream, an equatorward shift of the Ferrel cell and the subtropical jet, and a similar shift of the descending portions of the contracting Hadley cells. These changes in wind patterns may be responsible for the changes in precipitation and temperatures.

We can conclude that these four events caused a significant change in the atmospheric circulation of the planet, leading to a stronger winter circulation that drove more heat poleward through an increasingly steep temperature gradient. As discussed previously (ch. 11-16), the effects were most pronounced in

---

[164] Vinós, op. cit., p.106, and references within its chapter 6.

the Northern Hemisphere due to the stronger and more variable winter circulation toward the Arctic. These events lasted so long and had such a profound effect that by the time they ended and atmospheric circulation returned to normal, the planet's climate had already shifted to a different state. In the next chapter, we will examine the evidence supporting the hypothesis that changes in solar activity caused these events.

Box 20. Are present temperatures warmer than in the past 125,000 years?

The current global warming is a departure from the general cooling trend of the Neoglacial period. Undoubtedly, the massive amount of CO_2 released into the atmosphere by human activities is contributing to this warming. Most scientists agree that it is the primary cause of the warming. However, modern global warming began 180 years ago, long before the acceleration in emissions that has occurred since 1960. In addition, the increase in temperature does not show the expected acceleration for the exponential increase in atmospheric CO_2 that is currently occurring.

Many people are alarmed by the constant stream of alarming news about climate science, and a key question is how unusual is the current abrupt climate event. We've reached atmospheric CO_2 levels not seen on Earth in the last 3 million years since the mid-Pliocene warm period, which are 60% higher than during the Holocene Climatic Optimum. If CO_2 is indeed the primary driver of climate, as many believe, then the current temperature, after so much warming, should be somewhere between those two warm periods. Some Holocene temperature reconstructions support this, and it is expressed in the IPCC's 6th Assessment Report as *"surface temperatures of the past decade were probably warmer than when the long cooling trend began around 6500 years ago."*[165]

The claim that current surface temperatures are warmer than during the Holocene Climatic Optimum is unreliable. This conclusion is based on a comparison of a proxy Holocene temperature reconstruction with instrumental datasets. Such a comparison deserves several important criticisms. Proxies do not record temperature changes directly but result from biological or geological processes that respond to temperature changes. Converting proxy data into temperature changes involves many uncertainties and unproven assumptions. Combining marine and terrestrial proxies to compare temperature changes is also problematic because they don't change in the same way. Proxy and instrumental temperatures are fundamentally different and should not be compared quantitatively. Furthermore, the construction of a proxy collection or temperature dataset involves many human decisions that are prone to unintentional cognitive biases. Is there any other evidence that tells us whether the Holocene Climatic Optimum was warmer or colder than today? Yes, we have two: glaciers and trees.

---

[165] Gulev, S.K., et al., 2021. Climate change 2021: The physical science basis. 6th AR IPCC. p.378. doi.org/10.1017/9781009157896.004

The Holocene Climatic Optimum was the period in the last 100,000 years when glaciers were at their smallest, while the Little Ice Age was the period in the last 7,000 years when glaciers were at their largest. Between 8,000 and 4,000 years ago, glaciers were generally smaller than they are today in most mid to high-latitude regions of the Northern Hemisphere. The IPCC's 6th Assessment Report acknowledges that most glaciers in the Northern Hemisphere are now larger than before but points out that they have a relatively long time to adjust. However, 80% of the world's glaciers are very small, with an area of 1 km^2 (0.4 sq miles) or less, and glaciers are affected by the average annual temperature and precipitation at their surface rather than by global warming.

Tropical glaciers have experienced the greatest shrinkage since 1980, although warming has been less intense in these areas. Glaciers in mid-high latitudes, where warming has been more intense, have not retreated as much.[166] It's important to note that glacier shrinkage is not solely due to temperature increases; anthropogenic accumulations of black carbon and debris are also contributing factors. These non-climate factors are likely to exacerbate the current shrinkage. The extratropical Northern Hemisphere has experienced the most warming from modern global warming. The presence of several glaciers and small permanent ice patches there that didn't exist during the Holocene Climatic Optimum is strong evidence that the past was warmer than the present.

Another way to determine whether the Holocene Climatic Optimum was warmer than today is to look at biology. Trees do not grow above a certain height above sea level, known as the treeline. Temperature is the main factor that determines where the treeline is. As a result, the treeline has moved higher in many places over the past century, especially in the extratropical Northern Hemisphere, where winter warming has been particularly intense.

Numerous studies have shown that during the Holocene Climatic Optimum, the treeline in the Italian Alps, Swiss Central Alps (fig. B20), Pyrenees, Swedish Scandinavia, and British Columbia was much higher than today.

Figure B20. Treeline fluctuations in the central Swiss Alps during the Holocene. Altitude in meters above sea level. The present-day treeline in the central Swiss Alps is 150-200 m below the Holocene Climatic Optimum treeline.[167]

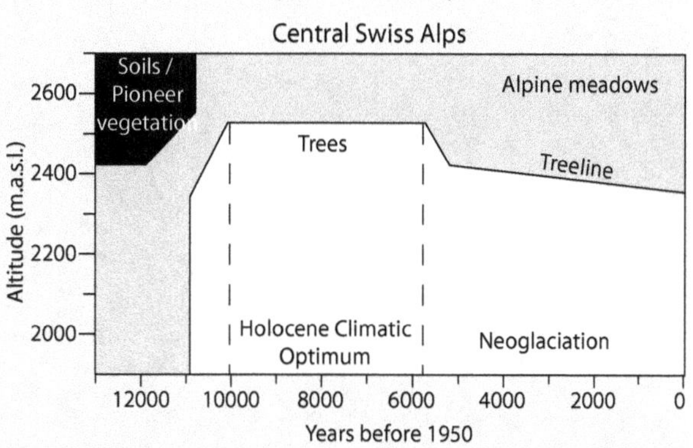

[166] Li, Y.J., et al., 2019. Adv. Clim. Change Res. 10 (4), pp.203–213. doi.org/10.1016/j.accre.2020.03.003

[167] Tinner, W. & Theurillat, J.P., 2003. Arct. Antarct. Alp. Res. 35 (2), pp.158–169. doi.org/10.1657/1523-0430(2003)035[0158:ULEAFO]2.0.CO;2

The Northern Hemisphere has experienced the most climate warming in recent decades. Studies have shown that many deciduous tree species in this hemisphere have reached thermal equilibrium, meaning they can't grow at higher altitudes because it's too cold.[168] However, during the Holocene Climatic Optimum, these same tree species were able to grow well beyond their current limits. This makes it clear that the planet cannot be warmer now than it was then, regardless of any proxy reconstructions or homogenization of temperature data. If the planet were warmer today, either the tree species would be out of thermal equilibrium, or their equilibrium altitude would be higher than it was during the Holocene Climatic Optimum.

In summary

During the Holocene, the climate was unstable and there were numerous abrupt climatic events, about two per millennium, that could not have been caused by changes in CO_2 levels. Some of these events had an almost periodic distribution and involved significant atmospheric reorganization, resulting in a steeper temperature gradient, reduced tropical zones, and expanded polar regions. This, in turn, increased poleward heat transport, leading to global cooling and significant changes in precipitation patterns. The current abrupt climate event is only the latest in a long chain. Despite the warming that has occurred, it's clear from glaciological and biological evidence that the Holocene Climatic Optimum was warmer than the present. This evidence is more reliable than uncertain reconstructions.

---

[168] Randin, C.F., et al., 2013. Glob. Ecol. Biogeogr. 22 (8), pp.913–923.
doi.org/10.1111/geb.12040

CHAPTER 23
PAST SOLAR ACTIVITY AND CLIMATE

Scientists have used radiocarbon dating to reconstruct past solar activity and have found evidence that some of the most important abrupt climatic events in history occurred during long and deep solar grand minimums, particularly the Spörer type. These solar minimums are part of a 2,500-year solar-climatic periodicity known as the Bray cycle, the most important climatic cycle at frequencies below 10,000 years. The importance of this solar cycle in climate change is further supported by evidence that the largest declines in human population have occurred during periods of prolonged low solar activity.

It's important to note that if solar activity has had such an effect on climate in the past, it is likely to have an appreciable impact on climate today. However, climate models do not currently include a significant effect of solar activity on climate, and many scientists reject this possibility.

Radiocarbon dating and past solar activity

Radiocarbon dating has been a scientific breakthrough. In addition to allowing archaeological dating, it has allowed scientists to determine past levels of solar activity. The method relies on the arrival of high-energy cosmic rays, which produce a radioactive isotope called ^{14}C in the atmosphere. This ^{14}C combines with oxygen to produce a small amount of radioactive $^{14}CO_2$ in the atmosphere, while the vast majority of CO_2 is non-radioactive $^{12}CO_2$. Plants use both types of CO_2, and the carbon atoms from these molecules end up in the bodies of all living organisms. Over time, ^{14}C radioactively decays to ^{12}C at a constant rate. Therefore, the older an organic sample is, the less ^{14}C it contains. Radiocarbon dating is based on determining the $^{14}C/^{12}C$ ratio of ancient organic remains.

Radiocarbon dating is not a perfect method of measuring time because the atmospheric $^{14}C/^{12}C$ ratio is not constant. This ratio is affected by changes in the amount of CO_2 in the atmosphere, which affects the amount of ^{12}C. However, during the last 11,000 years up to 1850, changes in CO_2 were minimal (ch. 21), so the large ^{12}C term varies very little during the Holocene, making it easy to adjust.

On the other hand, the production of ^{14}C is affected by changes in the solar and terrestrial magnetic fields, which affect the number of cosmic rays reaching the Earth. The Earth's magnetic field changes slowly over thousands of years, but the Sun's magnetic field is in constant flux. When the Sun is less active, its magnetic field weakens, allowing more cosmic rays to reach the Earth and produce more ^{14}C. This, in turn, increases the $^{14}C/^{12}C$ ratio, making older samples appear younger because they contain more ^{14}C.

Conversely, when solar activity is high, the radiocarbon clock runs slower because the $^{14}C/^{12}C$ ratio decreases. Since the radiocarbon clock runs irregularly, radiocarbon dates do not correspond linearly to calendar dates.

Radiocarbon scientists use a calibration curve to convert radiocarbon dates to calibrated or calendar dates when determining ancient dates. This process began in the 1950s, and the starting year for the calibrated calendar is 1950.

Therefore, many graphs refer to the time before 1950, which is considered the present in most paleo-dating studies. The calibration curve is a reliable scientific result, shown in figure 37a, independent of climate studies.

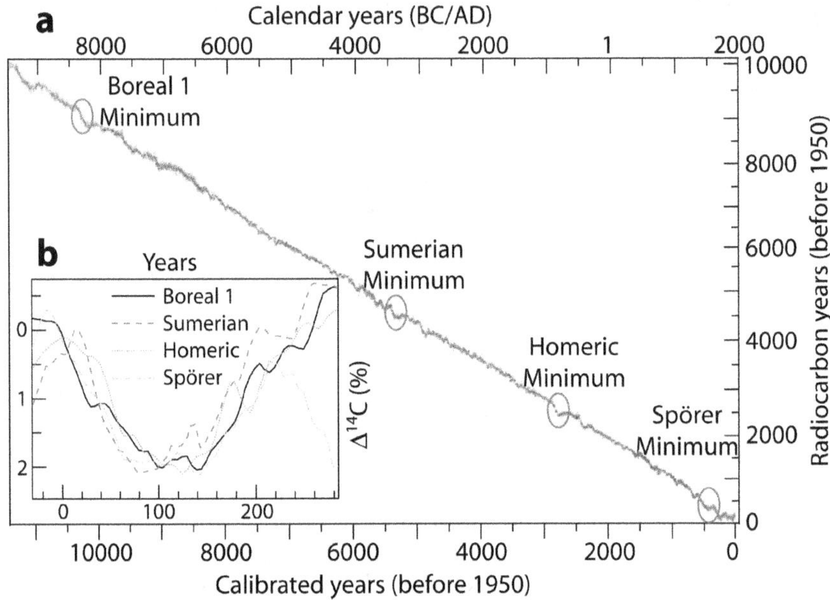

Figure 37. Radiocarbon dating and past solar activity. a) Radiocarbon calibration curve allowing conversion of radiocarbon dates to calibrated or calendar dates.[169] Four of the largest deviations from linearity are indicated by ovals with the name of the corresponding solar grand minimum. b) Superposition of the four Spörer-type minimums to compare their duration and effect on ^{14}C levels.

The curve in figure 37a shows long periods of low solar activity, called solar grand minimums, which appear as dents. Among these, there is a particular type of solar grand minimum called the Spörer type, which lasts for 200 years and has the lowest solar activity, causing a 2% increase in [14]C. There are only four Spörer-type minimums in the Holocene, marked with ovals in figure 37a and shown in figure 37b. These periods are precisely dated and correspond to the four two-century periods of lowest solar activity during the Holocene. What were the climatic conditions during these periods?

A 2500-year solar and climate cycle

To examine the relationship between solar activity and climate, we can combine figures 36 (ch. 22) and 37a into figure 38. The result is quite telling. The four periods of lowest solar activity during the Holocene coincide with four of the largest and most well-known abrupt climate events of the Holocene. This evidence is undeniable, and the direction of causality cannot be doubted since events on Earth do not influence solar activity. It is unsurprising that many paleoclimatologists are convinced that solar activity, despite its small

[169] Reimer, P.J., et al., 2013. Radiocarbon, 55 (4), pp.1869–1887.
    doi.org/10.2458/azu_js_rc.55.16947

energy changes, has an important effect on climate since the evidence clearly supports it. In fact, the authors of a study on Holocene climate change have called for an in-depth, multi-disciplinary assessment of the potential for solar modulation of climate on centennial scales.[170] However, most climate scientists deny this clear paleoclimatic evidence. A recent modeling study of the potential effects of a solar grand minimum occurring in the 21st century concluded that it would cause a temperature difference of only 0.3°C (0.5 °F), and thus global warming would continue, albeit at a slower rate.[171] This is in stark contrast to what the paleoclimate evidence indicates.

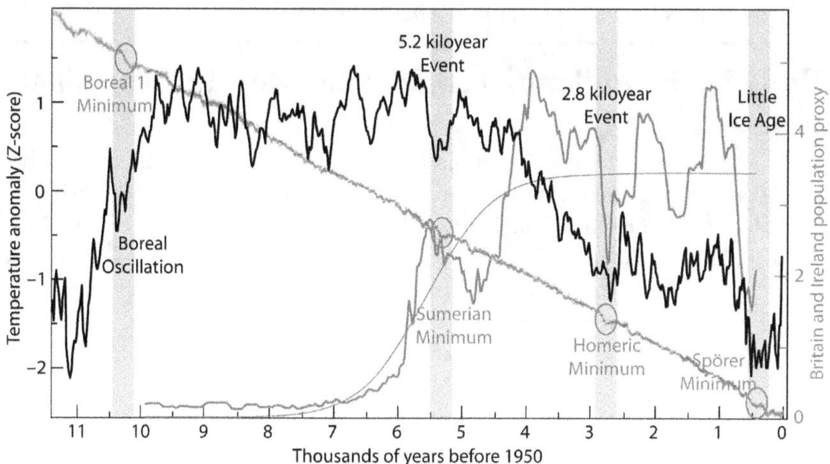

Figure 38. Solar-climate correspondence. Four major abrupt climate events (gray bars), indicated by a Holocene temperature reconstruction (black curve, ch. 21), coincide with the four Spörer-type solar grand minimums (gray ovals). A human population proxy for the British Isles (gray curve) indicates population declines during these and other cooling periods.[172]

In the previous chapter, we discussed the four major abrupt climate events that may have occurred according to a possible 2500-year irregular periodicity. This observation was first made by Roger Bray in 1968, who linked a 2500-year cycle in solar activity to climate.[173] As a result, this cycle has recently been named the Bray cycle.[174]

Most climatologists dispute the idea that solar activity has a substantial effect on climate. They argue that the 11-year solar cycle has only a small effect on climate and that there is no accepted mechanism for a stronger solar effect. Furthermore, the existence of longer solar cycles is problematic. A solar model

---

[170] Rohling, E.J., et al., 2002. Clim. Dyn. 18 (7), pp.587–593. doi.org/10.1007/s00382-001-0194-8

[171] Arsenovic, P., et al., 2018. Atmospheric Chem. Phys. 18 (5), pp.3469–3483. doi.org/10.5194/acp-18-3469-2018

[172] Bevan, A., et al., 2017. PNAS, 114 (49), pp.E10524–E10531. doi.org/10.1073/pnas.1709190114

[173] Bray, J.R., 1968. Nature, 220, pp.672–674. doi.org/10.1038/220672a0

[174] Vinós, J., 2022. Climate of the Past, Present and Future: A scientific debate. Critical Science Press.

has been developed to account for the known 11-year cycle, but longer cycles have no known cause. Some hypotheses have been proposed to explain longer cycles, suggesting that planetary orbits may affect solar activity through various mechanisms, but there is no evidence to support them.

The evidence suggests that the longer Spörer-type solar minimums have a greater impact on climate than the shorter Maunder-type minimums, which last about 70 years. This implies that the effect of low solar activity on climate accumulates over time, becoming stronger the longer the minimum lasts. This may explain why the 11-year solar cycle has a relatively small effect since solar activity is below average for only about five years.

Box 21. The millennial solar cycle over the past 2,000 years

Analysis of past solar activity reveals a distinct millennial frequency, based primarily on the most frequent occurrence of Maunder-type solar grand minimums. The Maunder Minimum, identified with newly invented telescopes, occurred between 1645 and 1715 and was the last solar grand minimum. It coincided with the Little Ice Age, characterized by lower temperatures and the largest glacier advances of the Holocene. Despite some warm summers in Europe, China, and North America, the period also saw some of the coldest winters on record.

The 1000-year solar cycle is named after John Eddy, the astronomer who revived interest in the Maunder Minimum in the 1970s. This cycle is highly irregular, prominent in the ^{14}C record during the early Holocene and the last 2000 years, but less so in between. The Bray and Eddy cycles are nearly in phase, meaning that their minimums occur close together in time every 5,000 years. The last time this happened was during the Little Ice Age, which resulted in a series of three solar grand minimums in less than 500 years, coinciding with the coldest period of the Holocene. This coincidence won't happen again for another 4,500 years, and when it does, it will have a good chance of triggering the next glaciation if it hasn't already started.

The evidence linking solar activity to significant climatic changes suggests that the Eddy Cycle has played a significant role over the past 2000 years. Figure B21 illustrates this, showing the ^{14}C record as a proxy for solar activity, with a 1000-year sinusoidal frequency obtained from the data by band-pass filtering. The figure also shows a climate proxy, the amount of iceberg discharge in the North Atlantic, as measured by the content of petrological tracers in benthic cores.[175] These tracers are transported by icebergs and released as they melt. During cold periods with higher winter snowfall, coastal glaciers advance more and release more icebergs, increasing the amount of tracers. While the two curves are not always in perfect agreement, their overall correlation is too strong to be dismissed as a coincidence. Any increase in iceberg activity, indicating colder temperatures, corresponds to a decrease in solar activity. Therefore, the observed relationship

---

[175] Bond, G., et al., 2001. Science, 294 (5549), pp.2130–2136.
  doi.org/10.1126/science.1065680

implies that solar activity has been the primary driver of climate on a centennial timescale over the past 2000 years.

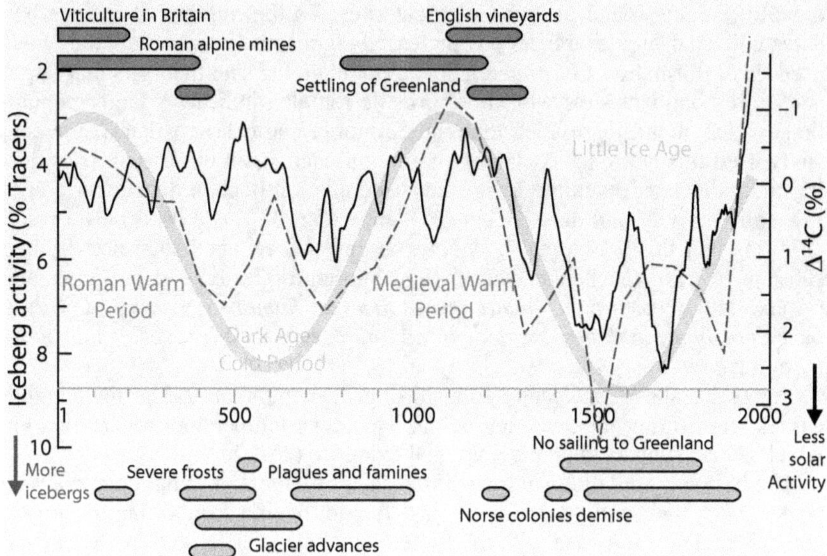

Figure B21. The millennial solar-climate cycle over the past 2000 years. Variation in [14]C (black curve), a proxy for solar activity, is compared to iceberg activity in the North Atlantic (dashed gray curve), a climate proxy. The light gray sine curve shows the millennial frequency of ^{14}C variability obtained by band filtering the data. It defines two warm and two cold periods, supported by a large amount of evidence, some of which are represented by dark and medium gray bars (see main text).

The climate of the last two millennia is characterized by four phases: the Roman Warm Period (ending around 400), the Dark Ages Cold Period (bipartite, with an early phase around 500 and a late phase around 700), the Medieval Warm Period (centered around 1100), and the Little Ice Age (beginning around 1300). This scheme, with its millennial quasi-periodicity, is well supported by a wealth of historical, biological, geological, and climatic evidence. A recent publication presents some of this evidence in the form of the colored bars shown in figure B21, where dark gray indicates warm and light gray indicates cold.[176]

However, this scheme poses a problem for some climate scientists because the present phase is supposed to be warm regardless of emissions due to increased solar activity since the end of the Little Ice Age. This contradicts climate models and defuses the climate emergency, even if rising CO_2 levels contribute to the observed warming.

---

[176] Moffa-Sánchez, P. & Hall, I.R., 2017. Nat. Commun. 8 (1), p.1726.
    doi.org/10.1038/s41467-017-01884-8

Solar-climate impact on human societies

Archaeologists use radiocarbon dating to determine the age of wood and organic remains found at archaeological sites. As the number of radiocarbon dates collected by researchers has increased significantly over time, they have used this information to estimate ancient populations. The theory is that larger and more abundant sites with more organic remains indicate a larger population, which in turn provides material for more radiocarbon dates. A recent study used this approach to assess population changes in the British Isles during the Holocene, revealing a significant population decline during periods of low solar activity and abrupt climate change (fig. 38, gray curve). The result confirms that this was a real phenomenon and that reduced solar activity and subsequent climate change caused food shortages, leading to suffering and misery. The authors note *"multiple instances of human population downturn over the Holocene that coincide with periodic episodes of reduced solar activity and climate reorganization."*[177]

Figure 38 shows a clear relationship between population and temperature curves, confirming the accuracy of this Holocene temperature reconstruction, which is also supported by glaciological evidence (ch. 21).

The human population study also reveals a millennial population cycle with peaks every thousand years for the last four millennia, supporting a climatic quasi-periodicity in phase with the millennial cycle of solar activity. However, most climatologists reject the idea of a millennial climate cycle because historical events such as the Roman and Medieval warm periods, the Dark Ages cold period, and the Little Ice Age cannot be adequately explained by the Enhanced CO_2 Effect climate hypothesis. Some scientists view these events as regional anomalies with little global impact. Accepting the millennial periodicity would imply that warming should occur even without human emissions, which contradicts climate models. To accept this climate periodicity would be to admit that the models are fundamentally wrong. This is unlikely to be admitted no matter how much evidence there is for a millennial climate cycle, solar or otherwise.

In summary

The variability of solar activity and climate is enormous. Focusing on the longest type of solar grand minimum shows that the four times it occurred during the Holocene coincided with four of the most prominent abrupt climate events. These events follow an irregular Bray cycle of about 2,500 years. In addition, the Eddy cycle, which occurs due to less severe solar grand minimums, follows a millennial frequency. Together, these cycles account for much of the Holocene climate variability. The periodicity caused by these solar cycles has significantly affected human populations during periods of prolonged low solar activity, as evidenced by the archaeological and historical records. The evidence suggests that if solar activity has affected climate in the past, it must be doing so now, even if we do not fully understand how. This evidence is inconsistent with our climate models and casts doubt on the existence of a climate crisis.

---

[177] Bevan, A., et al., 2017. PNAS, 114 (49), pp.E10524–E10531.
   doi.org/10.1073/pnas.1709190114

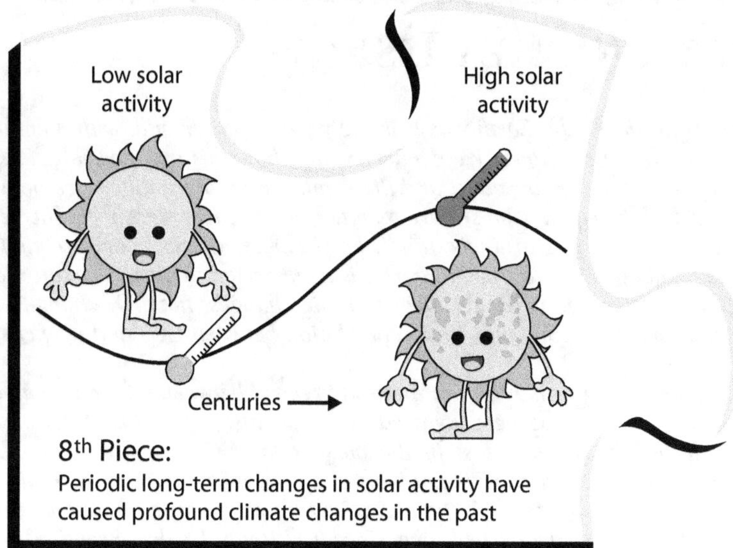

8th Piece:
Periodic long-term changes in solar activity have caused profound climate changes in the past

SECTION 6 KEY ISSUES

At certain times, the Earth was extremely warm and humid, with palm trees and crocodiles at the poles. We cannot explain how this was possible, as CO_2 levels could not have caused it, and the small temperature difference between the poles and the equator must have resulted in little heat transport. The key to solving this mystery is to understand that the poles act as cooling radiators, and the planet gets colder the more heat is transported to them. The late Cenozoic (current) ice age is the result of tectonic changes, not CO_2 changes, and the last 50 million years show little correlation between CO_2 and temperature changes.

However, the agreement in the Antarctic record between CO_2 and temperature during the Pleistocene convinced most scientists of a close relationship. This relationship is being lost in the present, as CO_2 levels are increasing enormously while Antarctica is not warming. The relationship is also missing for the last 11,000 years, with the two changing in opposite directions. Rising CO_2 levels and concurrent warming since 1975 may be the exception rather than the rule. Glaciological and biological evidence confirms that the Holocene Climatic Optimum was warmer than today, despite claims to the contrary.

In addition, the Holocene witnessed many abrupt climate events unrelated to CO_2. Four of the major ones, with a quasi-periodicity of 2,500 years, involved strong atmospheric reorganization, resulting in a steeper temperature gradient, reduced tropical zones, and expanded polar regions. This, in turn, increased poleward heat transport, leading to global cooling and significant changes in precipitation patterns. Surprisingly, these four events coincide with the only four solar grand minimums of a more profound type, lasting 200 years. Their climatic impact was accompanied by a strong effect on human populations at the time. If solar activity had such a strong climatic effect in the past, we cannot rule out its contribution to present climate change.

Section 7. Volcanoes

CHAPTER 24
VOLCANOES AND CLIMATE CHANGE

In 1815, Mt. Tambora exploded in the largest volcanic eruption in a thousand years. This caused a major weather disturbance in the Northern Hemisphere the following year. However, instrumental measurements and climate proxies have shown that the temperature impact of large eruptions is relatively modest and short-lived. On average, temperatures decrease by only 0.2-0.3°C for 3-4 years. Climate models often overestimate the impact of volcanic eruptions and do not account for the rapid recovery observed in the years following an eruption. Based on the available evidence, volcanic activity does not appear to play a major role as a climate forcing on centennial and longer time scales.

The Year Without a Summer

The largest volcanic eruption ever recorded in human history occurred in April 1815 on the Indonesian island of Sumbawa. The eruption of Mt. Tambora caused the deaths of 10,000 people, wiped out the island's vegetation, and killed at least 50,000 more from disease and starvation in the aftermath. The volcanic plume reached 43 kilometers into the stratosphere.

In the Northern Hemisphere, there was nothing unusual about the weather during the year of the volcanic eruption, except for some stunning sunsets. The following year, however, things took a dramatic turn. By early June 1816, more than a year after the eruption, reports of strange weather patterns began to surface. Late spring was either extremely dry or extremely wet in different parts of the Northern Hemisphere, and temperatures plummeted due to icy winds from the north. Every summer month brought heavy frosts in places that hadn't seen them before, and in other areas, the sky seemed to be constantly overcast. This combination of extreme conditions caused crop failures in many countries, leading to food shortages and malnutrition in Ireland, France, England, China, and the United States. The Year Without a Summer also sparked typhus epidemics in parts of Europe and cholera in India, as the monsoon rains arrived late and in torrents, causing devastating flooding in China's Yangtze Valley.

People could not connect the strange weather patterns to the volcanic eruption a year earlier, especially since most never even heard of it. It wasn't until the 20th century that scientists made the connection between volcanic eruptions and hemispheric weather. The eruption of Mt. Tambora was a massive event — even bigger than the eruption of Mt. Samalas in 1257. Nothing like it had happened in over a thousand years.

The weather effects of the Mt. Tambora eruption were considerable but short-lived. Records from the logbooks of the English East India Company and climate proxies show that temperatures in the Northern Hemisphere dropped by 0.5°C (0.9 °F) in 1816 but then recovered to previous levels in just six years. However, some climate models simulate a much larger and longer-lasting climate response to the eruption, indicating that they do not accurately reproduce the actual climatic effects of volcanic eruptions.[178]

Effect of volcanic eruptions on climate

Different volcanic eruptions release different amounts of gases. While water vapor makes up 50-90% of the emitted gas, the remaining gases vary from volcano to volcano. Carbon dioxide (CO_2) can make up 1-40%, sulfur dioxide (SO_2) 1-25%, hydrogen sulfide (H_2S) 1-10%, hydrochloric acid (HCl) 1-10%, and several other minor gases. H_2S oxidizes rapidly to SO_2, which becomes sulfate in the troposphere and clumps into aerosols. These aerosols affect cloud formation and are removed relatively quickly as acid rain. During explosive eruptions, which occur about every two years, SO_2 is also transported into the stratosphere, where most of it is of volcanic origin. Over several weeks to months, stratospheric SO_2 is converted to sulfate, leading to stratospheric dehydration and an increase in stratospheric aerosols that peaks about three months after the eruption and persists for about four years.

The increase in stratospheric sulfate aerosols has three important climatic effects. The first and most widely recognized is that they scatter incoming solar radiation, resulting in surface cooling. The second effect is that they absorb incoming near-infrared and outgoing longwave infrared radiation, leading to stratospheric warming. This warming affects the atmospheric circulation and leads to warm winters in the Northern Hemisphere for 1-2 years after a stratospheric eruption. The third effect is the very effective destruction of ozone caused by altered chemistry and perturbed warming rates. Ozone depletion affects the stratosphere, causing widespread atmospheric changes and increased solar UV radiation reaching the surface.

The climatic effects of volcanoes are complicated because they depend on the amount of SO_2 they inject into the stratosphere, their latitude, and the time of year they occur. These factors affect the dispersion of the stratospheric cloud, which in turn can have a global or hemispheric effect or no effect at all. For example, the 1980 eruption of Mt. St. Helens was a powerful volcanic eruption with no significant climatic impact.

Currently, volcanic CO_2 emissions account for only about 1% of anthropogenic emissions. However, in the distant past, higher levels of volcanic activity, especially from large igneous provinces, could have produced large amounts of CO_2. There is an ongoing debate as to whether the mass extinctions that sometimes occurred during these times were caused by SO_2 and cooling or CO_2 and warming, with the latter hypothesis being the most popular today.

The evidence from recent volcanic eruptions suggests that their effect on temperature is typically short-term, lasting a few years at most. Temperature records from eruptions in the 19th and 20th centuries show a temperature decrease of 0.2-0.3°C over 3-4 years. Proxies from the Northern Hemisphere and Europe also support this. Figure 39 shows the temperature change in the Northern Hemisphere for eight (bars) and 34 (lines) major eruptions in recent centuries.

---

[178] Brohan, P., et al., 2012. Clim. Past, 8 (5), pp.1551-1563.
     doi.org/10.5194/cp-8-1551-2012

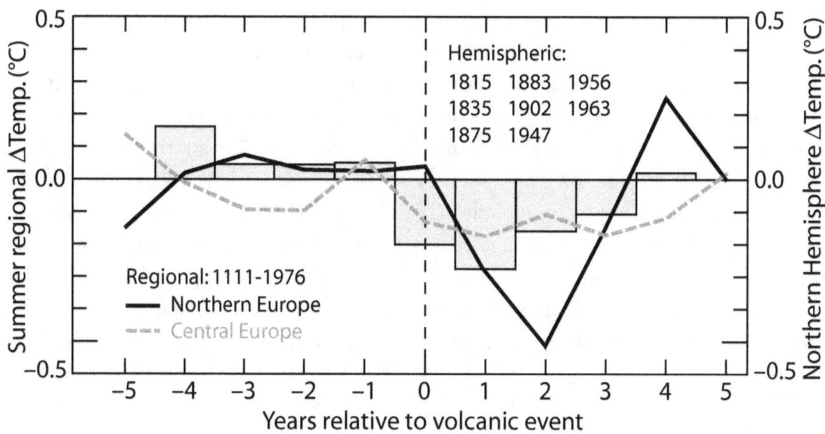

Figure 39. Regional and hemispheric temperature effects of volcanic eruptions. Gray bars, right scale, Northern Hemisphere temperature change in the years around eight major eruptions. Black and gray curves, left scale, estimate from tree-ring records of temperature change in Northern Europe (black curve) and Central Europe (dashed gray curve) during summers around 34 major volcanic eruptions from 1111-1976.[179]

The available evidence suggests that even the largest volcanic eruptions have a relatively modest and temporary negative effect on temperature. After about five years, the sulfate aerosols disappear from the stratosphere, and the radiative and chemical effects of the eruption end. Only delayed dynamic effects are possible at this time. It is difficult to explain why recovery from such events is so rapid. Sulfate aerosols cause a significant energy deficit at the surface, leading to a temperature drop. Meanwhile, the warmer stratosphere radiates more energy into space. When the situation ends, the planet is left with an energy deficit resulting from these effects. It is unclear where the energy for the subsequent rapid warming comes from. Climate models struggle to explain this phenomenon, probably because it remains unknown to scientists. This uncertainty may explain why the temperature recovery is slower in models where it is mediated by the ocean.

Box 22. Volcanic activity and the glacial cycle

In the 1970s, scientists speculated that volcanic eruptions might be a cause of glaciation because of their cooling effect. In the 1990s, however, it became clear that the relationship was the opposite of what was expected. In fact, there was a clear anti-correlation between volcanism and glaciation. This means that during the transition from cold glacial periods to warm interglacial periods, there were pulses of high volcanic activity.[180]

---

[179] Data from: Self, S., et al., 1981. J. Volcanol. Geotherm. Res. 11 (1), pp.41–60, doi.org/10.1016/0377-0273(81)90074-3 and Esper, J., et al., 2013. Bull Volcanol. 75, pp.1–14. doi.org/10.1007/s00445-013-0736-z

[180] Glazner, A.F., et al., 1999. Geophys. Res. Lett. 26 (12), pp.1759–1762. doi.org/10.1029/1999GL900333

In 2013, a study was conducted on the frequency of volcanic activity in the Pacific Ring of Fire, the site of many of the largest volcanic eruptions in history. The study used a large dataset of volcanic layer dates from multiple coring sites in volcanic zones along the Ring of Fire, spanning one million years. The results of the frequency analysis showed a statistically significant peak in volcanic eruption activity only at the 41,000-year periodicity associated with the Earth's tilt cycle, one of the Milankovitch frequencies (fig. B22).[181] This frequency of volcanic activity supports the hypothesis that changes in axial tilt (obliquity), rather than axial wobble (precession), are the main driver of the glacial cycle (box 19, ch. 21) because changes in volcanic activity are associated with increased crustal motion due to rapid melting of ice sheets as a glaciation ends and an interglacial begins.

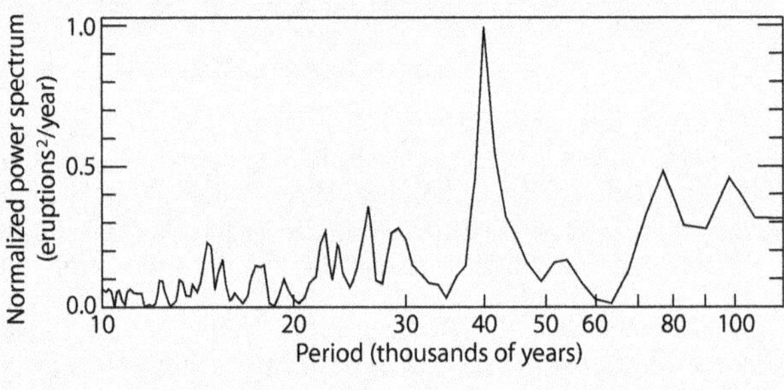

Figure B22. Volcanic activity and the glacial cycle. Changes in the frequency of volcanic eruptions correlate with periodic changes in the tilt of the Earth's axis. A power spectrum shows the fraction of a signal found at each frequency.

Ice core and volcanic data confirm that the most intense volcanic activity in the past 100,000 years occurred between 13,000 and 7,000 years ago (fig. 42, ch. 26). During this time, the rate of volcanic activity was between 2 and 6 times higher than background levels. However, this period also coincided with the greatest warming of the last 100,000 years. Therefore, this evidence challenges the notion that high volcanic activity leads to long-term cooling beyond the few years of temperature drop after each major eruption.

Some studies also suggest that as much as 40 ppm, or about half of the CO_2 increase during deglaciation, could be due to increased volcanic activity.[182] This implies that volcanic activity must be causing some warming beyond the short-term cooling.

In chapter 26, we will examine the hypothesis that the Little Ice Age resulted from unusual cooling caused by powerful volcanic eruptions. The fact that volcanic eruptions have only a small, short-term effect on temperature and are more a consequence than a cause of climate change makes it surprising that this theory is so widely accepted. The only plausible explanation is that we have few natural causes for climate change and are quick to dismiss the clear, evidence-based argument that the Little Ice Age was caused by long periods of low solar activity.

---

[181] Kutterolf, S., et al., 2013. Geology, 41 (2), pp.227–230. doi.org/10.1130/G33419.1
[182] Huybers, P. & Langmuir, C., 2009. Earth Planet. Sci. Lett. 286 (3-4), pp.479–491. doi.org/10.1016/j.epsl.2009.07.014

The 2022 Hunga Tonga Volcano Eruption

On January 15, 2022, the Hunga Tonga undersea volcano eruption was an exceptionally rare event. Submarine eruptions rarely produce plumes that reach the stratosphere, but the powerful explosion – the strongest in three decades – occurred in shallow water and injected a staggering 146 million tons of vaporized seawater into the stratosphere. Given the dryness of the stratosphere (fig. 9, ch. 7), this event caused an abrupt 13% increase in stratospheric water vapor. This increase was primarily concentrated in the Southern Hemisphere at altitudes between 26 and 34 km (16-21 miles). This additional water vapor quickly circled the globe as it was slowly transported poleward and downward by the Brewer-Dobson circulation. Most of this water will likely remain in the Southern Hemisphere, gradually exiting the stratosphere over the next few years.

The atmospheric radiation budget is particularly sensitive to changes in GHGs in the upper troposphere and stratosphere. Even small changes in stratospheric ozone or water content can lead to significant radiative forcing. This is because most outgoing longwave radiation is unlikely to be intercepted at this altitude due to the very low water vapor content. Thus, small changes in GHGs have a greater impact here than in the moist lower troposphere, which is highly opaque to infrared radiation. The increase in stratospheric water vapor should have a warming effect on the global average surface temperature.[183]

Similar to the eruption of Mt. Tambora in 1815, when some important climatic effects occurred 14 months later, some unusual changes were detected in the summer of 2023, 17 months after the Hunga Tonga eruption. In particular, Antarctic sea ice was very low due to strong northerly winds pushing the ice toward the coast. Also, the Antarctic ozone hole began to form significantly earlier in the year. In the Northern Hemisphere, the North Atlantic Oscillation (pressure difference between the Icelandic Low and the Azores High) was unusually negative (small pressure difference), and the North Atlantic surface temperature was unusually high. Globally, July was the warmest month ever recorded and September showed the largest temperature anomaly.

Whether these changes are primarily due to the Hunga Tonga volcanic eruption remains to be determined. Still, they are consistent with what might be expected, particularly the global warming and ozone hole effect, since stratospheric water vapor is involved in ozone depletion. However, any effects from the eruption should be temporary and disappear as the injected water vapor leaves the stratosphere over the next few years.

In summary

While volcanic eruptions can significantly impact weather, there is little evidence that they cause substantial climate change. In fact, the evidence suggests the opposite. Glacial-scale climate changes profoundly affect the frequency of volcanic eruptions. While some models exaggerate the impact of volcanic eruptions, it's hard to argue that they are a major driver of climate change on centennial and longer time scales.

---

[183] Jenkins, S., et al., 2023. Nat. Clim. Change, 13 (2), pp.127-129.
doi.org/10.1038/s41558-022-01568-2

CHAPTER 25
VOLCANOES AND MERIDIONAL HEAT TRANSPORT

When a large tropical volcano erupts, it affects the coupled system formed by the atmosphere and the ocean. One effect is that it can cause an El Niño-like response in the equatorial Pacific. Another effect is that it can lead to a warmer winter in northern Europe and North America, which may seem counterintuitive. This is because the eruption strengthens the polar vortex, which reduces the arrival of cold air masses from the poles. In addition, the eruption weakens the stratospheric circulation toward the poles by strengthening the adverse winds.

All of these dynamic effects combine to reduce the amount of heat transported toward the poles, counteracting the cooling effects of the volcano and leading to a faster recovery. However, between two and three decades after the eruption, some cooling can be observed in the Northern Hemisphere as a mysterious echo of the volcano's explosion.

The dynamic effects of volcanic eruptions

When a volcano erupts, excess sulfate can remain in the stratosphere for a few years. This sulfate has radiative and chemical effects that cause changes in the stratosphere and the Earth's surface, triggering a complex response between the atmosphere and the ocean. This response can last longer than the presence of sulfate in the stratosphere. The dynamic effects of a volcanic eruption are caused by changes in surface and lower stratospheric temperature patterns and gradients, which affect the transport of heat toward the poles. For this discussion, we will focus only on strong tropical eruptions, as the effects of extra-tropical eruptions are generally smaller and different.

When a volcano erupts, it causes more cooling at lower latitudes, which reduces the surface temperature gradient from the equator to the poles. In the stratosphere, however, the opposite occurs: heating is stronger at lower latitudes, increasing the temperature gradient. This causes the zonal winds (easterlies or westerlies) to become stronger and reduces the poleward heat transport by the Brewer-Dobson circulation.

These dynamic effects of volcanoes are mediated by meridional transport and have two main consequences. First, a volcanic eruption triggers a response similar to El Niño in the Pacific. Second, it strengthens the boreal polar vortex, leading to a positive phase of the North Atlantic Oscillation.

When stratospheric transport is reduced, some poleward heat transport must be redirected. The Brewer-Dobson circulation moves heat from the equator to the winter pole (fig. 23, ch. 14). But when this circulation is reduced, it can affect how heat accumulates and is distributed in the equatorial ocean. This, in turn, increases the likelihood of an El Niño event following a volcanic eruption since El Niño is an effective way to remove heat from the equatorial Pacific (ch. 18).

As the stratospheric zonal wind circulation strengthens, atmospheric waves are unable to reach the stratosphere. This leads to a stronger polar vortex (box 7, ch. 11; box 12, ch. 15). In winter, a closed and strong jet stream traps the very cold Arctic air. This leads to warm winter weather in northern Europe and North America. As a result, strong cooling usually occurs only after the winter, as was the case after the eruption of Mt. Tambora (ch. 24).

Volcanic eruptions have dynamic effects on the climate system related to changes in the transport of heat to the poles. These effects help explain why the planet can recover rapidly from a strong eruption in just a few years. As we saw in the previous chapter, the radiative effects of an eruption cause a significant loss of energy from the climate system. If this loss were not compensated for, each eruption would continue to cool the planet. However, we know that is not the case, so the dynamic effects at work must compensate for the energy loss.

One of the ways the planet conserves energy after a volcanic eruption is by reducing the transport of heat to the poles. This is accomplished by slowing down the stratospheric circulation and strengthening the polar vortex. In winter, the polar regions act as cooling radiators, so directing less heat toward them is an effective way to compensate for the energy loss caused by the eruption.

It's worth noting that the 1991 eruption of Mt. Pinatubo was followed by a decrease in outgoing longwave radiation from the Arctic of about 2 W/m^2 for five years (fig. 72, ch. 43), indicating that this mechanism is indeed at work. However, climate models do not currently include these transport effects, which means that they tend to overestimate the cooling effect of volcanic eruptions and point to the ocean as the source of the compensating energy.

The delayed effects of volcanic eruptions

A puzzling climate echo occurs two to three decades after a volcanic eruption. To illustrate this phenomenon, we can use a 1500-year Northern Hemisphere proxy temperature reconstruction that is annually resolved and calibrated to avoid loss of low-frequency variance.[184] The four strongest volcanic eruptions of this period, which were not followed by another very strong volcanic eruption for the next 60 years, occurred in 682, 1258, 1458, and 1815 AD. Figure 40 shows the reconstructed temperature from -10 to +60 years for each eruption after detrending and expressed as anomalies from the detrended mean. Note that the effect of the eruption is reduced and begins several years before the event due to the decadal smoothing applied by the authors of the study, which was necessary to preserve the low-frequency variance.

The delayed response to volcanic eruptions is an interesting phenomenon that has been observed. A second cooling phase occurs with a delay of about 20 years (for the 1458 and 1815 eruptions) to 30 years (for the 682 and 1258 eruptions). There are different theories about this delayed effect, with some scientists believing that it is due to an oceanic memory of the eruption, while others think that a bidecadal oscillation in the Atlantic Meridional Circulation may cause it.[185]

---

[184] Hegerl, G.C., et al., 2007. J. Clim. 20 (4), pp.650–666. doi.org/10.1175/JCLI4011.1
[185] Swingedouw, D., et al., 2015. Nat. Commun. 6 (1), p.6545.
    doi.org/10.1038/ncomms7545

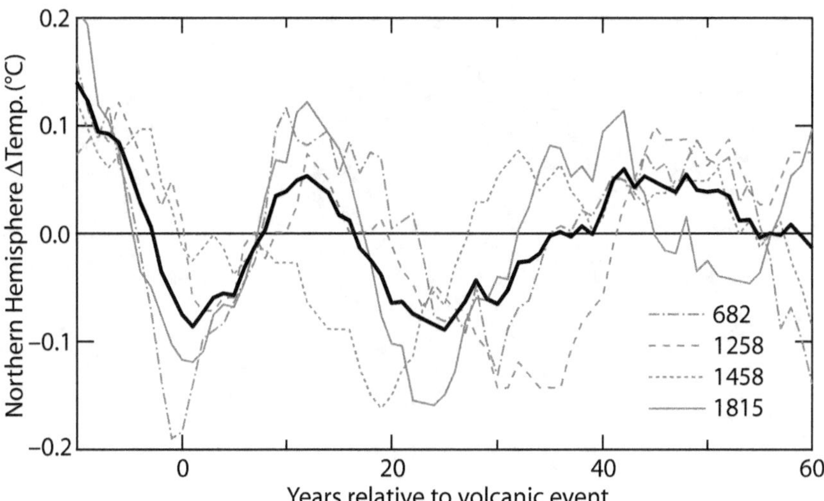

Figure 40. Delayed response to volcanic eruptions. Proxy reconstruction of the mean annual temperature from 30-90°N, smoothed over a decade, from 10 years before to 60 years after the indicated volcanic eruptions (gray curves). The thick black curve is the mean of the four gray curves.

However, the idea that the bidecadal oscillation is responsible is probably incorrect since the reduction in the Brewer-Dobson circulation and the strengthening of the polar vortex caused by the eruption imply a reduction in heat transport. This would lead to some warming that only partially compensates for the increased cooling due to the radiative effects of the eruption. The cooling effect after 20-30 years suggests an opposite effect on transport, which could be explained by the dynamic effects of the eruption resetting the 50-70 year multidecadal oceanic oscillations discussed in chapter 19. Thus, 20-30 years later, we may be in the opposite phase (increased transport and cooling) of the oceanic oscillations, and the detected effect is exactly what we would expect from these oscillations. This explanation does not require any special mechanism or oceanic memory.

In summary

The dynamic effects of a strong tropical volcanic eruption, such as an El Niño event, a decrease in stratospheric transport, and a strengthening of the polar vortex, have been observed. These effects can be explained by a reduction in poleward heat transport caused by altered temperature gradient patterns resulting from the radiative effects of stratospheric aerosols. Changes in transport can explain why the cooling caused by an eruption is less than expected and why the recovery is faster than expected. The observed echo cooling that occurs 2-3 decades later can be attributed to the effect of the eruption on transport through multidecadal ocean oscillations.

CHAPTER 26
VOLCANIC CONTRIBUTION TO THE LITTLE ICE AGE

The Little Ice Age lasted 550 years, from the 14th to the 18th century. During this period, glaciers expanded globally, reaching their largest size in 7000 years. The cooling was most pronounced in the North Atlantic region and was not continuous, with occasional warming periods lasting decades. Some of the worst famines and pandemics in recorded history also marked the Little Ice Age. Scientists are still debating its cause, as greenhouse gases have been ruled out. Paleoclimatic evidence suggests that the Sun was the primary cause, while many scientists believe that volcanoes were primarily responsible. Both factors likely contributed to the Little Ice Age, but there were periods of several centuries when volcanic activity was minimal.

The Little Ice Age

The Little Ice Age was the coldest period of the Holocene. Numerous biological, historical, and climate proxy records document this global cooling, although it was most pronounced in the Northern Hemisphere, particularly in the Atlantic region. Despite its regional prominence, glaciers in most areas of the world experienced major advances between the 14th and 18th centuries (fig. 35a, ch. 21), reaching their largest size in 7,000 years. The Little Ice Age was not continuously cold; there were periods of several decades during which warming and glacier shrinkage occurred, followed by a return to colder times.

The definition of the Little Ice Age is inconsistent, leading to some confusion. Scientists have not agreed on when it began, leading to different periods being studied under the same name. Climate proxies typically show a temperature peak during the Medieval Warm Period, between 1100 and 1150, followed by a decline. In the 13th century, three major volcanic eruptions occurred within 56 years: in 1230, 1257 (the great eruption of Mt. Samalas), and 1286. By 1300, the glaciers of the Alps were expanding rapidly, and this is the earliest date chosen by scientists as the beginning of the Little Ice Age.[186]

The largest famine ever recorded in Europe was the Great Famine, which occurred between 1315 and 1317. It was caused by bad weather that led to catastrophic crop failures for two consecutive years. This was followed thirty years later by the Black Death. Deteriorating weather is often followed by famine, which can weaken the immune systems of populations and contribute to pandemics. Food shortages can sometimes lead to population migrations and wars, resulting in a scenario known as the Four Horsemen. The Tapestry of the Apocalypse of Angers, France, woven over several years beginning in 1377, depicts how people felt about the consequences of the Little Ice Age.

---

[186] Matthews, J.A. & Briffa, K.R., 2005. Geogr. Ann. A, 87 (1), pp.17–36. doi.org/10.1111/j.0435-3676.2005.00242.x

An unexplained, abrupt and profound climatic change

The Little Ice Age is the most profound past climate change in thousands of years, and it is very recent, yet its causes remain poorly understood. The fact that scientists do not agree on its causes is a major problem, especially since we pretend to understand the causes of current climate change. If we don't understand the Little Ice Age, we cannot fully understand the nature of climate change.

The Little Ice Age was the result of a cooling period that occurred between 1100 and 1500, during which Northern Hemisphere temperatures may have decreased by 0.8°C (1.5 °F, fig. 41). Interestingly, the Law Dome Antarctic ice core shows that the level of CO_2 in the atmosphere was the same in 1100 and 1500 (282 ppm) and did not change by more than 3 ppm between these two dates. This suggests that the contribution of CO_2 levels to the Little Ice Age can be ruled out. In fact, this evidence contradicts the supposed relationship between CO_2 and temperature, as it shows that there was deep cooling for 400 years without any effect on CO_2 levels. Similarly, during the previous interglacial period, temperatures decreased for 8,000 years without a decrease in CO_2 levels (fig. 56, ch. 35). This means that temperatures can change for hundreds or even thousands of years without a corresponding change in CO_2 levels. Thus, it is not obvious that a temperature change can be attributed to a change in CO_2 if they change simultaneously, which casts doubt on the role of CO_2 as the main driver of climate change.

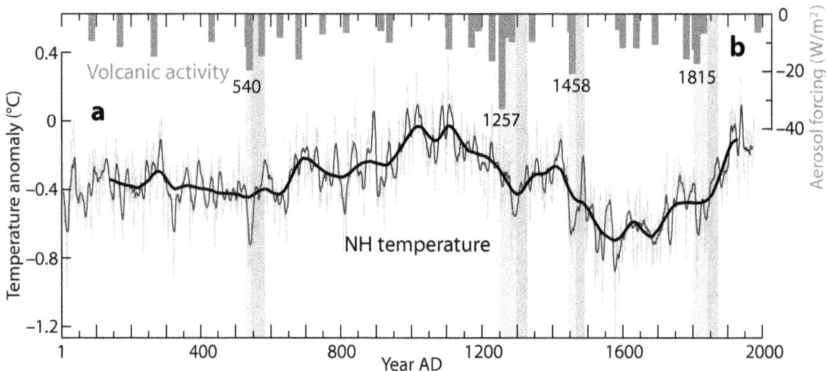

Figure 41. The effect of volcanic forcing on temperature during the last 2000 years. a) Reconstruction of the proxy temperature AD 1-1979 (light gray line) with a 10-year moving average and a slow component (thick black line). Light gray bars indicate temperature drops after the four major volcanic eruptions. medium gray bars indicate subsequent temperature recovery. b) Aerosol forcing of major volcanic eruptions from sulfate concentrations in Greenland and Antarctic ice cores.[187]

From 1200 to 1850, volcanic activity increased compared to the previous thousand years, leading some scientists to believe that the Little Ice Age was

---

[187] Data for the figure from Moberg, A., et al., 2005. Nature, 433 (7026), pp.613–617. doi.org/10.1038/nature03265 and Sigl, M., et al., 2015. Nature, 523 (7562), pp.543–549. doi.org/10.1038/nature14565

due to this increased activity.[188] However, there is no evidence of significant long-term cooling following recent volcanic eruptions (ch. 24 & 25). Figure 41 shows that short-term cooling followed the largest eruptions (light gray bars) but was immediately followed by warming (medium gray bars), as discussed in the previous chapter. In all cases, the previous centennial temperature trend, whether positive or negative, continued unchanged after the volcanic effects of a few years or decades. For example, the cooling effect of the eruptions of 1809, 1815 (Mt. Tambora), 1822, 1835, and 1843, one of the largest groups of strong eruptions in 2000 years, was followed by warming from the late 1840s onward, just a few years later (fig. 79, ch. 46). It is hard to believe that a few eruptions of lesser intensity between 1460 and 1780 were responsible for the cold conditions during 300 of the 550 years of the Little Ice Age.

Some scientists believe that the lowest solar activity in thousands of years caused the Little Ice Age. This hypothesis is supported by paleoclimatic evidence that shows profound cooling and climate change during every 200-year Spörer-type solar grand minimum in the Holocene (ch. 23). Cooling has also occurred many other times when there has been a 70-year Maunder-type solar grand minimum. The Little Ice Age coincided with one Spörer-type solar grand minimum and two Maunder-type solar grand minimums. Given this evidence, why isn't there a widespread belief among scientists that low solar activity was the main cause of the Little Ice Age?

The main problem with the impact of solar activity on the Little Ice Age is that it is unclear exactly how it affects climate because the change in total solar irradiance is too small to have a significant effect. Studies using climate models to examine the solar forcing during the Little Ice Age and its contribution to a proxy-derived climate signal conclude that the climate response to solar forcing is weak.[189] However, if the derived forcing is incorrect because the Sun acts on climate through indirect mechanisms, the conclusions of these studies may be inaccurate. This possibility is supported by paleoclimatic evidence showing that Spörer-type solar grand minimums have a strong effect on climate. Therefore, we should be cautious in accepting the conclusion that solar activity was not the main cause of the Little Ice Age.

The cause of the Little Ice Age is still uncertain and remains an unsolved problem that is an embarrassment to climate science. CO_2 is not supported as its cause by any evidence or theory, volcanic eruptions have weak support from both evidence and theory, and low solar activity is supported by evidence but not theory. It is troubling that we claim to be able to predict the climate of future centuries, yet we cannot accurately explain the climate of 300 years ago.

The Little Ice Age had very low volcanic activity

A comparison of the level of volcanic activity during the Little Ice Age with that of the rest of the Holocene is conspicuously absent from any climate study that suggests volcanic activity as the cause of the Little Ice Age. These studies would probably be unconvincing if such a comparison were made.

---

[188] Crowley, T.J., et al., 2008. PAGES news, 16 (2), pp.22–23.
   doi.org/10.22498/pages.16.2.22
[189] Hegerl, G.C., et al., 2003. Geophys. Res. Lett. 30 (5), 1242.
   doi.org/10.1029/2002GL016635

Analysis of a Greenland ice core has allowed the quantification of sulfate deposition from past volcanic activity.[190] The highest concentration of major volcanic signals in the last 110,000 years was found to have occurred between 13,000 and 7,000 years ago, consistent with evidence that periods of crustal adjustment during glacial-interglacial transitions (characterized by intense warming) are typically the most volcanically active (box 22, ch. 24).

Figure 42 shows the volcanic data collected in this study, with the data summed over 100-year periods (the last point corresponds to 1851-1950). The analysis shows that the level of volcanic activity during the warm Holocene Climatic Optimum was 2-4 times higher than during the cooler Neoglacial. Surprisingly, the long-term correlation between volcanic activity and temperature during the Holocene is positive instead of the widely accepted negative correlation. This is only possible if the long-term climatic effects of volcanic eruptions are negligible and temperatures recover within a few years after the eruption, as supported by the evidence in chapter 24. The data shown suggest that volcanic eruptions do not have a centennial cooling effect. There is also no evidence of a warming effect, as shown in figure 41. Thus, it can be concluded that volcanic activity is unlikely to have a significant centennial climatic effect.

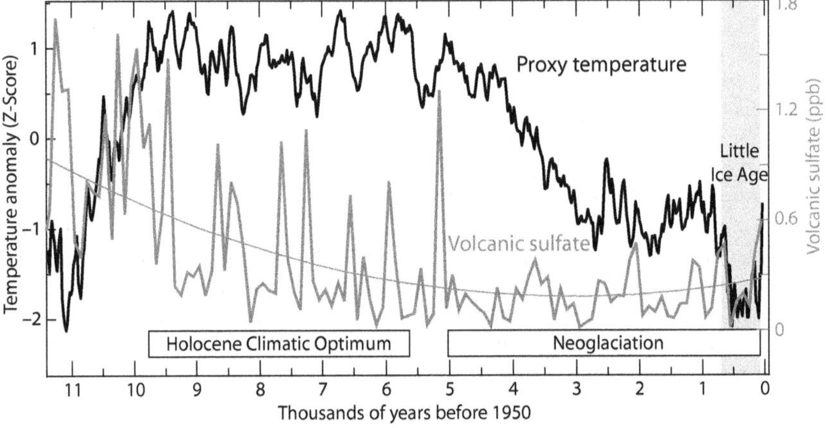

Figure 42. Holocene volcanic activity. A global temperature reconstruction (black line, see ch. 21, fig. 35) is compared with the centennial volcanic sulfate record from a Greenland ice core (gray line). The gray bar marks the Little Ice Age.

What further discredits the Little Ice Age volcanic hypothesis is that volcanic activity was extremely low for most of this period. In fact, the level of activity was even lower than the already low average volcanic activity during the Neoglacial period.

The level of volcanic activity was higher than average during two periods: before the Little Ice Age (1165-1345) and after 1765. It seems implausible to attribute the prolonged cold of the Little Ice Age for the 420 years between these periods to a double powerful eruption in 1452 and 1458. If that were the

---

[190] Zielinski, G.A., et al., 1996. Quat. Res. 45 (2), pp.109–118.
    doi.org/10.1006/qres.1996.0013

case, why didn't the rest of the Holocene also experience extreme cold, given that the level of volcanic activity was much higher?

The lack of evidence linking volcanic activity to long-term climate effects has not stopped some scientists from defending it as a cause of the Little Ice Age. Some suggest that volcanic eruptions can initiate a strong centennial sea ice feedback and cause important changes in the Atlantic Meridional Overturning Circulation – although models differ on whether it would increase or decrease – but only when solar activity is low.[191]

It is clear that the Little Ice Age was a consequence of low solar activity, although the exact mechanism remains unknown. This book presents a hypothesis that attributes solar-induced cooling to changes in poleward heat transport. The increased volcanic activity between 1165-1345 and 1765-1845 helps to explain why the cooling began around 1200 when solar activity was still high and why the Little Ice Age persisted for several decades after the end of the Maunder Minimum.

In Summary

The Little Ice Age, which lasted from 1300 to 1845, was global in extent but with notable regional variations. However, scientists have not reached a consensus on its global extent, duration, and underlying causes. This lack of agreement is embarrassing, given that this is the most recent period of major climate change. CO_2 levels remained unchanged for nearly 500 years after 1100, so it couldn't have been a contributing factor. Moreover, the four centuries between 1350 and 1750 had some of the lowest volcanic activity in the Holocene, meaning that only the initial cooling and the last 80 years of the Little Ice Age can be linked to volcanic activity. However, for most of the time between 1270 and 1720, solar activity was very low, and paleoclimatic evidence supports that extended periods of minimal solar activity coincide with periods of intense cooling. But this is problematic because most scientists don't believe that solar activity could affect the climate to such an extent. Moreover, the prospect of solar activity contributing to the current warming has been dismissed without a thorough understanding of the issue.

---

[191] Slawinska, J. & Robock, A., 2018. J. Clim. 31 (6), pp.2145–2167.
doi.org/10.1175/JCLI-D-16-0498.1

Section 7 Key Issues

The temperature effects of large eruptions are relatively modest and short-lived. Climate models often overestimate their impact and do not reproduce the observed rapid recovery. Based on the available evidence, volcanic activity does not appear to play a major role as a climate forcing on centennial and longer time scales.

Strong tropical volcanic eruptions show dynamic effects such as an El Niño event, a decrease in stratospheric transport, and a strengthening of the polar vortex. They probably represent a reduction in poleward heat transport as a result of an altered temperature gradient due to stratospheric aerosols. This may explain why the cooling is less and the recovery is faster than expected. A mysterious echo cooling observed 2-3 decades later can be attributed to the effect of the eruption on transport through multidecadal ocean oscillations.

Attempts to explain the Little Ice Age in terms of volcanic activity ignore that the four centuries between 1350 and 1750 had some of the lowest volcanic activity in the Holocene.

Section 8. The Sun

CHAPTER 27
THE EFFECTS OF A SOLAR GRAND MINIMUM

Over the past 6,000 years, the largest abrupt climate changes have occurred during three distinct periods. Interestingly, these periods coincide with the only three major 200-year solar grand minimums that have occurred during this time. This evidence suggests that the Sun is the primary driver of Holocene climate change, second only to the slow changes resulting from Earth's orbital variations over thousands of years.

Climate proxies provide valuable insights into the significant changes in atmospheric circulation, oceanic transport, precipitation, and temperature that occur during two centuries of low solar activity. The observed changes indicate an increase in the temperature gradient between the equator and the poles, resulting in increased poleward heat transport. This, in turn, causes the polar regions to expand and the tropics to contract. Unfortunately, most climatologists ignore this evidence of profound climate change caused by reduced solar activity. This is because they mistakenly consider solar variability to be a weak climate forcing based on faulty reasoning and because of the relatively small impact of the 11-year solar cycle.

A small change in irradiance

Many scientists reject the idea that solar activity is a significant contributor to global warming based on two main arguments. The first is that changes in solar irradiance are too small to cause the substantial change in the energy flux needed to cause climate change. The second argument is that the trends in temperature and solar activity have not coincided in recent decades.

However, the reasoning behind dismissing the relevant contribution of solar activity to global warming is flawed because of two implicit assumptions that haven't been proven to be true. The first assumption is that changes in total solar irradiance are the only way that solar activity affects climate. The second assumption is that the effect of solar activity on climate must result in a direct linear change in surface temperatures.

The first assumption is probably incorrect. Studies have shown that changes in the UV part of the solar spectrum mediate the effect of solar activity on climate.[192] Although solar UV energy is a small fraction of the total energy supplied by the Sun, it varies thirty times more with the solar cycle than the rest of the solar spectrum. In addition, most of this energy is delivered to the stratospheric ozone layer, where it has an enormous impact (box 1, ch. 2).

The second assumption is also likely to be wrong. The effect of solar activity on climate acts in a non-linear way on the atmospheric circulation (as ex-

---

[192] Gray, L.J., et al., 2010. Rev. Geophys. 48 (4), RG4001.
  doi.org/10.1029/2009RG000282

plained below). Therefore, any effect on temperature is secondary and complicated by various factors that affect temperature.

The dismissal of solar activity as a major climate determinant based on faulty reasoning is premature and unwarranted. We have evidence that it is the second most important climate determinant during the Holocene after slow-acting orbital changes. This evidence establishes the three Spörer-type solar grand minimums in the last 5500 years as the cause of the three major abrupt climatic events of the period. It also provides a wealth of information about how solar activity affects climate. However, this information has been ignored until recently, when for the first time, the evidence for what happened to the global climate during several major Spörer-type solar minimums was brought together in a single publication.[193] These events occur once every 2,500 years when the Sun enters a very unusual 200-year period of very low solar activity (fig. 37b, ch. 23).

Figure 43a shows a reconstruction of solar activity with the three abrupt climatic events highlighted by light gray bars, all of which coincide with Spörer-type solar grand minimums.[194] While other solar grand minimums have occurred in the past 6000 years, none have shown the same ^{14}C changes as those shown in figure 37b (ch. 23). At intervals of 5000 years, the troughs of the 2500-year Bray solar cycle and the 1000-year Eddy solar cycle coincide (box 21, ch. 23), resulting in a cluster of several solar grand minimums that induce a major climatic effect. Such an event occurred 5500 years ago and again 500 years ago.

This chapter graphically presents only a small fraction of the evidence for the climatic effects of a Spörer-type solar grand minimum. There is a wealth of evidence collected by scientists, some of which will be discussed. Despite the availability of this evidence, many scientists have chosen to ignore it. The reason is that the detailed mechanism of how the climate changes during low solar activity is unknown and, therefore, cannot be programmed into climate models. It is disturbing that the evidence that the models do not reflect is being ignored.

Information from atmospheric proxies

The previous chapter examined how sulfate deposition in ice cores provides insight into volcanic activity. Other ions present in ice cores reveal other climatic aspects. For example, an increase in the deposition of salts such as potassium is currently associated with winter atmospheric conditions. This occurs when the north polar vortex expands and the meridional circulation increases, leading to colder and windier conditions. The Siberian High-pressure system also strengthens, bringing frigid conditions to much of the Northern Hemisphere.[195] During the three major solar-driven climate events of the last 6000 years, the levels of these salts spiked (fig. 43d), indicating a major strengthening of the polar atmospheric circulation, resulting in extremely cold and windy winter conditions in the Northern Hemisphere.

---

[193] Vinós, J., 2022. Climate of the Past, Present and Future: A scientific debate. Critical Science Press. pp.67–88

[194] Wu, C.J., et al., 2018. Astron. Astrophys. 615, p.A93.
doi.org/10.1051/0004-6361/201731892

[195] Mayewski, P.A., et al., 2004. Quat. Res. 62 (3), pp.243–255.
doi.org/10.1016/j.yqres.2004.07.001

Other proxies that reflect wind strength at high latitudes, such as those obtained in Iceland, confirm that there was a southward shift of the polar front during these times, indicating an expansion of the Arctic region.[196] The North Atlantic Oscillation is a measure of the pressure difference between the Azores High and the Icelandic Low. When the difference is small the oscillation is in the negative phase, leading to a more wavy jet stream, more frequent and longer atmospheric blocking, and more cold Arctic air spilling over the midlatitudes. Reconstructions of the North Atlantic Oscillation during the Holocene show that it had very negative values during the Little Ice Age and the other two abrupt climate events studied.

During solar grand minimum events, the North Atlantic Oscillation often remains in a negative phase throughout the winter, which can intensify the winter climate effects in the region. This makes the region particularly sensitive to the effects of low solar activity.

Information from oceanic proxies

Chapter 17 discussed the importance of the Atlantic Meridional Overturning Circulation for heat transport to the Arctic. Several studies indicate that deep water formation in the North Atlantic was significantly reduced during the three abrupt climate events of interest. However, it is debatable whether the overturning circulation was significantly reduced as a consequence. There is evidence that salinity (fig. 43c) and water temperature below the thermocline – the transition layer between warmer water above and cooler water below – increased south of Iceland during these events.[197] This suggests that a compensatory mechanism involving an increased contribution of warm, salty water from the subtropical gyre at the expense of cooler, fresher water from the subpolar gyre drove the Atlantic Meridional Overturning Circulation. Thus, the circulation continued to transport warm water toward the Arctic and did not stop.

During the periods of low solar activity discussed in this chapter, there was an extraordinary growth of Norwegian coastal glaciers, which required an increase in winter precipitation. This increase in precipitation was made possible by the influx of more warm water during very cold winter conditions. The inflow of Atlantic water into the Nordic seas was discussed in chapter 17 (fig. 27) in the context of oceanic heat transport. The warm water inflow has been reconstructed over the Holocene and shows a strong 2,500-year periodicity, with a significant increase during periods of prolonged low solar activity, including the three periods considered in this chapter.[198]

---

[196] Jackson, M.G., et al., 2005. Geology, 33 (6), pp.509–512. doi.org/10.1130/G21489.1
[197] Thornalley, D.J., et al., 2009. Nature, 457 (7230), pp.711–714.
     doi.org/10.1038/nature07717
[198] Giraudeau, J., et al., 2010. Quat. Sci. Rev. 29 (9-10), pp.1276–1287.
     doi.org/10.1016/j.quascirev.2010.02.014

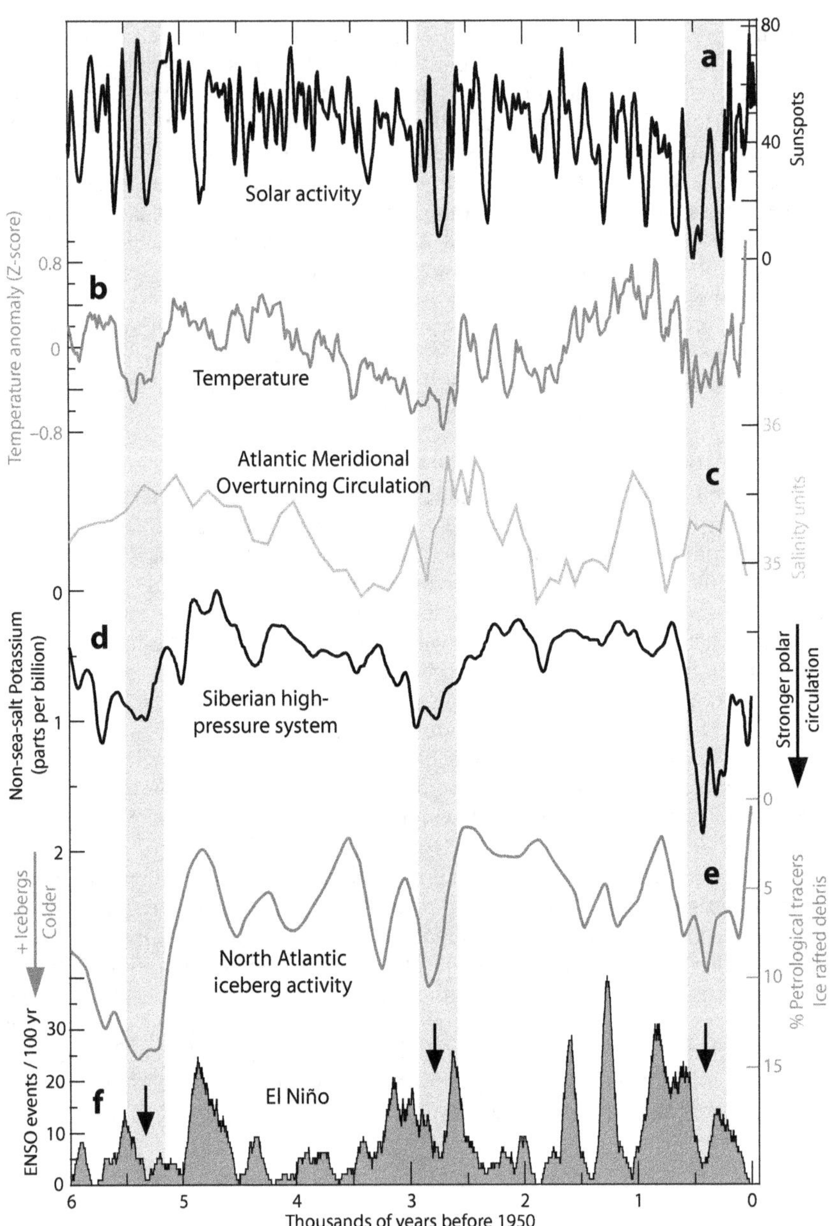

Figure 43. Climatic effects of a long solar grand minimum. a) Reconstruction of solar activity. b) Detrended proxy global temperature reconstruction. c) Salinity changes as a proxy for the strength of the Atlantic Meridional Overturning Circulation. d) Potassium content (inverted) in a Greenland ice core as a proxy for the polar atmospheric circulation and the strength of the Siberian High-Pressure System. e) Petrological tracers (inverted) as a proxy for iceberg activity in the North Atlantic. f) Frequency of strong El Niño events. Arrows indicate El Niño suppression during periods of very low solar activity.

In chapter 18, we discussed how the El Niño-Southern Oscillation contributes to poleward heat transport and showed its variability throughout the Holocene (fig. B15, ch. 18). Now we will focus on the frequency of strong Niño events over the last 6000 years (fig. 43f).[199] As the planet cooled during the Neoglaciation, the frequency of Niño events also increased, contributing to the cooling. Consistent with this interpretation, there was an increase in the frequency of Niño events during the cooling that led to the three events we are considering. However, when the Spörer-type solar grand minimum reached very low solar activity values, El Niño activity was suppressed (fig. 43f, arrows), and the equatorial Pacific shifted to a predominantly Niña condition. This type of semi-permanent La Niña condition has been observed during both warm periods, such as the Holocene Climatic Optimum, and cold periods, such as the central part of the Little Ice Age. This may indicate that since the planet's temperature is not changing, there is no need to adjust the intensity of poleward heat transport.

Information from hydrological proxies

Precipitation proxies provide information that is limited to specific regions because precipitation can vary widely from place to place. Even nearby locations can experience opposite changes, as seen after the eruption of Mt. Tambora when some parts of Europe became very wet while others became very dry (ch. 24). However, as noted above, the evidence suggests that winter precipitation in Norway increased during all three periods considered, leading to glacier growth.[200] Millennial-scale peaks in precipitation at these times have also been recorded in midlatitude locations such as Ireland and California.

Scientists have obtained a high-resolution record of Asian monsoon strength by analyzing a cave stalagmite from China.[201] This record reveals episodes of monsoon weakness or drought during the three solar grand minimums studied. Many scientists have traditionally linked most of the centennial and millennial variability of the Asian and Indian monsoons to solar variability.

Information about temperature changes

In previous chapters (ch. 21-23 & 26), we used a Holocene temperature reconstruction based on 73 global proxies. After detrending the reconstruction for the last 6000 years to remove the overall cooling trend of the Neoglaciation, we find that the three periods we studied show the largest cooling over thousands of years (fig. 43b). This finding contradicts the conventional wisdom that solar variations have little effect on temperature. In fact, the paleoclimate evidence suggests that solar activity, rather than CO_2, is the primary determinant of climate on centennial and millennial time scales.

The Indo-Pacific Warm Pool is the world's largest body of warm water, located in the tropical Indian and Pacific Oceans. It provides an excellent location for measuring the temperature of the planet over time. By analyzing a sub-

[199] Moy, C.M., et al., 2002. Nature, 420 (6912), pp.162–165.
doi.org/10.1038/nature01194
[200] Matthews, J.A., et al., 2005. Quat. Sci. Rev. 24 (1-2), pp.67–90.
doi.org/10.1016/j.quascirev.2004.07.003
[201] Wang, Y., et al., 2005. Science, 308 (5723), pp.854–857.
doi.org/10.1126/science.1106296

surface ocean temperature proxy in this region, researchers produce a reconstruction of the tropical ocean temperature during the Holocene similar to the global reconstruction used in this book.[202] This analysis also identifies the same three periods considered here as being significantly colder on a millennial scale in the tropical ocean. In addition, another study focused on ocean surface temperatures in the same region revealed a clear 2500-year periodicity in temperature during the Holocene. This periodicity was found to be related to solar activity.[203]

By analyzing benthic cores from the North Atlantic, researchers were able to construct a record of iceberg activity that showed remarkable agreement with the solar activity record (box 21, ch. 23).[204] This proxy record has been very useful in linking each centennial cold period in the North Atlantic to cooling and abrupt climate changes outside the region. This record is also consistent with the other evidence we have presented. The three periods of abrupt climate change that coincide with the three Spörer-type solar grand minimums show the greatest millennial iceberg activity in the North Atlantic (fig. 43e).

A long-term atmospheric reorganization

Thanks to numerous climate proxies, we can reconstruct the climatic changes that occur when solar activity becomes very low over many decades. Although we still do not fully understand how this happens, we do know what happens. Surprisingly, the changes inferred from the proxies are inconsistent with the global effect that a reduction in solar energy at the surface should have. Instead, they indicate a profound reorganization of the atmospheric circulation, mainly affecting the Northern Hemisphere. This reorganization is a slow process that becomes more profound the longer solar activity remains low. However, it begins to reverse as soon as solar activity returns to normal.

When the North Atlantic Oscillation becomes persistently negative, it indicates a weakened Arctic polar vortex. This leads to an increase in the exchange of air masses between the Arctic and the midlatitudes, resulting in very cold winters in the midlatitudes of the Northern Hemisphere.

In addition, the poleward transport of heat, moisture, and salts increases considerably. This, in turn, leads to a much more intense polar atmospheric circulation. The polar front also shifts southward, causing the Arctic to expand.

The latitudinal temperature gradient, the most important climatic determinant, becomes more pronounced. The southward shift of the jet stream and the subtropical jet causes a contraction of the Hadley cells. As a result of this contraction, the dryness produced by the downward branch of the Hadley cells also shifts southward. This leads to increased precipitation in the midlatitudes and a weakening of the monsoon.

As more heat from the ocean and atmosphere is directed toward the Arctic due to a steeper temperature gradient, the planet begins to cool as it loses energy. But the cooling is uneven. The Southern Hemisphere experiences less

---

[202] Rosenthal, Y., et al., 2013. Science, 342 (6158), pp.617–621.
   doi.org/10.1126/science.1240837
[203] Khider, D., et al., 2014. Paleoceanography, 29 (3), pp.143–159.
   doi.org/10.1002/2013PA002534
[204] Bond, G., et al., 2001. Science, 294 (5549), pp.2130–2136.
   doi.org/10.1126/science.1065680

cooling, the heat content of the tropical oceans decreases, and the high latitudes of the Northern Hemisphere cool more. The North Atlantic region, the main pathway for heat transport to the poles, experiences the greatest cooling and shows a large increase in iceberg activity. The energy deficit in the Northern Hemisphere causes more heat to be transported across the equator from the Southern Hemisphere. This is achieved by shifting the Intertropical Convergence Zone several latitudes southward so that the atmosphere will transport heat from the Southern Hemisphere to the Northern Hemisphere (fig. 82, ch.47), the opposite of the current situation.[205]

Solar-induced atmospheric reorganization is a gradual process that accumulates over time. The longer it continues, the more profound the changes and their effects. When solar activity returns to normal, it takes a similar amount of time to reverse the changes, which explains why the mid-19th century, with solar activity levels similar to those of the 20th century, had a different climate (fig. 79, ch.46). It is reasonable to assume that higher than normal solar activity should have the opposite effect, causing warming through atmospheric reorganization. Between 1935 and 2005, there was the longest period of above-average solar activity in at least the last 600 years, coinciding with the highest rate of warming in at least 600 years.

Studies show that the Hadley circulation has been expanding poleward at a rate of 0.5° latitude/decade since at least 1979, largely due to natural causes.[206] It seems possible that the climate change that occurred between 1850 and the start of our massive emissions around 1950 was due to the slow recovery from the solar-induced Little Ice Age.

The available evidence indicates that the effect of changes in solar activity on climate is neither adequately accounted for in climate theory nor properly reproduced in climate models. Therefore, our understanding of climate is inadequate to justify significant societal changes in response to observed climate changes since their cause is uncertain.

In summary

Evidence from climate proxies supports that prolonged periods of very low solar activity can cause a gradual reorganization of the atmosphere. This reorganization leads to a contraction of the tropics and an expansion of the polar regions. As a result, the temperature difference between the equator and the poles becomes more pronounced, driving more heat toward the poles. This increased heat loss in the polar regions, especially the Arctic, leads to global cooling, which is more pronounced in the northern midlatitudes. However, climate models cannot reproduce this effect because its mechanism is unknown, and the instrumental evidence for the much smaller effect of the 11-year solar cycle is unconvincing. As a result, most scientists ignore the evidence for solar activity as the primary driver of Holocene climate change. This renders our understanding of climate inadequate and unreliable.

---

[205] Sachs, J.P., et al., 2009. Nat. Geosci. 2 (7), pp.519–525. doi.org/10.1038/NGEO554
[206] Staten, P.W., et al., 2018. Nat. Clim. Change, 8 (9), pp.768–775.
   doi.org/10.1038/s41558-018-0246-2

CHAPTER 28
KNOWN EFFECTS OF THE SOLAR CYCLE

Although the change in solar energy caused by the solar cycle is too small to detect a climatic effect above the noise, we have consistently observed a temperature change in the global surface temperature and the heat budget of the tropical oceans that is four times larger than expected. The pattern of temperature changes is highly irregular, with higher warming in the mid to high latitudes of the Northern Hemisphere. Interestingly, this solar cycle temperature pattern is similar to the temperature pattern observed in recent decades as a result of global warming.

The warming caused by increased solar activity implies changes in poleward heat transport by the tropospheric and stratospheric circulation. As solar activity increases, there is a decrease in poleward heat transport, leading to heat accumulation at mid-high latitudes, cooling of the Arctic, and increased heat content of the tropical oceans. The observed changes in the latitudinal temperature gradient support this interpretation.

The solar cycle should have no detectable effect on climate

The Sun provides 99.9% of the energy to the climate system, and this energy – known as total solar irradiance – changes by only 0.1% over an 11-year solar cycle (ch. 2). This change is so small that its effects should be undetectable in the noise of climate variables. The predicted surface warming from a 1.1 W/m^2 increase is only 0.025°C, below the detection limit of 0.1°C.[207]

However, the temperature data consistently show that the solar signal in global temperature is 0.1°C, which is four times larger than we would expect from the energy change alone. This observation provides the first indication that the Sun's effect on climate is not due to surface changes in total solar irradiance.

The effect of the solar cycle on surface temperature

When there is a small increase in energy from the Sun, we would normally expect a small, evenly distributed change in surface temperature based on the amount of insolation. However, this is not what happens in reality. Instead, the change in surface temperature is highly uneven, with some regions cooling despite the increase in solar energy. These differences can only be explained by significant dynamical changes in the atmosphere and oceans, even though solar output varies by only 0.1%.

Figure 44a shows a map of surface temperature changes on a 5x5° grid attributed to the 11-year solar cycle between the 1996 solar minimum and the

---

[207] Wigley, T.M.L. & Raper, S.C.B., 1990. Geophys. Res. Lett. 17 (12), pp.2169–2172. doi.org/10.1029/GL017i012p02169

2002 solar maximum.[208] Although the global mean surface temperature increases by only 0.1°C during this cycle, it increases by 0.4°C (and over 1°C in certain areas) at 60°N latitude. The response pattern in figure 44a is determined with a lag of 1 month. This means that the climatic response to solar changes is very rapid, and this small lag is more likely for an atmospheric response than an oceanic one.

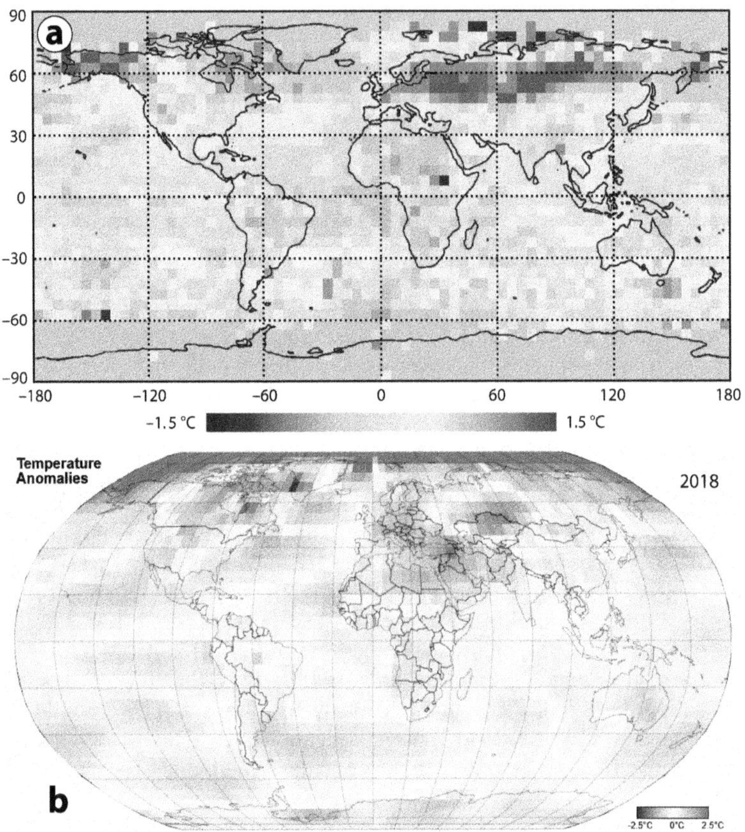

Figure 44. Surface temperature changes due to the solar cycle and global warming. a) Changes due to the solar cycle from 1996 to 2002. b) Global temperature anomaly for 2018.

When solar activity is high, the pattern of increased surface warming in the extratropics of the Northern Hemisphere and less warming in the tropics and Southern Hemisphere is similar to the pattern of recent global warming attributed primarily to increased CO_2 levels. Figure 44b shows a map of global surface temperature anomalies for 2018 on a 5x5° grid based on the 1991-2020 mean.[209]

---

[208] Lean, J.L., 2017. Sun-climate connections. In: Oxford Research Encyclopedia of Climate Science. doi.org/10.1093/acrefore/9780190228620.013.9

[209] Figure 44b from NOAA's global mapping tool.
www.ncei.noaa.gov/access/monitoring/climate-at-a-glance/global/mapping/2018

The greatest warming has occurred in the high latitudes of the Northern Hemisphere, with warming also occurring in association with the Gulf Stream and the Kuroshio Current. A pattern of four zones of increased warming separated by about 7,000 km (4,300 miles) in the Atlantic, Indian, and Pacific Oceans and the Tasman Sea is observed in the 30-60°S band of the Southern Hemisphere. This pattern is thought to be the result of an atmospheric wave.[210] While solar-induced climate change and global warming share these features, they do not indicate the result of the global distribution of a radiative effect produced by a change in solar activity or CO_2. Rather, they suggest a response to both forcings mediated by similar dynamic changes in the atmosphere and oceans.

This should be the second indication that the Sun's effect on climate is not the result of the effect of changes in total solar irradiance at the surface. Instead, it appears to be mediated by changes in the coupled atmosphere-ocean system. Therefore, to understand the solar effect on climate, we should focus on the atmosphere rather than the surface.

Effect of the solar cycle on the latitudinal temperature gradient

Figure 45 shows the temperature changes due to the solar cycle at the surface (as in fig. 44a) and in the lower stratosphere, calculated per circle of latitude (zonal mean). The clear relationship between the two curves indicates an atmospheric origin for the surface curve as well. The information shown in figure 45 is crucial to understanding the solar effect on climate. When solar activity is high, surface warming increases with increasing latitude, peaking at 0.4°C at 60°N. However, the Arctic experiences cooling when solar activity is high. Note the black curve in figure 45, which shows a sharp transition between strong warming at 60°N and cooling at 75°N. Between these two latitudes lies the polar vortex, which, when strong, severely restricts heat exchange. With increased solar activity, the latitudinal temperature gradient between the equator and 60°N is reduced, leading to a reduction in poleward heat transport.

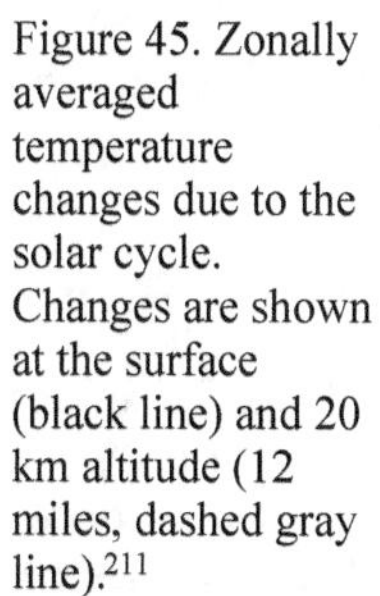
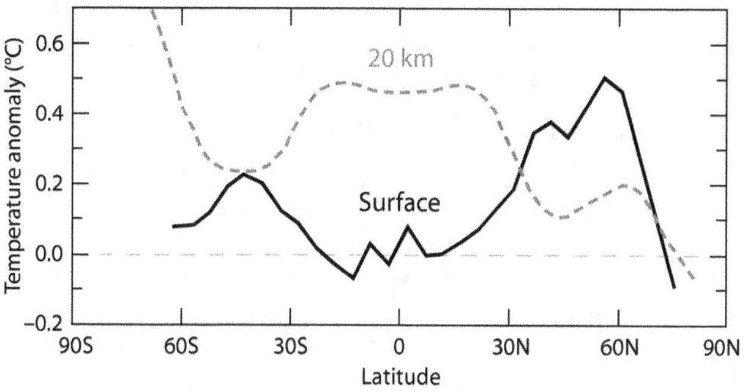

Figure 45. Zonally averaged temperature changes due to the solar cycle. Changes are shown at the surface (black line) and 20 km altitude (12 miles, dashed gray line).[211]

[210] Chiswell, S.M., 2021. Nat. Comm. 12 (1), p.4779.
doi.org/10.1038/s41467-021-25160-y
[211] Figure data from Lean, J.L., 2017. Sun-climate connections. In: Oxford Research Encyclopedia of Climate Science. doi.org/10.1093/acrefore/9780190228620.013.9

In the lower stratosphere, the situation is essentially the opposite of that in the surface (fig. 45, dashed gray line). The tropical stratosphere experiences significant warming due to increased ozone and UV radiation caused by increased solar activity. However, most of the ozone is transported to the high latitudes of the Northern Hemisphere by the Brewer-Dobson circulation, making it the region with the highest ozone concentration. This leads us to expect more warming there, but the lower warming observed in the high latitudes of the lower stratosphere of the Northern Hemisphere is puzzling. This phenomenon can only be explained by a decrease in meridional transport with an increase in solar activity. The decrease in stratospheric transport is consistent with the increase in the latitudinal temperature gradient shown by the dashed line in figure 45, which would have the effect of increasing the speed of the zonal (east-west) winds that oppose meridional transport. The two profiles shown in figure 45 reinforce the notion that increasing solar activity decreases poleward transport in both the lower stratosphere and the troposphere.

The heat budget of the tropical ocean

In box 14 (ch. 17), we discussed the influence of the solar cycle on oceanic heat transport. It is worth repeating the main findings of scientific studies, which show that changes in the heat budget of the tropical oceans have an 11-year frequency that is synchronized with the solar cycle.[212] The challenge, however, is that the changes in solar energy are too small by a factor of five to ten to be responsible for the observed temperature variations, as are all other solar effects.

If increased solar radiation is not the cause of ocean warming, then it must be due to the warming of the underlying ocean by the tropical atmosphere. The only way the atmosphere can do this is by reducing the amount of sensible and latent heat that flows from the ocean surface to the atmosphere. This is because, apart from shortwave energy from the Sun, the net energy flux in the tropics (and most of the planet) is from the ocean to the atmosphere. Since the atmosphere is primarily responsible for the planet's heat transport, the 11-year cycle in the tropical ocean heat budget is likely caused by a decrease in poleward heat transport. More solar heat remains in the tropical ocean during high solar activity because the atmosphere transports less of it. A decrease in wind speed could lead to the observed warming and decrease in transport, reducing both sensible heat transfer and evaporation (latent heat transfer).

Dynamic effects of the solar cycle

Dynamic changes in atmospheric circulation have been found to occur with the solar cycle.[213] As a result of increased solar radiation, the tropospheric circulation is altered in several ways. These changes include expansion of the Hadley cell, poleward movement of subtropical jets, stabilization of the polar vortex, reduced waviness of the jet stream, trapping of polar air, and reduced frequency of winter blocking events. The pressure gradient between the Icelandic Low and the Azores High also increases, resulting in the positive phase

---

[212] White, W.B., et al., 2003. J. Geophys. Res. Oceans, 108 (C8) 3248. doi.org/10.1029/2002JC001396

[213] Lean, J.L., 2017. Sun-climate connections. In: Oxford Research Encyclopedia of Climate Science. doi.org/10.1093/acrefore/9780190228620.013.9

of the North Atlantic Oscillation. These effects are well documented in the scientific literature. They may sound familiar, as they are the opposite of the larger scale effects shown in the previous chapter using several climate proxies during times of greatly reduced solar activity due to a grand minimum.

The known dynamical effects have one thing in common: they alter the transport of heat to the poles. During periods of high solar activity, meridional transport decreases, leading to warming in the midlatitudes and cooling in the Arctic. Conversely, during periods of low solar activity, meridional transport increases, leading to cooling in the midlatitudes and warming in the Arctic. However, meridional transport is not only influenced by solar activity. Other factors such as volcanic eruptions (ch. 25), the Quasi-Biennial Oscillation (ch. 14), the oceanic multidecadal oscillations (ch. 19), and the El Niño-Southern Oscillation (ch. 18) also play an important role. As a result, changes in transport and their climatic effects, including surface temperature, are not expected to coincide with solar activity in the short term because these other factors also influence them. Only on centennial scales, when all other factors tend to average out, do changes in transport become primarily dependent on solar activity, which exhibits centennial and millennial cycles that substantially alter solar activity over decades and even centuries, overriding all other factors and inducing long-term warming or cooling.

Reproduction of the solar cycle effects in climate models

Until recently, climate models did not include the stratosphere, lacked a realistic Quasi-Biennial Oscillation, and could not account for the complex response of ozone to solar radiation. As a result, they have been unable to reproduce the effects of the solar cycle or understand the impact of a solar grand minimum. Climate models typically show weaker and delayed temperature responses to changes in solar irradiance. Not all models account for solar-induced ozone changes, the polar vortex, and surface changes. Furthermore, models often produce unrealistic warming of the Arctic in response to increased solar activity, contrary to the observed cooling. More worryingly, the dynamic response of the stratosphere to changes in solar activity is missing from most models.[214] The main hypothesis of solar forcing on climate, discussed in the next chapter, is based on this response. While the models qualitatively capture some solar effects on climate, we can conclude that they are most likely not reproduced by the correct mechanism.

In summary

Although the energy changes involved are small, the climate system shows a distinctive pattern of surface temperature change in response to increased solar cycle activity. This pattern is reminiscent of the warming observed during the last decades. The effect of increased solar activity is minimal in the tropics and weaker in the Southern Hemisphere compared to the Northern Hemisphere. The observed temperature changes are consistent with the observed strengthening of the polar vortex, which should reduce poleward heat transport. These temperature changes should also reduce heat transport by decreasing the temperature gradient between the equator and 60°N, the latitude of

---

[214] Misios, S., et al., 2016. Q. J. R. Meteorol. Soc. 142 (695), pp.928–941.
doi.org/10.1002/qj.2695

maximum solar cycle heating. In the lower stratosphere, the observed changes suggest a decrease in the poleward heat transport circulation. Changes in the heat content of the tropical ocean imply a decrease in the ocean-atmosphere heat flux, consistent with the decrease in heat transport due to increased solar activity. The dynamic changes in atmospheric circulation resulting from changes in solar cycle activity are well known. They are of the same nature as those observed during the long solar grand minimums of the past, although the magnitude is different. Climate models show a weak reproduction of the observed effects in response to solar cycle changes, with a much larger lag, and typically lack the crucial dynamic response of the stratosphere.

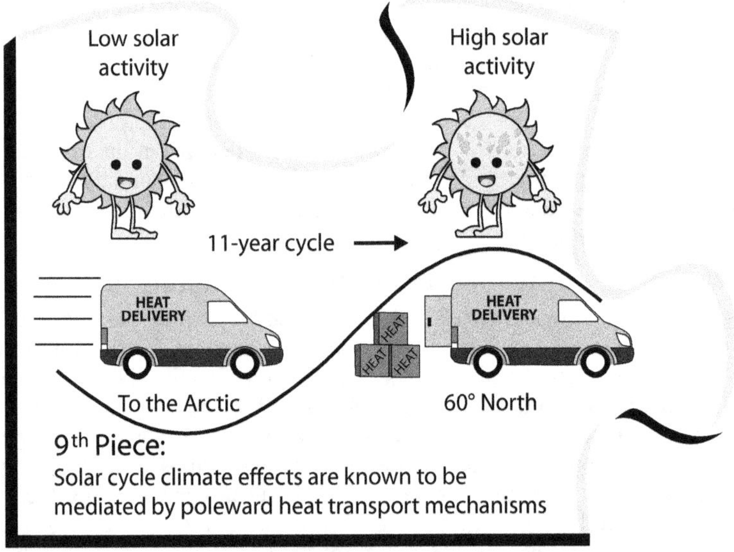

CHAPTER 29
THE STRATOSPHERIC PATHWAY

Changes in the tropical ozone layer caused by variations in solar activity can have substantial climatic effects at the surface. This occurs through a known signaling pathway involving changes in wind speed and atmospheric wave propagation in the stratosphere, ultimately affecting the strength of the polar vortex during winter. These changes in the polar vortex directly impact pressure patterns and tropospheric circulation, which in turn regulate the poleward transport of heat and the exchange of air masses between the Arctic and the midlatitudes. As a result, winter weather in the Northern Hemisphere is largely determined by these processes.

For the past 50 years, scientists have known that a certain type of atmospheric wave, called a planetary wave, could be responsible for changing the Earth's climate in response to solar activity. These waves provide the energy needed for climate changes in response to solar changes. However, because planetary waves are influenced by every factor that affects stratospheric conditions, the effects of solar activity on surface climate are not always consistent or easy to detect. Despite these challenges, the Sun's influence on climate is real, and its most obvious effect is reflected in the frequency of cold winters in the midlatitudes of the Northern Hemisphere.

A solar effect from the stratosphere to the surface

Scientists have studied whether and how sunspot changes affect the climate for over two hundred years. However, early attempts to understand this phenomenon were hampered by a lack of data and knowledge about the atmosphere. Scientists struggled to find evidence of the effect at the surface because it is secondary, highly variable, and nonlinear.

It wasn't until 1994 that scientists discovered a mechanism for how the Sun's variable activity affects the climate. By observing and modeling the changes that occur in the stratosphere in response to the solar cycle, they were able to identify the process.[215] This mechanism has since been repeatedly confirmed and observed in atmospheric reanalyses, which are climate data assimilation products.[216] Some climate models that more accurately represent the stratosphere and ozone chemistry have also replicated this mechanism to some extent.[217] After three decades of research, we now have a better understanding of how solar activity influences climate.

During the solar cycle, an increase in solar activity leads to a greater relative increase in UV radiation than total radiation. This increase in UV radiation has a major impact on the tropical ozone layer, increasing ozone production and raising temperatures by about 1.5°C (2.7 °F, fig. 46) due to increased absorp-

[215] Haigh, J.D., 1994. Nature, 370 (6490), pp.544–546. doi.org/10.1038/370544a0

[216] Mitchell, D.M., et al., 2015. Q. J. R. Meteorol. Soc. 141 (691), pp.2011–2031. doi.org/10.1002/qj.2492

[217] Kodera, K., et al., 2016. Atmospheric Chem. Phys. 16 (20), pp.12925–12944. doi.org/10.5194/acp-16-12925-2016

tion of UV radiation. Stratospheric ozone plays a critical role in this process, acting as a signal receiver and amplifying the effect.

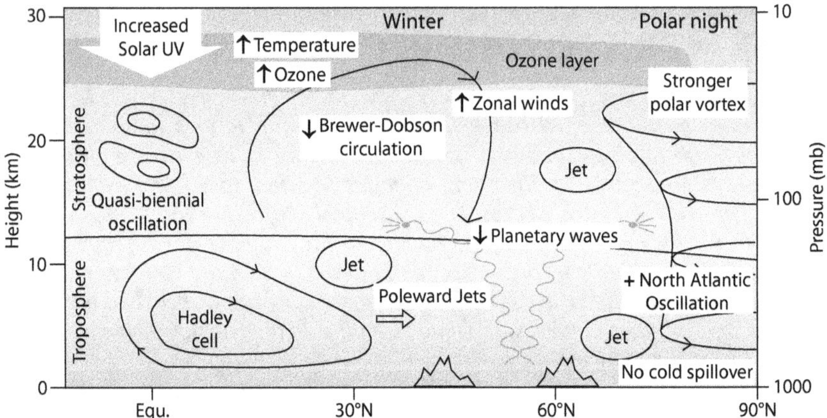

Figure 46. Top-down mechanism of the solar effect on climate. The effects of increased solar activity are shown in white boxes.

As the tropical ozone layer warms, it creates a larger temperature gradient between the tropical and polar stratospheres. This gradient directly affects the speed of east-west (zonal) winds, which depends on the size of the gradient. The steeper the gradient, the faster the winds.

Atmospheric waves are similar to ocean waves and travel through the troposphere, carrying energy and momentum and affecting weather. For a more detailed explanation, see box 10 (ch. 14). A special type of atmospheric wave, called a planetary wave, is created by temperature differences between land and sea and by winds blowing over large mountain ranges. Planetary waves are more common in the Northern Hemisphere. These waves can reach the stratosphere, but only if the zonal winds in the stratosphere are not too strong and blowing from the west. Otherwise, they are reflected back into the troposphere.

When planetary waves break up, they provide momentum and energy that contribute to the poleward transport of heat and ozone by the Brewer-Dobson circulation. They also weaken the polar vortex by reducing the speed of the winds that form it. This effect is particularly noticeable in the Northern Hemisphere.

As solar activity increases, zonal winds in the stratosphere accelerate, making it more difficult for planetary waves to reach the stratosphere. As a result, the polar vortex remains strong throughout the winter. Although the increase in solar activity is relatively small in terms of radiation, it profoundly affects the stratospheric circulation and its transport of heat and ozone poleward.

Through stratosphere-troposphere coupling (ch. 15), the solar signal is transmitted to the troposphere. The strength of the polar vortex determines the winter state of the North Atlantic Oscillation, which becomes anomalously positive during high solar activity. In addition, the position of the jet stream is affected by the strength of the vortex and moves poleward during high solar activity. This movement keeps cold Arctic air masses in the Arctic region, leading to warmer winters in the midlatitudes of the Northern Hemisphere.

In the tropics, changes in atmospheric circulation caused by a poleward shift of the jet stream and a decrease in the upward branch of the Brewer-Dobson circulation lead to an expansion of the Hadley circulation and a similar shift of the subtropical jet. These changes affect precipitation patterns and cause warming at 60°N, as less heat can be transported to the Arctic through a strengthened polar vortex.

When solar activity decreases, the opposite effects occur. This mechanism can explain all the solar cycle-related climate changes discussed in the previous chapter.

Recent studies have confirmed the role of the Northern Hemisphere polar vortex as a transmission system that produces changes in the tropospheric circulation in response to changes in solar activity.[218]

The role of planetary waves

It may seem surprising today, but back in 1974, an atmospheric physicist named Colin Hines was skeptical that the Sun could affect Earth's climate. He proposed that the only way solar variations could affect climate was by altering the propagation of planetary waves.[219] Interestingly, he suggested that this mechanism would be more relevant at mid and high latitudes during winter. Hines was the first to propose how the Sun influences the climate of our planet.

For the past 50 years, scientists have known that the small magnitude of changes in total solar irradiance does not preclude solar variability from affecting our planet's climate. It is a mistake to focus on changes in irradiance as the only solar forcing. Atmospheric waves play a critical role by moving energy through the atmosphere. The Sun does not directly provide the energy to change our climate; rather, the change in UV radiation in the ozone layer tips the balance between two different atmospheric circulation states. Planetary waves then provide the energy to shift the circulation.

When solar activity is high, less wave energy reaches the stratosphere, strengthening the polar vortex and reducing heat transport to the poles, resulting in energy conservation and warming. Conversely, more wave energy reaches the stratosphere when solar activity is low, weakening the polar vortex and increasing heat transport to the poles, resulting in energy loss and cooling.

Studying planetary waves in the stratosphere is challenging, and research on this topic is relatively new. However, even with these limitations, researchers have already found in the 55-75°N stratosphere that the amplitude of planetary waves decreases during solar maximums. Conversely, during solar minimums, changes in the meridional temperature gradient and vertical wind shear lead to an increase in the amplitude of planetary waves (fig. 67, ch. 42).[220] The effect of the solar cycle on these waves is substantial, explaining about 25% of the variability in their amplitude.

---

[218] Veretenenko, S., 2022. Atmosphere, 13 (7), p.1132. doi.org/10.3390/atmos13071132
[219] Hines, C.O., 1974. J. Atmos. Sci. 31 (2), pp.589–591.
   doi.org/10.1175/1520-0469(1974)031<0589:APMFTP>2.0.CO;2
[220] Powell Jr, A.M. & Jianjun, X., 2011. J. Atmos. Sol. Terr. Phys. 73 (7–8), pp.825–838.
   doi.org/10.1016/j.jastp.2011.02.001

Box 23. The first evidence of a solar effect on climate

Sunspots have been observed since ancient times, and due to the Sun's primary role in climate and weather patterns, many scientists have attempted to find a connection between sunspots and climate. William Herschel, the discoverer of Uranus and infrared radiation, began scientific research on the subject in 1801.

Despite the considerable amount of research devoted to the subject, the results have been inconclusive, leading to controversy in the field. Most of the effects found were either transient, statistically insignificant, or nonexistent, resulting in a negative perception of the Sun-climate connection. This perception persists to this day, and many scientists refuse to consider the possibility of a link, even when presented with evidence, due to a lack of confidence in the field.

The quest for the first concrete evidence of a solar influence on climate ended in 1987 when Karin Labitzke, a German researcher at the Free University of Berlin, discovered it in the polar stratosphere during the dark winter months.[221] This lack of sunlight, where the solar effect was first discovered, provides further evidence that changes in total irradiance are not the cause of the solar effect on climate.

In 1980, researchers discovered the Holton-Tan effect, which showed that the Quasi-Biennial Oscillation (a pattern of stratospheric winds above the equator) affects the global circulation in the stratosphere. This is discussed in more detail in box 11 (ch. 14). During winter, the effect of the Quasi-Biennial Oscillation can reach the pole and change the way planetary waves propagate. This is because the winds of the oscillation change between easterly and westerly phases every two years. When the winds are in their easterly phase, they change the zonal circulation pattern, allowing more planetary waves to reach the vortex, weakening it and raising the temperature inside. The opposite happens when the winds are in their westerly phase: the vortex is stronger and the temperatures inside are colder.

Karin Labitzke's insight was to notice that the solar effect on the stratosphere depends on the phase of the Quasi-Biennial Oscillation. During winters in the westerly phase of the oscillation (light gray data in fig. B23), high solar activity corresponds to higher polar stratospheric temperatures, while low solar activity corresponds to lower temperatures. Keep in mind that these differences are due to dynamic effects on the vortex in the absence of sunlight. However, in winters during the easterly phase of the oscillation (dark gray data in fig. B23), high solar activity corresponds to lower polar stratospheric temperatures, and low solar activity corresponds to higher temperatures.

At the time, many scientists were puzzled by the phenomenon because the mechanism discussed in this chapter was unknown and would be developed ten years later. They were confused that a change in the direction of the equatorial wind could completely reverse a solar effect. As a result, most scientists ignored it,

---

221 Labitzke, K., 1987. Geophys. Res. Lett. 14 (5), pp.535–537.
   doi.org/10.1029/GL014i005p00535

and today it is rarely mentioned in climate books, even those that focus on the atmosphere.

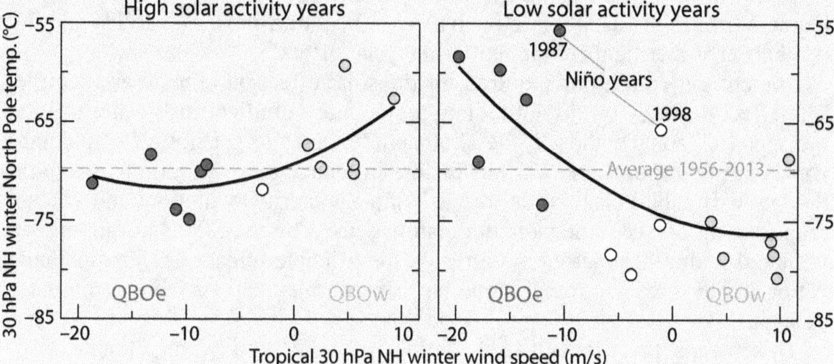

Figure B23. Effect of solar activity and the Quasi-Biennial Oscillation on winter polar stratospheric temperature. High solar activity years are shown in the left panel and low solar activity years in the right panel. Easterly years of the Quasi-Biennial Oscillation are shown in dark gray and westerly years in light gray. Two strong El Niño years are identified.

The solar effect on climate can be reversed because it depends on the conditions of planetary wave propagation in the stratosphere, which are affected by any factor that modifies temperature or wind speed there. One such factor is the Quasi-Biennial Oscillation that modulates the global circulation in the stratosphere through the Holton-Tan effect. When the oscillation reverses, so does the solar effect on climate, as the propagation of planetary waves in the stratosphere changes from being facilitated by the change in solar activity to being hindered. El Niño events also affect the conditions for planetary wave propagation in the stratosphere. Figure B23 shows the effect of the 1987 and 1998 El Niño events on winter polar stratospheric temperature. However, this composite behavior is too complex even for many scientists, so it is generally ignored.

Karin Labitzke spent the next 27 years working on her discovery. She identified the changes in the winter tropospheric circulation that occur as a result of the solar cycle and contributed significantly to the understanding of the stratospheric solar signaling pathway discussed in this chapter. Her discovery marked the end of a 185-year quest for definitive evidence of a solar effect on climate, but unfortunately, she died without receiving the recognition she deserved. Her extraordinary discovery was not well received at the time because the climate science community was embracing the Enhanced CO_2 Effect hypothesis of climate change and viewed any solar hypothesis as undesirable competition.

The inconsistency of the solar effect

The solar effect on climate is an unlikely outcome because the changes in solar radiation are too small. Three specific conditions must exist for the effect to be detectable. First, there must be an ozone layer for the solar signal to be received and cause temperature changes there. Second, the continents must not be located primarily in the tropics, as there would not be enough planetary wave activity outside of this area to produce an observable effect on the vortices. This is currently the case in the Southern Hemisphere. Finally, the planet must be in an ice age since the polar vortex requires very low winter polar temperatures to act as an effective barrier to heat transport. The ability to affect this barrier is a critical component of the solar effect.

The current conditions required for the solar effect on climate are complex. The effect depends on the latitudinal temperature gradient in the stratosphere, the speed of zonal winds in the stratosphere, and the generation of planetary waves in the troposphere. Various factors influence these conditions, including the Quasi-Biennial Oscillation, the El Niño-Southern Oscillation, and volcanic eruptions. In essence, the signaling pathway used by the Sun is not unique and is shared with other factors. As a result, the ultimate impact on climate change is not due to the solar signal alone but is a complex and variable combination of signals.

This means that predicting the effect of solar activity on climate is not straightforward, as it depends on a combination of factors. Therefore, we cannot forecast winter weather in the Northern Hemisphere based on solar activity alone. We must also take these other factors into account. In addition, summers are minimally affected by solar changes, which explains why some summers during the Little Ice Age were quite warm despite lower solar activity.

Some people predicted that global temperatures would drop because solar cycles 24 and 25 have less activity. However, many more people, including NASA, argue that the Sun has little effect on climate because this drop has not occurred. But the truth is not that simple. The solar effect on climate is real and important but highly unpredictable in the short term. It only becomes apparent when the number of cold winters in the Northern Hemisphere rises above the background noise of all the other factors influencing stratospheric conditions.

The climatic effect of low solar activity has already been observed, as the frequency of very cold winters in the midlatitudes of the Northern Hemisphere has increased in recent decades. However, this trend has puzzled many scientists who do not understand solar effects on climate, especially since climate models predict the opposite.[222] There is some inconclusive discussion that it could be due to the loss of sea ice in the Arctic, but we know it's the result of decreased solar activity because that's exactly what's expected and what happened during the Little Ice Age. Chapter 42 further discusses how the low activity of solar cycles 24 and 25 affects climate.

If solar activity remains low for a prolonged period, the percentage of cold winters would increase as the effect of all other factors would average out to zero. As a result, we would see actual global cooling, and the energy content of the climate system would decrease. However, I expect an increase in solar activity with solar cycle 26, so there is no need to worry about undue global cooling caused by the Sun if I am correct.

A word about ozone levels

The solar effect on climate depends on the ozone layer. However, human emissions have caused a significant decrease in ozone levels due to increased

---

[222] Cohen, J., et al., 2020. Nat. Clim. Change, 10 (1), pp.20–29.
   doi.org/10.1038/s41558-019-0662-y

chlorine- and bromine-containing substances in the stratosphere. This ozone depletion could potentially affect the solar effect on climate. This question has not been thoroughly investigated because most scientists consider the solar effect on climate to be irrelevant. It is a complex issue to determine what effect the loss of ozone might have on the solar signal. I cannot predict whether it would lead to a decrease or an increase in the climate's response to solar changes, although I suspect the former. However, the latter is a more worrying possibility, as it could make the climate more sensitive to changes in solar activity.

In summary

The solar effect on climate is a complex process. Stratospheric ozone plays a critical role in receiving and amplifying the solar signal, while planetary waves – not changes in solar irradiance – provide the energy to produce climate effects. The transmission system is the polar vortex, which converts the energy into changes in the winter atmospheric circulation. This pathway is shared by other factors that also influence the propagation of planetary waves, such as the Quasi-Biennial Oscillation, El Niño, and volcanic eruptions. The effect of solar activity on climate is inconsistent and difficult to identify because of these shared factors. In the short term, changes in solar activity influence the frequency of cold winters in the Northern Hemisphere. In the long term, they can alter the energy content of the climate system and lead to profound climate changes.

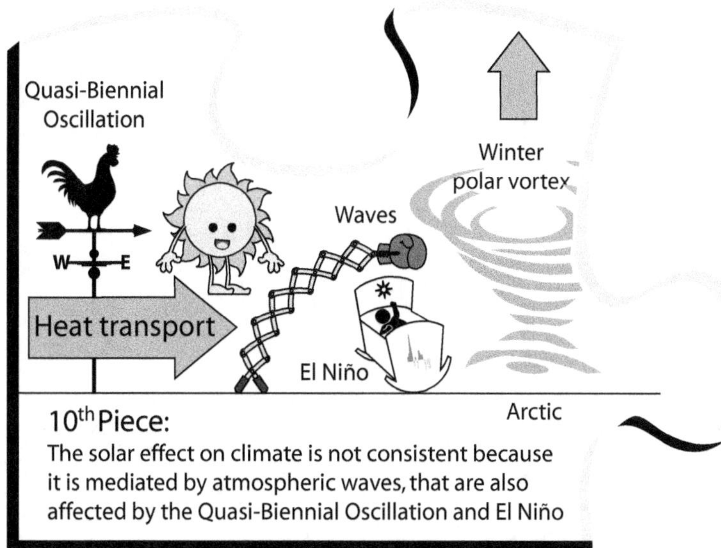

CHAPTER 30
EARTH ROTATION AND THE SOLAR CYCLE

Twice a year, the global atmospheric circulation is rearranged to direct more heat toward the wintering pole as a result of the increased temperature gradient between the equator and that pole. This causes the Earth to speed up its rotation, shortening the days by a fraction of a millisecond. This change in rotation depends on solar activity. Winters of low solar activity experience a greater increase in rotation speed than winters of high solar activity. The result is an 11-year cycle in the Earth's winter rotation speed changes due to the solar effect on global atmospheric circulation. Climatologists continue to ignore this five-decade-old finding and its implications.

The semi-annual change in the Earth's rotation

The Earth's climate system is strongly influenced by the seasons. As a pole moves from facing the Sun (summer) to facing away from it (winter), there's a big increase in atmospheric circulation and heat transport toward that pole. The planet's rotation responds to these changes because the atmosphere carries angular momentum (rotational inertia) and the total momentum must be conserved. As a result, a semi-annual pattern is clearly visible in the rotation rate. Scientists use precise radio astronomical calculations to measure these changes and detect microsecond changes in day length. We have already discussed the seasonal changes in the Earth's rotation rate in chapter 16 (fig. 24), where you can find a more detailed explanation.

The seasonal (semi-annual) change in the Earth's rotation rate is known to be the result of the exchange of momentum between the atmosphere and the solid Earth. Scientists have known for many years that the seasonal variation in day length reflects changes in the zonal circulation of the atmosphere.[223] Meanwhile, interannual variations are caused by other atmospheric phenomena. The biennial component of day length corresponds to changes in the Quasi-Biennial Oscillation, while the 3-4 year component corresponds to the El Niño-Southern Oscillation signal.

The effect of solar activity on the Earth's rotation

The Earth's rotation is affected by solar activity. This phenomenon has been measured since the 1960s and documented in scientific publications every decade. The effect has never been disproved but is still ignored by most scientists. More recent studies using 50 years of data continue to confirm this effect.[224]

Figure 47a shows changes in day length (ΔLOD) in milliseconds for two years: 2014, a year of high solar activity, and 2017, a year of low solar activity.

---

[223] Lambeck, K. & Cazenave, A., 1973. Geophys. J. Int. 32 (1), pp.79–93.
  doi.org/10.1111/j.1365-246X.1973.tb06521.x

[224] Le Mouël, J.L., et al., 2010. Geophys. Res. Lett. 37 (15), L15307.
  doi.org/10.1029/2010GL043185

As winter sets in in the Northern Hemisphere, atmospheric circulation intensifies to transport more heat toward the Arctic. As a result, the Earth spins faster and the day becomes a fraction of a millisecond shorter (indicated by gray arrows). However, during the year of low solar activity, the atmospheric circulation increases even more, indicating greater heat transport toward the poles. Chapter 42 provides additional evidence for this connection.

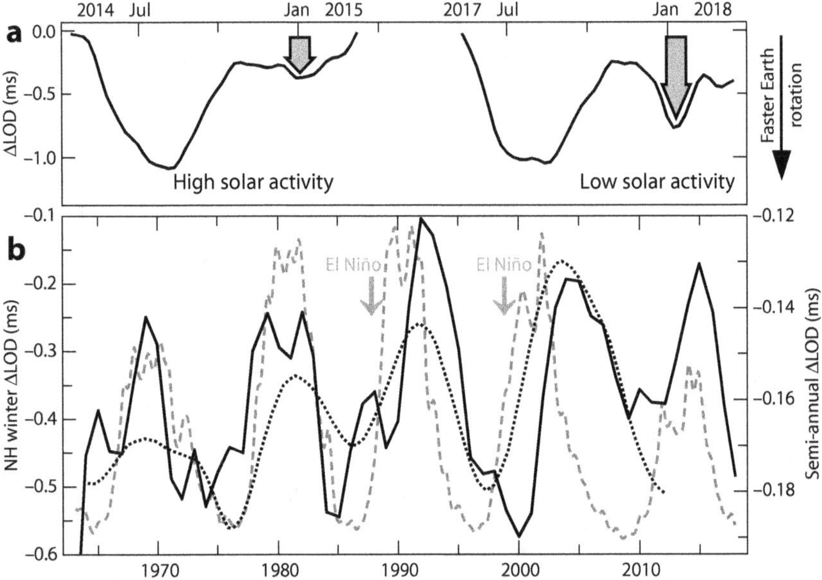

Figure 47. Solar activity affects Earth's rotation. a) Changes in day length in milliseconds for 2014 and 2017. Dark gray arrows indicate spin acceleration during Northern Hemisphere winter. b) Acceleration of the Northern Hemisphere winter spin (black line, three-year average). The dashed gray line is the solar cycle, while the dotted line (right scale) is from one of several studies confirming this effect.[225] Gray arrows indicate El Niño events.

The black solid curve in figure 47b represents the acceleration of Earth's rotation for the Northern Hemisphere winters. Over 56 years of data, this curve shows an 11-year cycle in winter atmospheric circulation changes synchronized with the solar cycle (fig. 47b, dashed gray curve). Winters with low solar activity show a greater rotation acceleration than winters with high solar activity, resulting in about 0.35 milliseconds shorter days.

While there is a clear correlation between changes in the Earth's rotation during winter and solar activity, it is known that other atmospheric phenomena can also affect this variable. As seen in figure 47b, the effects of the 1987 and 1998 El Niño events are visible (indicated by light gray arrows). This anomaly is also evident in the polar stratospheric temperatures discussed in chapter 29 (fig. B23, box 23), as they are related. Because of these additional factors, we

---

[225] Barlyaeva, T., et al., 2014. Ann. Geophys. 32 (7), pp.761-771.
    doi.org/10.5194/angeo-32-761-2014

don't expect a perfect fit between winter changes in Earth's rotation and solar activity.

The meaning of the solar cycle-Earth rotation relationship

Changes in the Earth's zonal circulation are reflected in the seasonal variation of day length because the zonal winds are responsible for the transfer of angular momentum between the atmosphere and the solid Earth. In 1976, a study found a relationship between multidecadal changes in day length and climate changes.[226] The trend in several climate indices was found to coincide with the trend in changes in day length. This study even predicted the warming trend that began immediately thereafter. The results of the study have been reproduced more recently using updated indices.[227]

The winter increase in atmospheric circulation is caused by an increase in the temperature gradient between the equator and the pole, where low solar radiation and strong radiative cooling cause the coldest temperatures in the hemisphere. This gradient change leads to an enhanced heat transport toward the pole, which is even stronger during low solar activity, resulting in a stronger increase in the Earth's rotation speed. We can confirm this interpretation because high solar activity is associated with Arctic cooling, while low solar activity is associated with Arctic warming (fig. 45, ch. 28; fig. B23, ch. 29). Low solar activity increases the frequency of cold winters because warm air reaching the Arctic rises above the cold air and pushes it toward the midlatitude continents. This exchange causes opposite temperature trends in the Arctic and midlatitude continents during winter, with one region warming and the other cooling.

The effect of solar activity on the Earth's rotation is often ignored or unknown by most climatologists. However, it is clear that changes in solar activity affect the Earth's rotation. We know this because solar activity cannot be affected by changes in the Earth's rotation, and the change in total solar irradiance is too small to cause a rotation change. Therefore, it must be the case that solar activity affects global atmospheric circulation in a way that causes changes in rotation.

However, this has some uncomfortable implications. Since solar activity affects the Earth's rotation rate, it is incorrect to claim that solar changes are too small to affect climate or that we fully understand how solar activity affects climate. It also suggests that climate models are missing a crucial piece of information that affects their reliability. Many climate scientists would rather ignore this inconvenient truth than admit that their field may be built on a weak foundation. To admit this would be to admit a profound ignorance of a major societal problem, which is unlikely.

In summary

Solar activity affects the Earth's rotational speed by changing the strength of the meridional circulation, which is responsible for transporting heat to the poles. In years of low solar activity, the atmospheric circulation increases more, causing the Earth to rotate faster and directing more heat to the Arctic in

---

[226] Lambeck, K. & Cazenave, A., 1976. Geophys. J. Int. 46 (3), pp.555–573. doi.org/10.1111/j.1365-246X.1976.tb01248.x
[227] Mazarella, A., 2013. Nat. Sci. 5 (1A), pp.149–155. doi.org/10.4236/ns.2013.51A023

winter. As a result, the Arctic experiences a warmer winter while the midlati-tudes become colder. Several studies provide evidence that solar activity modi-fies atmospheric circulation and changes heat transport on a hemispheric scale. However, this conclusion contradicts the current representation of Earth's climate in models and our understanding of climate change, suggesting that the solar effect is poorly understood.

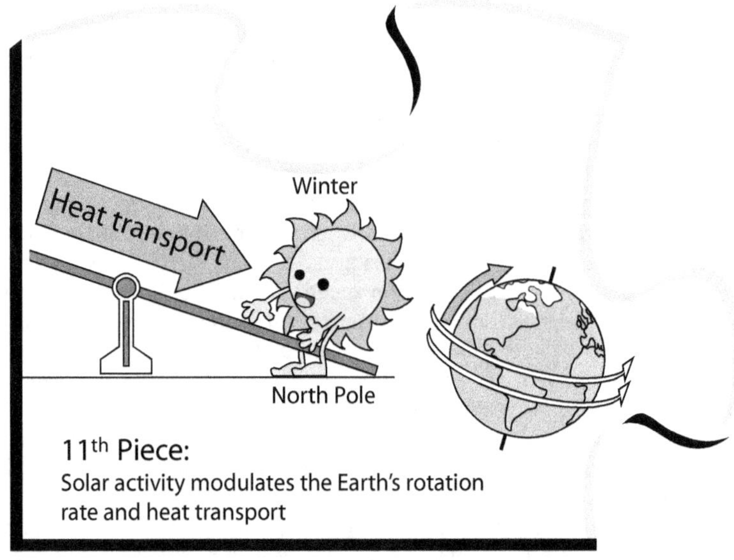

SECTION 8 KEY ISSUES

Over the past 6,000 years, three major abrupt climate changes coincide with three major 200-year solar grand minimums. Climate proxies indicate that extended periods of low solar activity cause atmospheric reorganization, shrinking the tropics and expanding the polar regions. This intensifies the latitudinal temperature gradient, driving more heat toward the poles. The increased heat loss, especially in the Arctic, leads to a pronounced global cooling, mainly affecting the northern midlatitudes. Despite this evidence, climate models cannot reproduce this effect because the mechanism is unknown, and the 11-year solar cycle has a much smaller effect.

The small energy change of the solar cycle affects the surface temperature and the heat budget of the tropical oceans four times more than expected. As solar activity increases, poleward heat transport decreases, leading to heat accumulation in the northern high latitudes, cooling of the Arctic, and increased heat content in the tropical oceans. The changes in atmospheric circulation dynamics during the solar cycle are similar to those observed during past solar grand minimums, although on a smaller scale.

Stratospheric ozone receives and amplifies the solar signal, and planetary waves provide the energy for climate effects. The polar vortex converts wave energy into winter atmospheric circulation changes jointly influenced by solar activity, the Quasi-Biennial Oscillation, El Niño, and volcanic eruptions. Short-term solar activity affects the frequency of cold winters in the Northern Hemisphere. Long-term changes can alter the energy content of the climate system, leading to profound climate changes.

Solar activity affects the Earth's rotation rate through changes in the meridional circulation, which is responsible for transporting heat to the poles. During years of low solar activity, the atmospheric circulation intensifies, causing the Earth to rotate faster and directing more heat to the Arctic in winter. As a result, the Arctic experiences warmer winters while the midlatitudes become colder.

PART III. THE WINTER GATEKEEPER HYPOTHESIS

SECTION 9. CLIMATE REGIMES AND SHIFTS

CHAPTER 31
1976 IS THE YEAR THE CLIMATE CHANGED

The most recent period of global warming began in 1976, after a cooling period between 1945 and 1975. In addition, CO_2 emissions before 1950 were not large enough to have a significant effect on Earth's temperature. In 1976, a sudden climate shift occurred in the Pacific Ocean at the time of important changes in the Earth's atmosphere. This event resulted in a state of enhanced zonal circulation, which likely reduced heat transport through the atmosphere toward the poles, affecting the global temperature trend. This climate event is evident in several climate-related variables, but climate models have been unable to reproduce it. Furthermore, the IPCC provides no explanation for this event or for the cooling period that occurred between 1945 and 1975.

The decades before and after 1976

Before 1940, atmospheric CO_2 levels rose slowly, with an annual increase of less than 0.5 ppm and no acceleration. Between 1940 and 1950, levels remained stable, but this was not known at the time because systematic measurements did not begin until 1958. In the early 1960s, it became clear that CO_2 levels were increasing. But despite the increase in CO_2, the planet's surface had been cooling since 1945. As a result, scientists believed that some other factor was having a greater effect on the Earth's temperature. Nevertheless, most atmospheric scientists were convinced that the increase in CO_2 would eventually lead to global warming. In 1967, the first climate model was developed based on this knowledge.

By 1975, CO_2 levels had increased by 15 ppm, or 5%, in just 17 years, indicating a significant acceleration in the rate of increase. However, despite this increase, no warming was detected, only cooling.

The turning point in the Earth's climate occurred in 1976 when measurements began to show a warming trend that became more pronounced in the early 1980s. For scientists, the turning point came in 1985, when data from the Vostok ice core confirmed the significant role of CO_2 in the Pleistocene climate. The agreement between the measurements and past data convinced most scientists of the so-called climate consensus.

Since 1976, the Earth's surface has experienced a warming trend without a significant cooling period, while CO_2 levels have increased exponentially. In the 1970s, CO_2 levels were growing at a rate of 1 ppm/year, but now they are rising at 2.5 ppm/year, an increase of 150%. However, this exponential increase in CO_2 levels has not had a significant effect on the rate of warming, which has not changed significantly over the past 45 years despite the large increase in the rate of CO_2 growth.

Although 1976 was not a notable year in terms of CO_2, it marked a breakpoint in the temperature trend.

What happened in 1976?

It took scientists 15 years to notice what happened to the climate in 1976. In 1991, a study was published showing a drastic shift in 40 environmental variables in the Pacific climate that year. These variables included air and water temperatures, the Southern Oscillation, chlorophyll, geese, salmon, crabs, glaciers, atmospheric dust, corals, carbon dioxide, winds, ice cover, and transport through the Bering Strait. The changes suggested that one of the largest ecosystems on Earth occasionally undergoes abrupt changes.[228]

The sudden changes in the Northern Hemisphere winter atmospheric circulation that occurred in 1976 resembled a weakened, quasi-permanent El Niño. The 1976 event began when the ocean-atmosphere system had not fully recovered from the 1976-77 El Niño, and the shift was described as a change in the background climate state.[229]

After this discovery, scientists looked more closely at previous climate and fishery data from the North Pacific. They found that the 1976 shift was not an isolated event but part of a larger 50-70-year climate oscillation they called the Pacific Decadal Oscillation. This oscillation had caused previous abrupt shifts.[230]

The events of 1976 and 1977 were not a gradual change as expected from an oscillation but a sudden shift. It began with the El Niño of 1976-77, visible in the Pacific Decadal Oscillation index (fig. 48a), and was followed by an increase in global surface temperature (fig. 48f). This was followed by a change in the atmosphere that caused a noticeable change in the planet's climate.

Frictional torque is the torque exerted on the surface by wind friction. Frictional torque anomalies are associated with sea-level pressure anomalies at high latitudes and contribute to subsequent mountain torque anomalies.[231] Mountain torque is the torque exerted by a pressure difference on either side of a mountain, and its changes are associated with changes in the zonal (east-west) wind circulation.

After the 1976-77 El Niño, a frictional torque anomaly (fig. 48c) appeared, which was compensated by a change in mountain torque of the opposite sign (fig. 48d) so that the total torque was maintained. However, due to the persistent changes in the torques, there was a persistent increase in the atmospheric angular momentum (fig. 48b). The change in the wind speed anomaly responsible for the torque changes (fig. 48e) has already been discussed in chapter 19 (fig. 31) regarding the role of ocean oscillations in poleward heat transport.

---

[228] Ebbesmeyer, C.C., et al., 1991. Proceedings of the Seventh PACLIM Workshop, April 1990. Interagency Ecological Studies Program Technical Report, 26 pp.115–126. hdl.handle.net/1834/22168

[229] Graham, N.E., 1994. Clim. Dynam. 10, pp.135–162. doi.org/10.1007/BF00210626

[230] Mantua, N.J., et al., 1997. Bull. Am. Meteorol. Soc. 78 (6), pp.1069–1080. doi.org/10.1175/1520-0477(1997)078<1069:APICOW>2.0.CO;2

[231] Weickmann, K., 2003. Mon. Weather Rev. 131 (11), pp.2608–2622. doi.org/10.1175/1520-0493(2003)131<2608:MTGFTA>2.0.CO;2

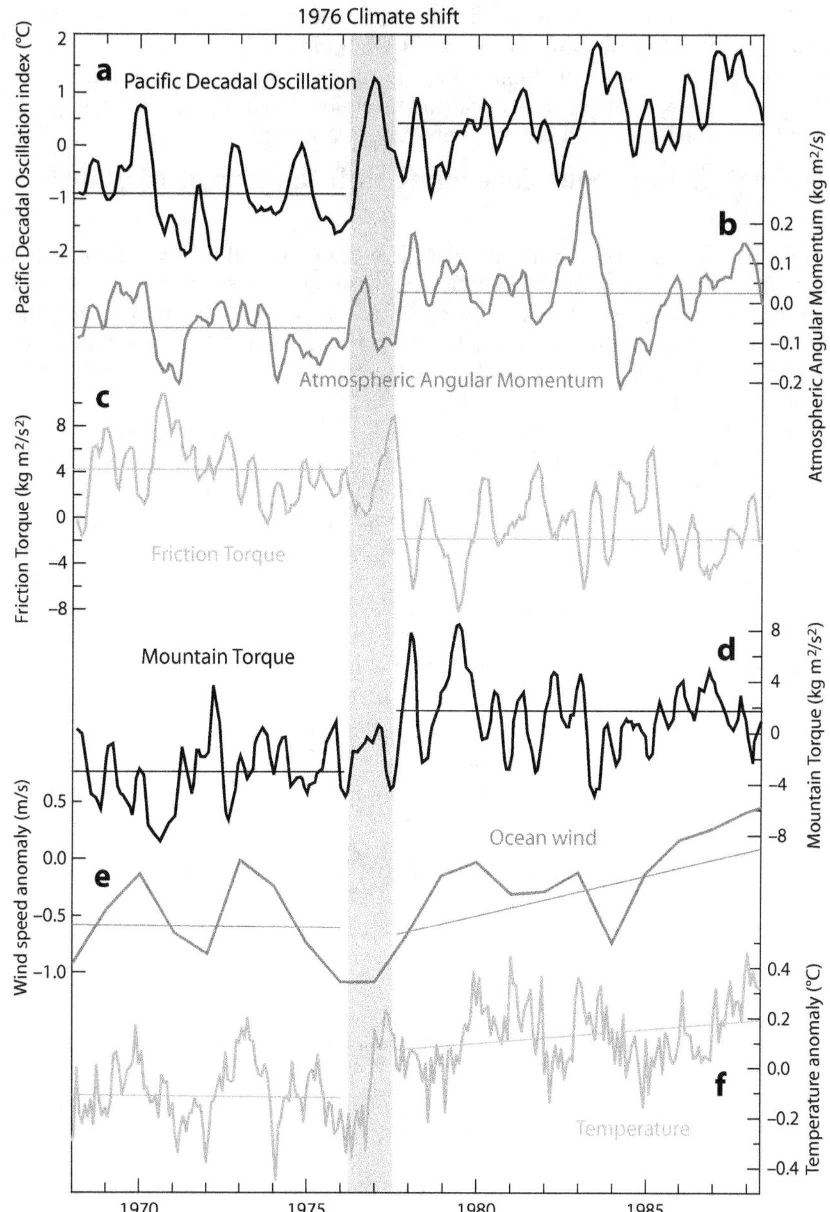

Figure 48. The climate shift in 1976. Changes in the indicated variables are shown as thick lines. The thin lines are the mean value on either side of the climate shift (indicated by the gray bar), except in e) and f), where the thin lines are the respective long-term trends.[232]

---

[232] Data for the figure from Marcus, S.L., et al., 2011. J. Geophys. Res. 116 D03107. doi.org/10.1029/2010JD015032 Yu, L., 2007. J. Clim. 20 (21), pp.5376–5390. doi.org/10.1175/2007JCLI1714.1 and UK MetOffice HadCRUT5 temperature dataset.

Thus, there was a sudden shift in the winter atmospheric circulation in 1976, resulting in a strengthening of the zonal wind circulation and a weakening of the meridional wind circulation, which is responsible for transporting heat to the poles. These changes imply that the meridional heat transport decreased at that time, coinciding with the onset of global warming.

Official silence about the climate shift that triggered global warming

By the time the 1976 Shift and Pacific Decadal Oscillation were identified, the IPCC had already published its First Assessment Report. The report supported the notion that the primary cause of global surface warming was the continuing increase in CO_2 from human emissions. Anything not human in origin was labeled "natural" or "internal variability" and given no role in the long-term temperature trend.

The IPCC reports offer no explanation for the cooling period from 1945 to 1975. Some scientists speculate that it may have been caused by the rapid increase in sulfur emissions from industry and the burning of fossil fuels, which have a cooling effect. These emissions peaked in the early 1970s when several countries introduced legislation to limit them. These scientists believe that by 1976, the warming effect of rising CO_2 levels exceeded the cooling effect of sulfur emissions. However, this contradicts climate models and IPCC reports, which suggest that anthropogenic forcing has always been positive, meaning that the cooling period from 1945 to 1975 could not have had a human cause.

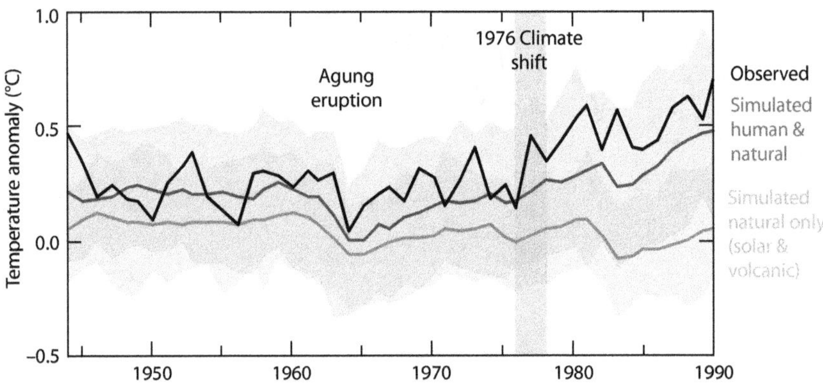

Figure 49. Global surface temperature changes and their proposed causes. This figure from the 6th Assessment Report has been cropped and annotated. Observed changes are shown in black, climate model simulations with the response to both human and natural factors are shown in dark gray, and natural-only factors are shown in medium gray.

Climate models have been unable to reproduce either the warming of the early 20th century or the cooling of the mid-20th century, whether using all forcings or only natural forcings. Figure 49 (from figure SPM.1 of the 6th Assessment Report) shows that the climate models (brown curve) simulate a slight warming in the late 1940s and 1950s, followed by a cooling effect due to the eruption of Mt. Agung in 1963 and then a continuous warming trend.[233] How-

---

[233] IPCC, 2021: Summary for Policymakers. doi.org/10.1017/9781009157896.001

ever, the 1976 climate breakpoint, which has been extensively studied by dozens of scientists and resulted in significant changes in the Pacific climate, is not reflected in the IPCC reports or climate models. According to climate models, the break in the temperature trend occurred at the 1963 eruption, which was 13 years before the 1976 shift.

The cooling period from 1945 to 1975 is well documented and was not caused by the eruption of Mt. Agung in 1963. Many people still remember the cold winters of the early 1970s, and this cooling period ended with an abrupt atmospheric climate shift in 1976. The inability of climate models to reproduce this cooling period or the 1976 climate shift raises questions about their ability to accurately predict future climate.

In Summary

In 1976, a global climate shift had a major impact on the Pacific Ocean and marked the beginning of the current warming trend. This shift has been studied extensively and is evident in many climate-related variables. The available evidence suggests that the shift was due to changes in atmospheric circulation, with an increase in zonal winds and a decrease in meridional winds, leading to a warming phase in the multidecadal oceanic oscillations due to reduced poleward heat transport. Despite its well-documented nature, the 1945-1975 cooling period and the 1976 climate shift are not explained by climate models or the IPCC, so they are ignored.

CHAPTER 32
CLIMATE REGIMES AND SHIFTS

Periodic abrupt climate changes occur on all time scales. On multidecadal scales, they are characterized by climate shifts that separate climate regimes lasting a few decades. Although there is substantial evidence for the existence of these climate regimes and shifts, they remain controversial. Since the cause of these events is unknown, they are referred to as intrinsic low-frequency climate variability. However, their manifestation is linked to changes in the global atmospheric circulation. These shifts must represent persistent changes in energy flow modes, probably indicating distinct poleward heat transport regimes.

Abrupt climate change on different time scales

Climate changes all the time, but sometimes it changes much faster than the normal rate of change. This type of climate change is called abrupt. By this standard, the current climate change can be classified as abrupt.

The realization that climate can change so abruptly as to be noticeable within a decade is a relatively recent discovery. It was made in the mid-1980s by studying ice cores from Greenland, which identified abrupt warming peaks of millennial frequency that occurred during the last glaciation. The warming rate calculated for these events is 1.4°C/decade (2.5 °F/decade) in northern Europe, which is ten times faster than today's rate.

On a different timescale, the transition from a glacial to an interglacial period, known as deglaciation, is a relatively rapid change on the tens of thousands of years timescale of the glacial cycle. During the last deglaciation, sea level rose at an average rate of 1.2 cm (half an inch) per year for 8,000 years, four times faster than today.

In chapter 22, when we reviewed the evidence for abrupt climatic events during the Holocene, we characterized them as periods of one or two centuries during which the climate changed much more rapidly than the millennial mean change.

In the previous chapter, we noted that there are cases of abrupt climatic changes, or climate shifts, that occur within one to a few years, separated by a few decades.

Climate can be likened to a fractal structure that behaves similarly on all time scales, with periods of rapid climate change interspersed with longer periods of less change. The longer the time frame, the greater the impact of abrupt changes.

Abrupt climate changes mark the beginning and end of distinct periods in which the climate system exhibits less variability in its energy content and fluxes. This can be observed in glacial and interglacial periods, which remain stable for some time before returning to their previous state. This property is known as metastability. For example, interglacials tend to remain stable for an average of 14,000 years, while Greenland's millennial events, known as interstadials, last for centuries. Abrupt Holocene events last only a century or two before reversing, while climate shifts separate distinct climate regimes that last several decades. It should be noted that climate shifts can also be expected to reverse.

Climate regimes and shifts

Prior to the discovery of the 1976 climate shift, ecologists had developed a theoretical concept to explain rapid transitions between alternative stable states, primarily in grazing ecosystems. These rapid transitions are known as regime shifts. In 1989, a study used this concept to explain the alternating sardine and anchovy regimes occurring simultaneously in the Pacific and other oceans, possibly in response to climate change.[234] The study had already identified a regime shift from anchovy to sardine in the mid-1970s (fig. 50). These fisheries studies ultimately led to the discovery of the 1976 climate shift and the Pacific Decadal Oscillation. Additional climate shifts were identified in 1925 and 1946, separating periods when long-term sea surface temperatures showed a predominantly warmer or cooler anomaly. These periods of reduced variability were called climate regimes.

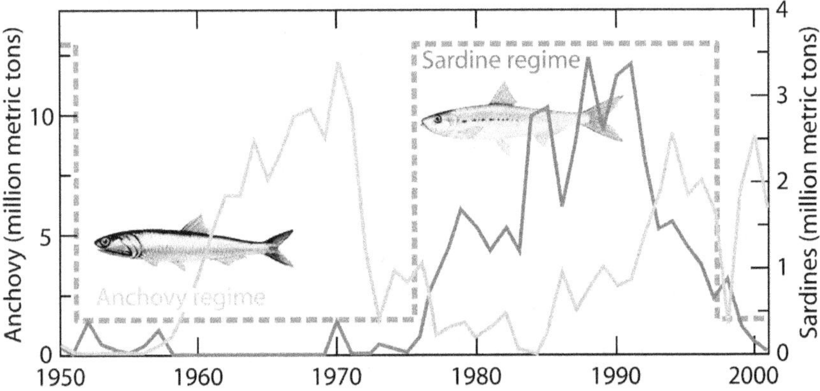

Figure 50. Alternation between a cold anchovy regime and a warm sardine regime in the Pacific Ocean. The data indicate the respective catches landed in Peru. The alternation of regimes, indicated by the gray dashed line, coincides with known climatic shifts.[235]

The existence of climate regimes and shifts is a topic of much debate among scientists. Weather and climate are highly variable, and to make sense of this complexity and predict future changes, researchers look for large-scale spatial patterns and trends in climate data over time. These patterns are known as intrinsic modes of climate variability, and scientists pay particular attention to teleconnections, which are detectable relationships between different modes of variability. An example of a low-frequency mode of variability is the Pacific Decadal Oscillation, which operates on an interdecadal time scale. Changes that occur on longer time scales tend to have larger effects because variability on smaller scales averages out over time.

---

[234] Lluch-Belda, D., et al., 1989. S. Afr. J. Mar. Sci. 8 (1), pp.195–205.
    doi.org/10.2989/02577618909504561
[235] Figure after Chavez, F.P., et al., 2003. Science, 299 (5604), pp.217–221.
    doi.org/10.1126/science.1075880

Many scientists believe that climate regimes and shifts are real because they cause significant changes in multiple variables in a short time.[236] However, other scientists who focus on only one variable may view these changes as background noise, which is abundant in the climate system. However, the application of noise-reduction techniques suggests that Pacific climate changes are more than just noise.[237]

Atmospheric modes related to heat transport

In the previous chapter, we discussed the global atmospheric effect of the 1976 climate shift. We also looked at the correspondence between changes in global temperature trends, the Pacific Decadal Oscillation, and the Atlantic Multidecadal Oscillation that we discussed in chapter 19 (fig. 31). Based on this information, we can conclude that we're dealing with a global atmospheric change that has the strongest impact on the Northern Hemisphere. This change leads to spatiotemporal variability, which manifests itself as different modes of low-frequency climate variability and their teleconnections.

It appears that the changes are caused by the influence of the atmosphere on the ocean, as indicated by variations in atmospheric angular momentum and sea level pressure. In addition, the changes occur over a short time (just a single winter), which is inconsistent with the long lags that typically occur in the communication between different ocean basins.

Climate is the result of energy flows acting on the matter of the climate system. Different climate regimes require different energy flows. Since these regimes are metastable and alternate, this suggests that there are also metastable modes of energy flow in the climate system. In chapter 19, we explored that multidecadal ocean oscillations are likely due to variations in the intensity of poleward heat transport. The existence of climate regimes and shifts that span at least one hemisphere supports this explanation.

Climate models are designed to represent energy fluxes within the climate system, but they struggle to represent low-frequency variability accurately and completely miss climate shifts. Given that variability in poleward heat transport is poorly understood and lacks a valid theory (ch. 12), this may be one of the major deficiencies of climate models.

Tipping points

In recent years, much attention has been paid to the possibility of tipping points in the climate system.[238] But do the abrupt changes described for each climate timescale meet the criteria for tipping points? The answer depends on how we define tipping points. If we define a tipping point as a large change resulting from small initial changes, then all of the abrupt changes mentioned at the beginning of this chapter qualify as tipping points. This means that the climate goes through a tipping point every few decades for reasons that have nothing to do with the increase in CO_2.

---

[236] Hare, S.R. & Mantua, N.J., 2000. Prog. Oceanogr. 47 (2-4), pp.103–145.
doi.org/10.1016/S0079-6611(00)00033-1

[237] Rodionov, S.N., 2006. Geophys. Res. Lett. 33 (12), L12707.
doi.org/10.1029/2006GL025904

[238] Lenton, T.M., et al., 2019. Nature, 575 (7784), pp.592–595.
doi.org/10.1038/d41586-019-03595-0

By the most widely accepted definition, abrupt climate changes are not tipping points because they do not have irreversible consequences once they cross a threshold. Rather, these changes can and do reverse themselves over time. Some scientists are concerned that human-induced changes in the atmosphere could push the climate past a critical threshold and onto a different trajectory, leading to more severe warming, but such fears are often based on limited evidence.[239] It is worth noting that catastrophism has always existed, and the tendency to catastrophize is a common cognitive distortion.

Let's consider some possible abrupt climate changes. Multidecadal climate changes occur periodically with little notice. Centennial abrupt climate events are more severe (fig. 38, ch. 23), but because they have a cooling effect, we are less vulnerable to them in our current situation after recent warming. Large abrupt warming events occurred during the last glaciation, but they require specific conditions that are not present outside of glaciations.

The Holocene is approaching the average duration of interglacials over the last 800,000 years. It's based on unfounded assumptions to think that we can avoid the next glaciation because of our high CO_2 emissions.[240] It's clear from the available evidence that the primary climate risk in our distant future is a return to glacial conditions. No interglacial period has lasted long after obliquity (the tilt of the Earth's axis) drops below 23°, which will happen in 3,400 years.

The changes in the climate system over the past 45 years do not indicate a substantial acceleration of the observed warming, and they could continue for a long time without reaching a tipping point. If scientists weren't constantly warning us about climate change, we might not worry much about it. Moreover, climate change tends to reverse course over time. As we discussed earlier, the climate cooled significantly for thousands of years 124,000 years ago and for hundreds of years after 1100 AD, despite no change in atmospheric CO_2 levels. Although we should not rule out the possibility of this happening again, many scientists dismiss it and only fear further warming if CO_2 emissions are not reduced.

In summary

The climate system exhibits abrupt changes at all time scales, resulting in distinct climate regimes that are long-lived and characterized by differences in energy content and fluxes. At multidecadal scales, for example, these changes are seen as multidecadal oceanic phases separated by climatic shifts. These abrupt changes are not significantly related to changes in atmospheric CO_2 levels but, instead, appear to be related to changes in heat transport. However, climate models cannot accurately reproduce low-frequency oscillations and do not reproduce climate shifts. Because abrupt climate shifts are reversible and their climatic timescale determines their intensity, they cannot be classified as tipping points in the usual sense.

---

[239] Steffen, W., et al., 2018. PNAS 115 (33), pp.8252–8259.
doi.org/10.1073/pnas.1810141115

[240] Vinós, J., 2022. Climate of the Past, Present and Future: A scientific debate. Critical Science Press. pp.239–253.

CHAPTER 33
1997, THE CLIMATE CHANGED AGAIN

The year 1998 is often cited as the beginning of a controversial "pause" in global warming. What many people don't know is that important and abrupt changes in several climate variables occurred soon after 1997. These changes made 1997 the most important climate shift in decades, affecting not only global temperatures but also Pacific Ocean climate patterns and global atmospheric circulation. This shift caused changes in tropical extent, cloudiness, wind speed, and the Earth's rotation rate. Even more puzzling were unexplained changes in the stratosphere, including a change in its cooling trend and a decrease in water vapor, suggesting an increase in poleward heat transport. Unfortunately, scientists have yet to fully understand these changes, and the significance of the 1997 climate shift remains unrecognized.

The Pause

In 1997-98, there was an El Niño event in the Pacific Ocean. Eight years later, a scientist reported in a newspaper that there had been no warming since El Niño and questioned whether human emissions were solely responsible for climate change.[241] This sparked controversy, with other scientists arguing that eight years was too short to draw conclusions. By 2012, however, many scientists acknowledged a pause in global warming and began investigating its cause. In 2014, two scientific journals jointly published a special issue dedicated to the pause, which included several articles with different explanations. Despite the many proposed explanations, there is still no consensus on the cause of the pause.[242]

The major El Niño event of 2015-16 marked the end of the global warming pause observed from 1998-2014. Subsequently, changes were made to several temperature datasets, transforming the pause into a period of continued global warming in surface temperature data. While some scientists argued in 2016 that the pause was real, it is no longer mentioned in publications.[243] As a result, the concept of a global warming pause is no longer recognized in mainstream climate science.

However, the global warming pause was only one of several effects of a major climate shift that occurred in 1997. Nevertheless, climate scientists have yet to fully explain or acknowledge many of these other effects because they do not meet their expectations.

---

[241] Carter, R.M., There IS a problem with global warming... it stopped in 1998. The Telegraph, 09 April 2006.

[242] Springer Nature (2014) Focus: Recent slowdown in global warming. www.nature.com/collections/sthnxgntvp

[243] Fyfe, J.C., et al., 2016. Nat. Clim. Change, 6 (3), pp.224–228. doi.org/10.1038/nclimate2938

Changes everywhere

Like the 1976 climate shift, the 1997 shift caused a phase change in the Pacific Decadal Oscillation, but with a different sign. The shift is most evident in the cumulative value of its index (Fig. 51a), where a peak is observed when cold anomalies of the Pacific surface temperature become more frequent than warm anomalies since 1997.[244] Researchers noticed the change in the Pacific Decadal Oscillation after 1997, although it showed some differences from the change in 1976 and was not simply a reversal.[245]

Because the Pacific Decadal Oscillation functions similarly to a long-lived El Niño-like pattern, we can observe comparable trends in the Cumulative Multivariate El Niño Index (fig. 51b, black curve).[246] The index shows that an increase in the occurrence of La Niña events followed the El Niño of 1997-98, a trend that has continued into the early 2020s.

The character of El Niño events has also changed, with an increase in the frequency of Central Pacific El Niño events and a decrease in Eastern Pacific El Niño events. These changes are evident in the oceanic regions of the equatorial Pacific, where the volume of water above 300 m depth with temperatures above 20°C (68 °F) showed a noticeable change. After 1998, the frequent negative anomalies associated with El Niño events in the eastern Pacific, which significantly reduced the volume of warm water, were no longer observed (fig. 51b, dashed gray curve).[247] These results indicate that not only the frequency of El Niño events has changed, but also their character.

The 1997 climate shift had a global impact on atmospheric circulation that extended beyond the Pacific. The Hadley cells responsible for weather patterns in the tropics intensified and expanded (fig. 51c).[248] The expansion of the Hadley cells resulted in a poleward shift of the subtropical jet, causing changes in precipitation patterns and atmospheric circulation.

Circulation changes have also affected cloudiness. In particular, low cloud cover in the extratropical regions of the Northern Hemisphere decreased markedly after 2000 (fig. 51d).[249] However, it's difficult to determine the effect of this decrease on radiative fluxes because cloud cover trends can vary with altitude, hemisphere, and region, and these variations can offset the effects of the changes.

Changes in wind speed may be responsible for the observed changes in cloud cover. In the late 20th century, winds over land had been slowing for decades, raising concerns about future wind energy production (fig. 51e). Meanwhile, ocean winds were increasing in speed. However, after the 1997 climate shift, land and ocean winds reversed their trends, and onshore winds began to

---

[244] ERSST v5 Pacific Decadal Oscillation annual data from NOAA, cumulative values were detrended.

[245] Litzow, M.A., 2006. ICES J. Mar. Sci. 63 (8), pp.1386–1396. doi.org/10.1016/j.icesjms.2006.06.003

[246] Multivariate ENSO Index v2 data from NOAA.

[247] Data from the TAO Project Office of NOAA.

[248] Nguyen, H., et al., 2013. J. Clim. 26 (10), pp.3357–3376. doi.org/10.1175/JCLI-D-12-00224.1

[249] Data from EUMETSAT CM SAF dataset. Dübal, H.R. & Vahrenholt, F., 2021. Atmosphere, 12 (10), p.1297. doi.org/10.3390/atmos12101297

increase in speed again, easing concerns.[250] Still, scientists remain puzzled by these changes. The initial trend and subsequent reversal were surprising and cannot be explained as a simple response to increased CO_2.

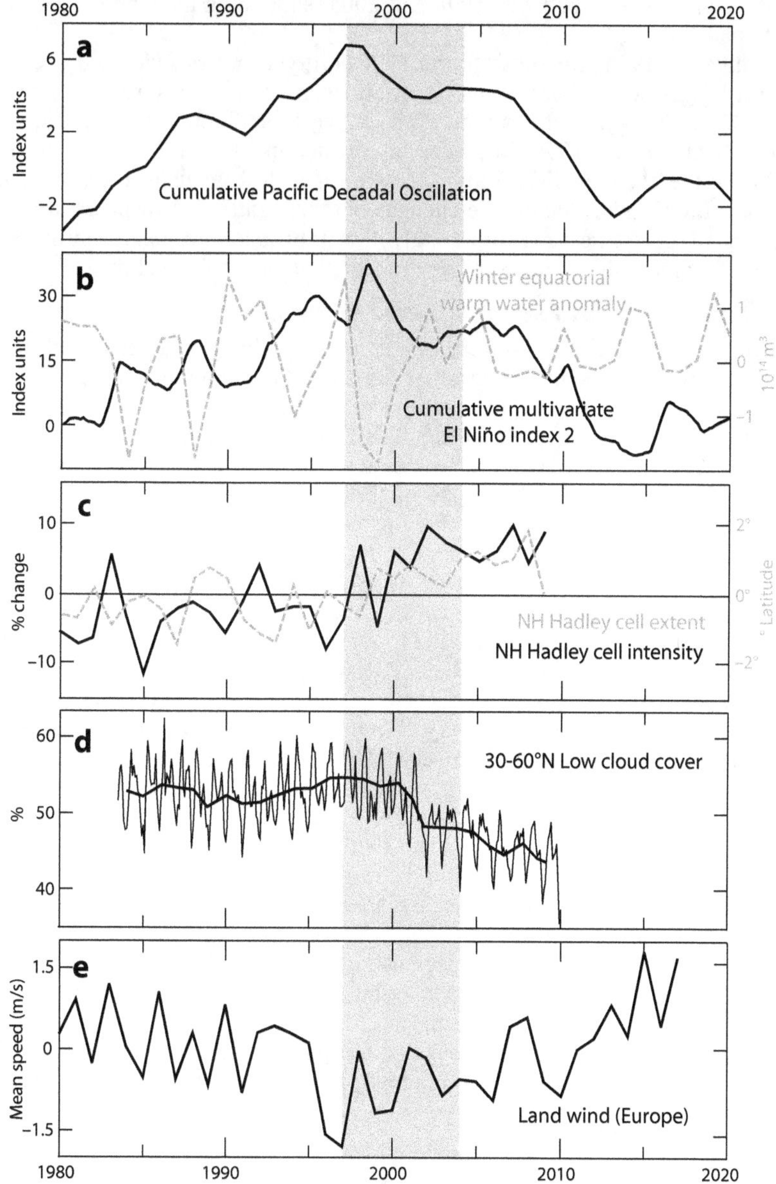

Figure 51. Some changes at the time of the 1997 climate shift. The gray bar marks the interval 1997-2004, when most of the observed abrupt changes occurred.

---

[250] Zeng, Z., et al., 2019. Nat. Clim. Change, 9 (12), pp.979–985.
doi.org/10.1038/s41558-019-0622-6

Similar to the 1976 climate shift, changes during the 1997 shift affected its atmospheric angular momentum. After 1997, there was a decrease in momentum similar in magnitude to the increase after 1976. As a result, the Earth accelerated its rotation between 1998 and 2004, shortening the length of the day by two milliseconds.

Unlike the 1976 climate shift, the 1997 shift was not as sudden. It occurred over a longer period, with some changes occurring over seven years between 1997 and 2004, while others took even longer. In addition, some climate variables affected by the 1976 shift were not affected in 1997.

While the changes discussed so far are difficult to explain in terms of anthropogenic climate change, the changes observed in the stratosphere are even more challenging. Perhaps surprisingly, the stratosphere responded strongly to the 1997 climate shift, underscoring the global nature of these shifts.

Difficult to explain stratospheric changes

After the 1997 climate shift, the most striking change was observed in the stratosphere. The cooling trend observed since the beginning of the satellite record in 1979 weakened considerably and even came to a complete halt in the lower stratosphere (fig. 52a & b).[251] The cooling of the stratosphere is attributed to increased CO_2 and water vapor levels caused by the enhanced greenhouse effect and is considered one of the fingerprints of human-induced climate change. Some have argued that this cooling is evidence that the warming is not caused by increased solar activity since such activity should have a warming effect on the stratosphere. However, this argument ignores the dynamic changes that solar activity can cause in the stratosphere.

Climate models have been unable to reproduce the observed changes in the stratosphere under current conditions. These models predicted that the cooling trend would continue, so the unexpected changes have puzzled scientists who consider the phenomenon a mystery.[252] Some scientists have suggested that the changes may be due to the recovery of ozone leading to a warming of the stratosphere.[253] However, this argument does not explain the clear breakpoint in the temperature trend observed in 1997 nor other sudden changes observed in the stratosphere at that time and shortly thereafter.

The Brewer-Dobson circulation is responsible for the slow transport of stratospheric air toward the poles. Because of this, changes in the circulation can take several years to become apparent in different regions of the stratosphere. In 2001, an unexpected decrease in the stratospheric water vapor content was observed, with values decreasing by 10% (fig. 52c).[254] This contradicts the assumption that increasing CO_2 levels should lead to an increase in stratospheric water vapor. Although stratospheric water vapor is much less abundant than its tropospheric counterpart, its radiative emissions have a strong

---

[251] Seidel, D.J., et al., 2016. J. Geophys. Res. Atmos. 121 (2), pp.664–681. doi.org/10.1002/2015JD024039

[252] Thompson, D.W., et al., 2012. Nature, 491 (7426), pp.692–697. doi.org/10.1038/nature11579

[253] Maycock, A.C., et al., 2018. Geophys. Res. Lett. 45 (18), pp.9919–9933. doi.org/10.1029/2018GL078035

[254] Data for figure 52c and d, from Randel, W. & Park, M., 2019. J. Geophys. Res. Atmos. 124 (13), pp.7018–7033. doi.org/10.1029/2019JD030648

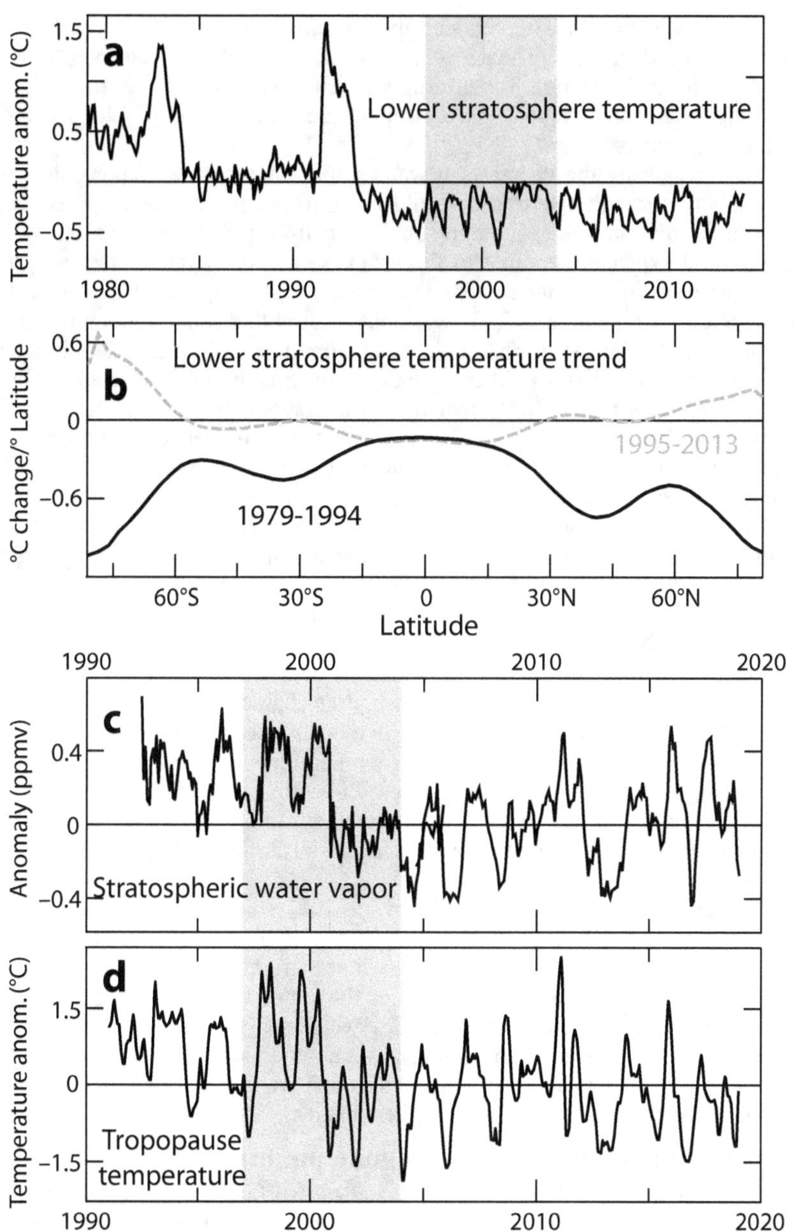

Figure 52. Changes in the stratosphere after the 1997 climate shift. The strong warming peaks in stratospheric temperature visible in the upper panel correspond to volcanic eruptions.

influence on the CO_2 effect because they are not masked at this altitude. Cooling of the stratosphere by water vapor should lead to surface warming, as discussed in chapter 7. The 10% decrease in stratospheric water vapor observed in

2001 was estimated to have reduced the warming effect of CO_2 by 25% over the following decade.[255] The 13% increase in stratospheric water vapor as a result of the 2022 Hunga Tonga eruption (ch. 24) should have a significant warming effect on global surface temperature, although it should disappear within a few years.

What is causing the decrease in water vapor? Most water vapor enters the stratosphere through the tropical cold-point tropopause, where the Brewer-Dobson circulation acts as a pump sucking air from the troposphere. Radiative cooling and expansion cooling of the air cause a very effective freeze-drying, removing most of its water content as it enters the stratosphere. The decrease in stratospheric water vapor in 2001 was consistent with a concurrent abrupt cooling of about 1°C at the tropical cold point tropopause (fig. 52d). These events have been explained as a sudden increase in the strength of the Brewer-Dobson circulation since this is the effect that would result from an increase in its pumping action.[256] This explanation is supported by evidence from changes in ozone and planetary wave activity, indicating that the Brewer-Dobson circulation experienced a significant increase early in the century.

An interesting finding, highly relevant to the thesis of this book, is that changes in solar activity can affect water vapor in the lower stratosphere.[257] The simplest explanation for this effect is that solar activity regulates the strength of the Brewer-Dobson circulation. This regulation is supported by the information presented in chapters 28 and 29, which provides evidence for the hypothesis discussed in section 11. In addition, if we compare the 1979-1995 lower stratospheric temperature trend, which is characterized by above-average solar activity, as shown in figure 52b with a black line, with the effect of a solar cycle maximum on lower stratospheric temperatures by latitude at 20 km (12 miles) height, as shown in figure 45 (ch. 28) with a gray dashed line, we observe a remarkable similarity, especially in the Northern Hemisphere. This similarity may imply that solar activity plays a role in the important trends in stratospheric temperature.

All changes in the stratosphere from the 1997 shift are consistent with an increase in poleward stratospheric heat transport by the Brewer-Dobson meridional circulation. The changes in the lower stratospheric temperature trend increase with latitude and are largest at the poles (fig. 52b), suggesting increased heat and ozone transport across a weakened vortex. However, the reason for these changes and their timing is unknown to scientists. Climate models do not help to understand poleward heat transport, and their simulations may confuse scientists about the nature of the changes.

How is it possible that this has gone unnoticed?

In this chapter, we have only scratched the surface of the vast amount of evidence for a major climate shift in 1997. In the next chapter, we will examine the evidence for changes in the Arctic following this shift. What is perplexing

---

[255] Solomon, S.et al., 2010. Science, 327 (5970), pp.1219–1223.
  doi.org/10.1126/science.1182488
[256] Randel, W.J., et al., 2006. J. Geophys. Res. Atmos. 111, D12312.
  doi.org/10.1029/2005JD006744
[257] Schieferdecker, T., et al., 2015. Atmos. Chem. Phys. 15 (17), pp.9851–9863.
  doi.org/10.5194/acp-15-9851-2015

is that scientists do not recognize this major climate shift as a global phenomenon that has affected many areas of the atmosphere-ocean system in a short period of time. When possible, they attribute the changes to the increasing impact of human activities on the climate. Alternatively, specific explanations are sought for each change rather than a common one. The inability to link the changes is worrying.

The fact that most climate scientists failed to notice such an abrupt change is incredible, given the attention paid to climate. It shows that when widespread belief in a scientific hypothesis is accompanied by discouragement of those who seek alternative explanations, the scientific method breaks down.

In summary

The changes that occurred after the 1997 climate shift were substantial and global in nature, affecting even the stratosphere. The Pacific Ocean, the surface temperature trend, and the lower stratospheric temperature trend experienced the strongest effects. Considering all these abrupt changes as part of a single phenomenon, it seems likely that there was a change in the global heat transport system, resulting in an abrupt increase in the magnitude of poleward transport. However, despite the magnitude of these changes, scientists and climate models cannot explain their cause as a result of increased atmospheric CO_2. As a result, this climate shift has gone unnoticed and is being ignored.

CHAPTER 34
ARCTIC WARMING IS NOT AMPLIFICATION

A hundred years ago, the Arctic experienced a period of intense warming. A similar phenomenon has occurred since 1997, but this time scientists have a different explanation. Climate models suggest that rising CO_2 levels amplify the warming effect in the Arctic. However, this Arctic amplification was not observed during two decades of global warming from 1976 to 1997. Instead, Arctic warming in the 21st century has occurred mostly in winter. It is primarily due to increased heat transport into the region rather than a change in the ice-albedo feedback. Arctic warming has been used to mask the fact that the rest of the planet is warming more slowly than before.

Arctic warming, a consequence of the 1997 shift

Chapter 11 showed us that atmospheric circulation and heat transport to the poles are much more robust in the winter hemisphere. In addition, the polar vortex restricts heat transport to the polar regions during this season. In chapter 16, we discovered that atmospheric wave activity increases during Northern Hemisphere winters. This activity weakens the polar vortex, creating blocking patterns in the jet stream that redirect storms toward the Arctic. These storms are the main source of heat during the Arctic winter.

The concept of "polar amplification" emerged from early climate models in 1975. It refers to the greater increase in surface temperature at higher latitudes, which in the models was initially caused by the retreat of snow boundaries and lower tropospheric thermal stability, which limited convective warming.[258] Over time, the term was changed to "Arctic amplification" because it became clear that Antarctica, except for the small peninsula, was not warming.

As shown in chapter 9 (fig. 13), it's clear that as the planet warms over millions of years, the tropics experience minimal warming. However, the warming increases as we move closer to the poles. For example, during the early Eocene, 50 million years ago, the polar regions had tropical conditions (fig. 32, ch. 20), while the tropics were not much warmer than they are today. In the long term, the evidence supports the concept of polar amplification.

About a century ago, the Arctic experienced an intense warming that was well documented by scientists and newspapers of the time (fig. 53 top). This early 20th-century Arctic warming is known to modern scientists (fig. 53 bottom).[259] But they don't have a clear explanation for it. At that time, human emissions were much lower and could not have caused such a strong warming

---

[258] Manabe, S. & Wetherald, R.T., 1975. J. Atmos. Sci. 32 (1), pp.3–15. doi.org/10.1175/1520-0469(1975)032<0003:TEODTC>2.0.CO;2
[259] Frauenfeld, O.W., et al., 2011. J. Geophys. Res. Atmos. 116, p.D08104. doi.org/10.1029/2010JD014918

effect. In addition, solar activity was low in the 1920s. All this suggests that we don't have a good understanding of why the Arctic is warming.

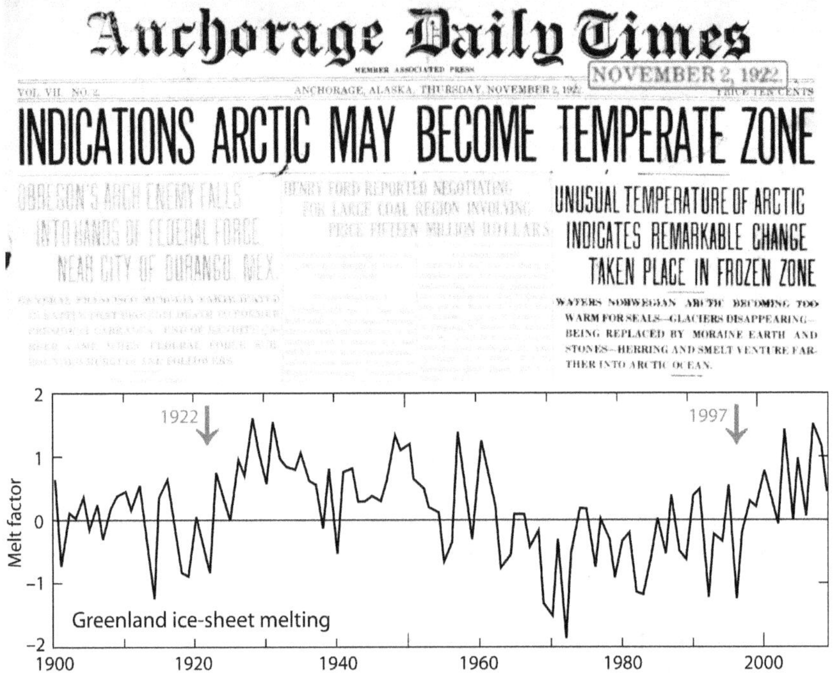

Figure 53. Arctic warming in the early 20th century. It was reported in newspapers in 1922 and can be seen in data from recent studies.

It's quite surprising that very little Arctic amplification had occurred by 1995, given all the global warming that took place from the late 1970s to the early 1990s. While we now tend to think of Arctic amplification as a natural consequence of global warming, the evidence suggests otherwise. In fact, they may be occurring independently, even though climate models suggest that Arctic amplification is an undelayed effect of global warming. As late as 1996, Arctic experts said that *"the relative lack of observed warming and relatively small ice retreat may indicate that [climate models] are overemphasizing the sensitivity of climate to high-latitude processes."*[260]

In 1997, as the rest of the planet experienced a slowdown in global warming, the Arctic began to warm at a much faster rate. Scientists were relieved to see the onset of Arctic amplification and were quick to attribute it to human CO_2 emissions. However, the climatic impact of the 1997 shift on the Arctic was sudden and abrupt, typical of natural changes, rather than the expected gradual response to continued increases in CO_2 levels.

---

[260] Curry, J.A., et al., 1996. J. Clim. 9 (8), pp.1731–1764.
    doi.org/10.1175/1520-0442(1996)009<1731:OOACAR>2.0.CO;2

Box 24. Dealing with the Pause

The onset of Arctic warming and the global warming Pause (ch. 33) occurred simultaneously due to the 1997 climate shift. This provided an opportunity to compensate for the lack of warming in other regions of the world in the temperature datasets by including more Arctic warming, which had previously been underrepresented due to the limited number of measurements available.

The UK Met Office Hadley Centre's HadCRUT3 temperature dataset was replaced in 2012 by version 4, which turned the Pause into a warming period by including additional stations from the high latitudes of the Northern Hemisphere, which was the only region experiencing warming at the time (fig. B24). In 2021, the update to HadCRUT5 eliminated the Pause by revising the method used to account for Arctic warming over large areas lacking observations. The authors stated that *"increased warming results from an improved representation of Arctic warming,"* and the revisions brought HadCRUT in line with other global surface temperature datasets.[261] It seems that if everybody does it…

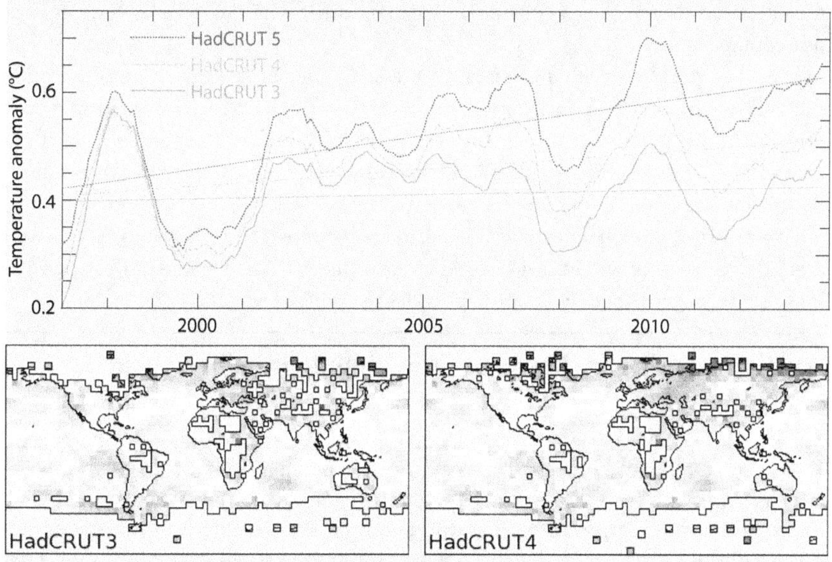

Figure B24. Removing the Pause in HadCRUT datasets.

Changing the way Arctic warming is represented is justified for geographical reasons. However, it is problematic because it makes the temperature dataset a mix of apples and oranges. The polar regions are unique in that their atmosphere becomes extremely dry in winter (see box 4, ch. 7). Moisture limits temperature change because water vapor releases latent heat as the air cools and absorbs it as it warms. With little water vapor, even a small amount of energy can cause large changes in air temperature. A single winter storm can raise Arctic temperatures by

[261] Morice, C.P., et al., 2021. J. Geophys. Res. Atmos. 126 (3), p.e2019JD032361.
    doi.org/10.1029/2019JD032361

20°C (36 °F) for more than a week (fig. B13, ch. 16). Using temperature anomalies exacerbates this problem because the Arctic anomalies are enormous, so the effect of Arctic warming on the global mean is greatly magnified, leading to an unrealistic result. It leads us to believe that the planet is warming significantly when, in fact, it is transferring a small fraction of its energy to the Arctic during the winter. After influencing the thermometer readings, this energy leaves the planet through radiative cooling.

The changes made to the temperature datasets are inadequate, as evidenced by the fact that prior to these revisions, the trends in surface and lower stratospheric temperatures were opposite. Stratospheric cooling as the surface warms is important because it is the expected behavior of greenhouse warming. Since 1997, however, the lower stratosphere has stopped cooling (ch. 33). With the changes in the datasets, the surface temperature record now shows warming. This creates a problem: you cannot have it both ways. Either there is little surface warming and the physical connection between the surface and the stratosphere remains intact, or the surface warming is enhanced and the physical connection is broken. Choosing the latter option requires a contrived explanation for why the lower stratosphere is not cooling as it should.

Another problem with the changes made to the temperature datasets is that they show the Arctic contributing more to warming in the 21st century, even though the planet is warming at the same rate. Given the accelerating rate of increase in CO_2 levels, this contradicts the Enhanced CO_2 Effect hypothesis since the planet should be increasing its rate of warming (ch. 46).

Moreover, if the changes had cooled the record instead of warming it, there is a good chance they would not have been introduced. This illustrates the scientific bias that has unfortunately become prevalent in climate science.

Arctic warming is the result of a change in winter heat transport

After 1997, there was a large increase in winter temperatures in the central Arctic, but summer temperatures remained unchanged (fig. 54a).[262] This is because the summer heat is primarily used to melt ice once the temperature exceeds the melting point. As a result, summer sea-ice extent in the Arctic declined sharply after 1997, raising concerns about an ice-free Arctic in the near future (fig. 54b).[263] Increased ice melt also occurred in Greenland, increasing freshwater flux to the Arctic Ocean and leading to a more negative ice sheet mass balance (fig. 54c).[264] However, most of the abrupt Arctic trends have stabilized somewhat after the first few years, much to the chagrin of those who predicted a rapid disappearance of Arctic sea ice.

---

[262] Data from Danish Meteorological Institute.
[263] Data from National Snow & Ice Data Center.
[264] Data from Dukhovskoy, D.S.,et al., 2019. J. Geophys. Res. Oceans, 124 (5), pp.3333–3360. doi.org/10.1029/2018JC014686 Mouginot, J., et al., 2019. PNAS, 116 (19), pp.9239–9244. doi.org/10.1073/pnas.1904242116

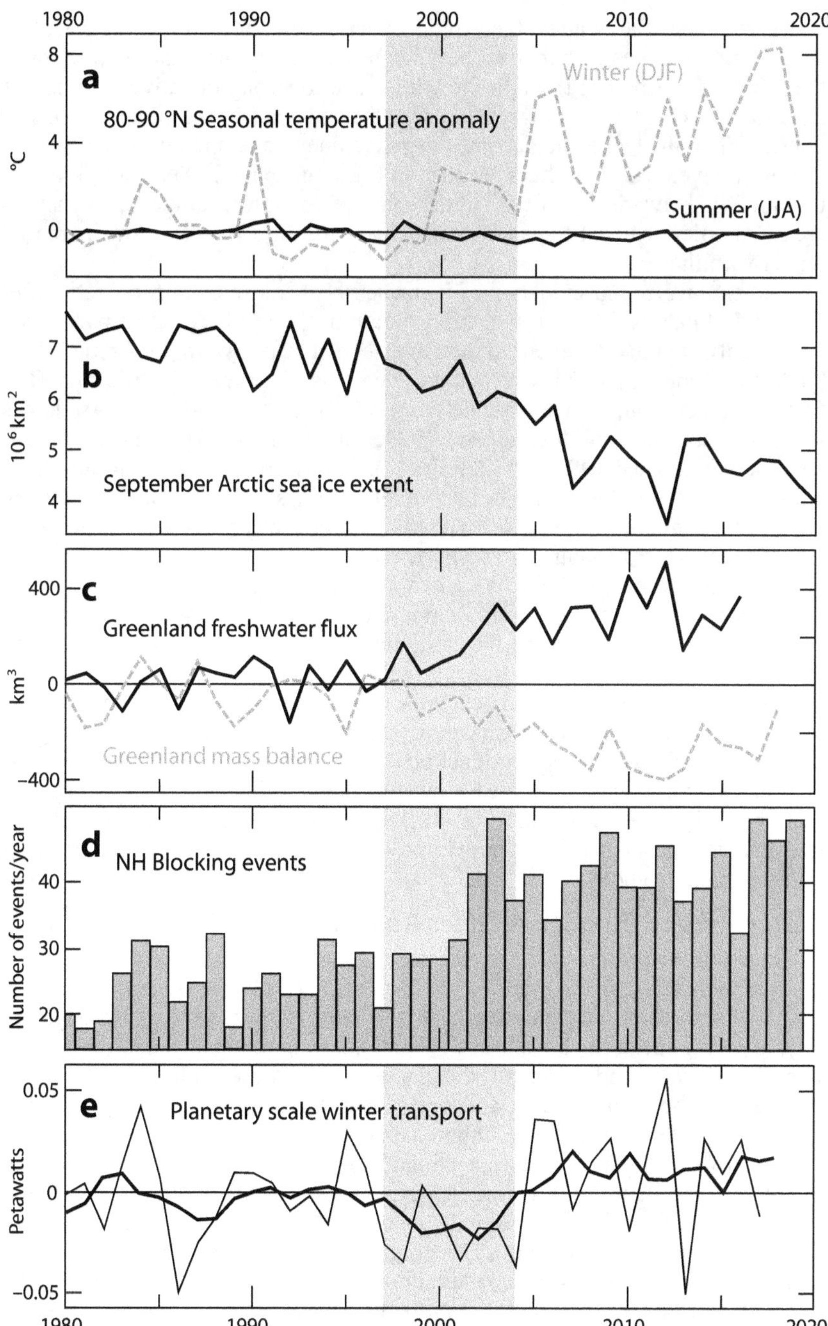

Figure 54. The 1997 Arctic Shift. The gray bar marks the interval 1997-2004, when most of the observed abrupt changes occurred.

During the polar winter, the Arctic experiences almost total darkness for several months, meaning that no heat is produced. Any heat that is present in the fall does not persist through the winter due to strong radiative cooling. As discussed in chapter 7, the greenhouse effect is weak in the polar regions in winter. When the lapse rate becomes negative due to a temperature inversion, it can even have a cooling effect, as seen in the stratosphere. Therefore, it's clear that the winter warming in the Arctic comes from lower latitudes, and an increase in poleward heat transport is necessary for the Arctic to show increased winter warming.

Studies have found evidence of increased heat transport to the Arctic after the 1997 climate shift. This is caused by atmospheric blocking events, which temporarily stop the zonal circulation and send storms toward the Arctic. Since 2000, the frequency of these blocking events has increased significantly (fig. 54d).[265] In addition, planetary-scale wintertime heat transport to the Arctic has also increased since 2000 (fig. 54e).[266] Ocean heat transport to the Arctic has also increased since 1997 (fig. 27, ch. 17). The decrease in albedo may have amplified the warming of the Arctic, but albedo does not matter in winter when there is no sunlight in the Arctic. Overall, it seems clear that the recent Arctic warming is due to a change in transport.

In recent decades, the Arctic has warmed more than the rest of the planet. This has led to a decrease in the temperature difference between the equator and the North Pole, reducing the latitudinal temperature gradient. However, instead of a decrease in heat transported along this gradient, an increase has been observed, contrary to expectations. This observation challenges the prevailing theory of meridional transport, which proposes that heat transport is driven by maximum entropy production (fig. 19, ch. 12). Instead, it shows that heat transport can actually increase even as the gradient decreases. It suggests that an unidentified factor plays a major role in determining meridional heat transport independent of the temperature gradient, supporting this book's new climate change hypothesis.

Unrecognized climate regimes and shifts of unknown origin

The evidence presented for Arctic warming and previously for the rest of the planet shows that climate regimes are distinct states of atmospheric circulation with different levels of poleward heat transport. Rather than changing gradually, these regimes can shift abruptly from one state to another. Similar to how interglacials vary and occur at irregular intervals, these multidecadal climate regimes can be unpredictable and occur at different intervals. They may not follow the perfect cycle that we might expect.

The problem is that the abrupt climate shifts described are not part of our scientific understanding of climate change. They don't fit the idea of a changing climate responding to gradually increasing CO_2 levels. Moreover, many changes originating from the 1997 climate shift have been attributed to increased human emissions. As a result, climate shifts and regimes often get little attention because they're seen as unforced variability or ignored altogether.

---

[265] Data from Lupo, A.R., 2021. Ann. N.Y. Acad. Sci. 1504 (1), pp.5–24. doi.org/10.1111/nyas.14557

[266] Data from Rydsaa, J.H., et al., 2021. Q. J. R. Meteorol. Soc. 147 (737), pp.2281–2292. doi.org/10.1002/qj.4022

What's more, climate models fail to produce these shifts, compounding the problem. Climate science relies heavily on these models (table 2, ch. 50), so it is not an option to acknowledge that they may miss crucial aspects of climate.

The inability to accurately identify climate shifts and regimes hinders our ability to study their causes and understand them as global rather than regional phenomena. Some scientists have proposed that they are related to synchronized chaos, but this is not widely accepted.[267] As we discussed in chapter 19, there is evidence that ocean oscillations had a different period during the Little Ice Age. Therefore, climate regimes are influenced by different conditions and factors that have changed over time. This means that their role in recent warming cannot be dismissed as cyclical behavior that evens out over time.

Box 25. When should we expect the next climate shift?

Scientists have identified four previous climate shifts in the Pacific Ocean, which occurred in 1925, 1946, 1976, and 1997. The variable intervals between these shifts (20-30 years) make it difficult to predict when the next one will occur.

Chapter 18 discussed the effect of solar activity on the El Niño-Southern Oscillation, while chapters 29 and 30 discussed its effect on winter polar stratospheric temperatures and planetary rotation, respectively. These phenomena are all related to heat transport, and I considered the possibility that external solar forcing could also play a role in triggering climate shifts. Interestingly, the four climate shifts identified in the Pacific Ocean during the 20th century occurred 1-3 years after a solar minimum. This suggests that the climate regimes of the past century have lasted 2-3 solar cycles.

In chapter 14 (box 11), we discussed the Holton-Tan effect, which links the phase of the tropical Quasi-Biennial Oscillation to the strength of the polar vortex through planetary wave propagation. Recent research has shown that this effect is most pronounced during solar minimums. However, the Holton-Tan effect weakened considerably during the period of reduced poleward transport between 1976 and 1997, which we will explore further in chapter 39. This suggests that the tropical-polar stratosphere and the stratosphere-troposphere couplings are stronger during solar minimum winters, making it an appropriate time for coordinated changes in the strength of poleward transport.

There is a highly speculative mechanism that could determine the appropriate solar minimum for an abrupt climate shift. It involves the 9.1-year frequency exhibited by the Atlantic and Pacific multidecadal oscillations, which some scientists believe originates in lunisolar tides.[268]

The difference in frequency between this 9.1-year lunar tidal cycle and the 11-year solar cycle causes them to shift from correlated to anticorrelated states, which could result in constructive and destructive interference that matches the multide-

[267] Tsonis, A.A., et al., 2007. Geophys. Res. Lett. 34, L13705. doi.org/10.1029/2007GL030288

[268] Muller, R.A., et al., 2013. J. Geophys. Res. Atmos. 118, 5280–5286. doi.org/10.1002/jgrd.50458

cadal oceanic periodicity. Figure B25b shows the quasi-decadal frequencies of the solar cycle and Atlantic Multidecadal Oscillation (AMO) and their changing correlation. The correlation curve has a similar appearance to the Pacific Decadal Oscillation curve.

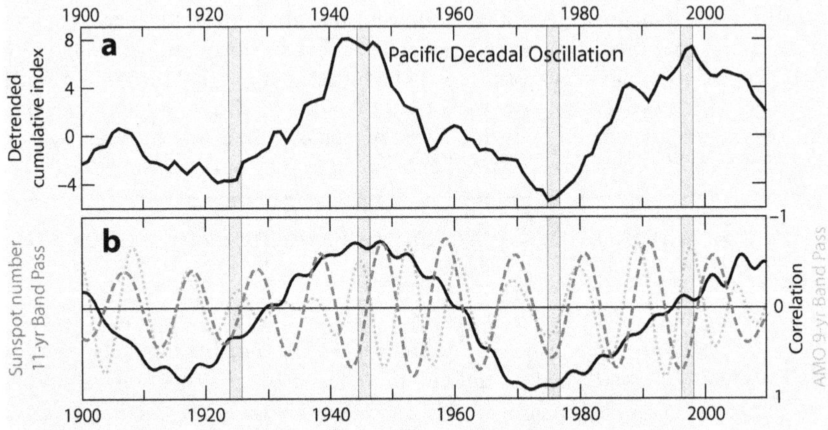

Figure B25. Conjecture on the timing of climate regimes and shifts by solar and lunar cycles. a) Cumulative Pacific Decadal Oscillation with identified climate shifts marked with gray bars. b) Band filtering of the sunspot data (dashed dark gray line) and the Atlantic Multidecadal Oscillation (dotted light gray line) showing only their 11- and 9-year frequencies, respectively. The black line is the correlation (inverted) between the two frequencies.

One might speculate that the periodic change in transport strength that leads to the observed climate shifts could be caused by the interaction between the effects of oceanic and atmospheric tides on the tropospheric component of the meridional transport and the effects of the solar cycle on the stratospheric component. This conjecture is supported by the presence of the 9- and 11-year frequencies in the Fourier analysis of the daily North Atlantic Oscillation data.[269]

According to this conjecture, if we assume its validity and the absence of any climate shift after the end of solar cycle 24, then we can expect the next climate shift to occur within the three years following the projected end of cycle 25, which gives us a date between 2031 and 2035. This shift could reduce poleward transport and cause Arctic cooling, contrary to our current expectations (fig. 93, ch. 51).

In summary

Arctic warming in the 21ˢᵗ century differs from the Arctic amplification predicted by climate models in response to global warming. It began 20 years later and was caused by a climate shift in 1997 that increased the transport of heat from lower latitudes to the Arctic. This abrupt change in heat transport caused both the Pause in global warming and the Arctic warming. However, global surface temperature datasets have not been able to accurately reflect

[269] Alvarez-Ramirez, J., et al., 2011. Adv. Space Res. 47 (4), pp.748–756.
doi.org/10.1016/j.asr.2010.09.030

these changes due to adjustments made to the relative weight of Arctic temperature anomalies in the global mean. The next climate change could occur between 2031 and 2035, and if it does, it could reduce Arctic warming despite expectations to the contrary.

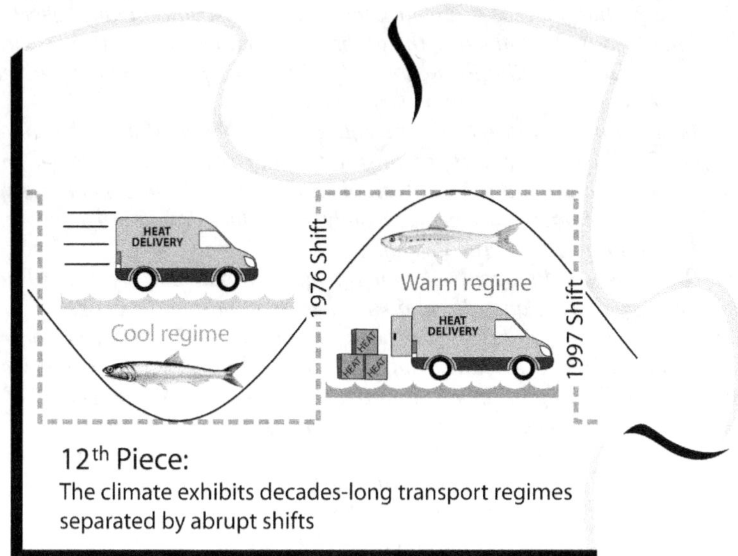

SECTION 9 KEY ISSUES

Recent global warming began in 1976 with a sudden climate shift in the Pacific Ocean that increased zonal atmospheric circulation and decreased poleward heat transport, affecting the global temperature trend. As a result, the multidecadal oceanic oscillations changed from a cold phase, which had led to the 1945-1975 cooling period, to a warm phase.

The abrupt climate shift of 1976 revealed the existence of multidecadal climate regimes separated by abrupt transitions. They result from changes in the global atmospheric circulation that establish distinct poleward heat transport regimes. Climate models cannot accurately reproduce low-frequency oscillations and do not reproduce climate shifts.

In 1997, a new shift occurred that affected global temperatures, Pacific Ocean climate patterns, and global atmospheric circulation. It caused changes in tropical extent, cloudiness, wind speed, and even the Earth's rotation rate. The stratosphere was particularly affected, changing its cooling and water vapor trends, indicating an increase in poleward heat transport. The 1997 climate shift remains unrecognized by most scientists.

One hundred years ago, the Arctic experienced an unexplained period of intense warming. The 1997 climate shift has led to a new period of intense Arctic warming. It has occurred mainly in winter. It has been confused with Arctic amplification due to feedback to increased CO_2, but the evidence supports that it is primarily due to increased heat transport into the region.

SECTION 10. THE NEED FOR A NEW THEORY

CHAPTER 35
EXPLAINING THE PRESENT OBSCURES THE PAST

The "Enhanced CO$_2$ Effect" hypothesis was originally a failed attempt in the 19th century to explain glaciations by changes in CO$_2$. Today, the hypothesis proposes that human emissions of warming greenhouse gases and cooling aerosols, with a small contribution from natural factors, explain climate change in recent decades. According to the hypothesis, most of the warming is due to feedbacks that respond to the warming caused by increased CO$_2$. This makes the climate very sensitive to warming or cooling from any source since almost twice as much warming occurs in response to the initial warming. While the hypothesis explains the recent warming since 1976 quite well, it doesn't explain past climate changes. The mid-20th century cooling, the early 20th-century warming, the Little Ice Age, the Holocene temperature evolution, abrupt climate events during the Holocene, and the early Eocene 50 million years ago all remain unexplained. We, therefore, need a theory of climate change that can explain all climate change, not just some of it.

About theories and hypotheses

A scientific theory is a plausible explanation based on our current, but still incomplete, scientific knowledge of the physical world. Acceptance of a theory does not make it correct, and rejection does not make it incorrect. For example, Bohr's incorrect theory that electrons orbit the nucleus of an atom was accepted even by Einstein. In contrast, Milankovitch's theory, which explains the glacial cycle through orbital variations, was initially rejected by most scientists for decades until it was finally accepted in 1976. Similarly, Darwin's theory of evolution fell out of favor in the late 19th century before being re-popularized by the modern synthesis in the 1940s. As scientific knowledge expands over time, theories may be modified or rejected. Although the central ideas of Darwin's theory of evolution and Milankovitch's orbital theory are still considered correct, they have undergone significant changes over time.

Scientific hypotheses are tentative proposals based on observations and supported by evidence. The day-to-day work of an experimental scientist (like me) is to make observations, develop hypotheses that fit the evidence, and design clever tests that will reveal if the hypothesis is wrong or support it. Because the reality is incredibly complex and our minds are limited, most hypotheses are wrong. As Thomas Henry Huxley said in 1870, the great tragedy of science — the slaying of a beautiful hypothesis by an ugly fact — is constantly being enacted before the eyes of scientists. However, if a hypothesis is not refuted by new evidence, over time it may be incorporated into larger theories or become a theory in its own right.

Climate model results are not evidence

Climatology is not an experimental science, which makes it difficult to test hypotheses. Without experimentation, hypotheses can only be tested with evidence that was not available when they were proposed. Sometimes new evidence can be found to support or contradict a hypothesis, such as the benthic cores that supported Milankovitch's theory in 1976, making alternative hypotheses very unlikely. However, this process can be slow and uncertain. Therefore, climatologists have developed computer models that incorporate much of what is known about climate. They use these models to test their hypotheses.

We will discuss climate models in section 15, but it is important to note that they are a product of the human mind and, despite their complexity, are subject to the limitations of human understanding. Climate models do not reflect reality, so they are not scientific evidence. This is something that climate scientists can forget and is worth repeating. <u>The results of climate models are not scientific evidence</u>. The results of climate models are only theoretical support within the theoretical framework used to build the program. Climate models will never give an answer that requires something that isn't in their code. Therefore, they are circular reasoning and do not prove anything except their internal consistency.

The Enhanced CO_2 Effect hypothesis

It is a hypothesis to propose that human emissions of cooling aerosols and warming greenhouse gases, with a small contribution from natural factors, can explain the climate changes observed in recent decades. In this hypothesis, the primary cause of climate change is the change in CO_2 levels, and most of the effect is thought to come from climatic feedbacks that cannot be measured. Therefore, it is correct to refer to it as the "Enhanced CO_2 Effect hypothesis."

In a sense, this is the only climate hypothesis that is being tested experimentally. As two leading climate scientists remarked in 1957, *"Human beings are now carrying out a large-scale geophysical experiment of a kind that could not have happened in the past nor be reproduced in the future."*[270] While the results of this experiment will ultimately determine whether the hypothesis is correct, we cannot afford to wait decades or centuries to find out, especially since the hypothesis has received a great deal of attention and is leading to significant societal and energy system changes based on the assumption that it is correct. Therefore, it is essential to test the hypothesis with newly available evidence that was not used to construct it.

Chapters 7 and 8 explain the difference between the Enhanced CO_2 Effect hypothesis and the greenhouse effect theory. These two concepts are often mistakenly used interchangeably. The greenhouse effect theory is based on the idea that an increase in greenhouse gas levels will cause the atmosphere to become more opaque to infrared radiation, resulting in some surface warming. However, the theory does not specify how much warming will occur, as this depends on various factors within the climate system that are not yet fully understood. The greenhouse effect theory is widely accepted by the scientific community, including those who are skeptical of the Enhanced CO_2 Effect hypothesis.

---

[270] Revelle, R. & Suess, H.E., 1957. Tellus, 9 (1), pp.18–27.
    doi.org/10.3402/tellusa.v9i1.9075

The Enhanced CO_2 Effect hypothesis is an extension of the greenhouse effect theory. It proposes that changes in atmospheric CO_2 have been the primary cause of climate change throughout the planet's geological epochs, including the present. However, if the Enhanced CO_2 Effect hypothesis proves to be false, it would not discredit the greenhouse effect theory.

The Enhanced CO_2 Effect hypothesis was originally developed to explain glacial periods, but Milankovitch's theory disproved it as their cause in 1976. In the 1960s, the hypothesis gained some support as an important cause of climate change due to rising CO_2 levels and early climate models. However, during the cooling period observed at that time, many scientists believed that other factors played a more important role. It was not until global warming became apparent and the first deep Antarctic ice core provided evidence of a strong correlation between temperature and CO_2 levels throughout the Pleistocene (fig. 34a, ch. 21) that the hypothesis gained widespread acceptance.

As noted above, the popularity of a hypothesis does not determine its accuracy. A scientific consensus for or against a hypothesis does not prove it right or wrong. Only evidence can support or refute a hypothesis. However, even if the evidence supports a hypothesis, it cannot be considered definitively correct because additional evidence may emerge that contradicts it. For example, centuries of evidence indicated that swans were all white, but the discovery of black swans in Australia in the 18th century disproved this notion.

The original version of the CO_2 hypothesis proposed that the greenhouse effect theory could explain glaciations. In the late 19th century, scientists were trying to understand why glaciations had occurred in the distant past, but they did not think that climate change was a significant issue at the time. In the 1930s, the hypothesis was revived in an attempt to explain the unusual warming that occurred in the early 20th century. Although this attempt failed, the hypothesis began to be considered relevant to explaining recent climate changes.

A hypothesis developed to explain recent warming by changes in CO_2

According to the greenhouse effect theory, changes in CO_2 levels by themselves have a small effect. For substantial warming or cooling to occur, the climate must be highly sensitive to these changes and respond strongly to the warming (or cooling) caused by the change in CO_2. This response includes water vapor, the planet's primary greenhouse gas, and changes in cloudiness and ice albedo. Together, these factors amplify the effect of CO_2, leading to the Enhanced CO_2 Effect hypothesis proposed in 1896 and developed in the 1960s.

The Enhanced CO_2 Effect hypothesis has gained popularity because it effectively explains all of the warming since 1976. This is due to the recruitment of other climatic factors as feedbacks to the CO_2 signal. However, the hypothesis has a drawback: it implies that the climate is highly sensitive to warming or cooling, whatever the cause. This is because the feedbacks respond positively to warming or cooling from any source, not to changes in CO_2 levels.

However, the climate is a stable system that has maintained life-supporting temperatures for over 500 million years, even in the face of asteroid impacts and entire erupting igneous provinces. This suggests a system dominated by negative feedbacks that stabilize it rather than positive feedbacks that destabilize it.

Our current understanding of the natural drivers of climate change is extraordinarily poor. The popularity of the Enhanced CO_2 Effect hypothesis does not help because it requires that natural factors produce only small changes in temperature for changes in CO_2 to explain climate change in a system that the hypothesis says is very sensitive to changes in temperature.

For example, changes in solar radiation are very small, and without considering the dynamic changes caused by solar activity (discussed in chapters 27-30), the solar effect can easily be dismissed as insignificant. Volcanic eruptions also pose a challenge, as models tend to exaggerate their cooling effects relative to observations (ch. 24). However, if volcanic eruptions were frequent enough in the past to raise CO_2 levels, they can be used to explain warming.

However, the value of a hypothesis is judged by what it cannot explain, not just what it can, since a hypothesis is usually consistent with the evidence from which it was created. The main problem with the Enhanced CO_2 Effect hypothesis is that by accounting for recent warming, it makes it impossible to explain past climate changes. This shows that the hypothesis is inadequate or incomplete in its current form.

Past climate change becomes inexplicable

We don't have to look far back in time to see this problem. As discussed in chapter 31, the Enhanced CO_2 Effect hypothesis cannot explain the cooling period between 1945 and 1975. During this period, there was a positive net anthropogenic forcing that should have led to warming, but instead, there was cooling. Figure 55, taken from the IPCC 6th Assessment Report, shows the combined effect on temperature of human emissions of greenhouse gases and aerosols from 1900-1990 (red line).[271] Rather than showing them separately, this modified figure shows their joint effect according to the average of multi-model simulations. It's worth noting that even natural causes, as hypothesized, cannot be responsible either, as they should have had a positive effect on temperature throughout the period, except for a few years following the eruption of Mt. Agung in 1963.

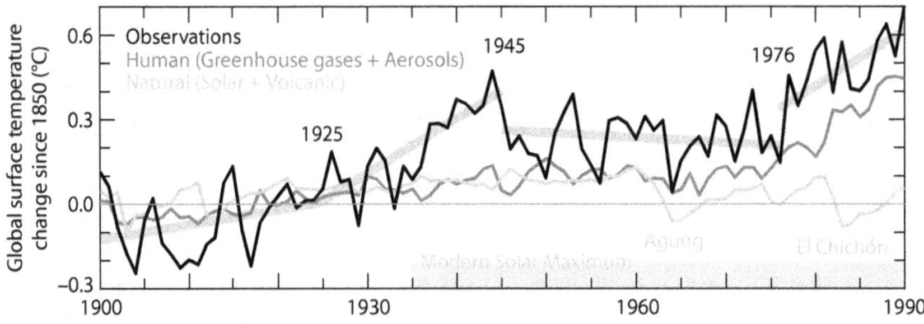

Figure 55. Global surface temperature change and its causes in the 6th Assessment Report. Observations (black line) are compared with climate model simulations of the effect on the temperature of human (medium gray line) and natural (light gray) forcings.

---

[271] Eyring, V., et al., 2021. Climate Change 2021: The Physical Science Basis. 6th AR IPCC. pp. 515–516. doi.org/10.1017/9781009157896.005

Trend lines for the observations (thick gray lines) correspond to periods separated by known climate shifts in the Pacific Ocean. The period of the Modern Solar Maximum and two volcanic eruptions are indicated.

According to the hypothesis and climate models, the net anthropogenic and natural forcings were small before the 1970s. This suggests that the only possible outcome was a gradual rise in temperature from the 1910s to the 1970s, with warming accelerating thereafter. However, the evidence shows that this did not happen. Instead, the global climate experienced an intense warming in the early 20th century and a clear cooling in the mid-20th century. Thus, the Enhanced CO_2 Effect hypothesis cannot explain climate change prior to 1976. Because our understanding of climate change is still limited, it is unwise to accept an incomplete or flawed hypothesis, especially when it is coupled with urgent calls for significant societal and energy system changes.

The Enhanced CO_2 Effect hypothesis faces a similar problem regarding the climate of the past 11,700 years, as discussed in chapter 21. Climate models project a gradual warming trend throughout the Holocene, failing to capture the Neoglacial cooling period that is well documented by global glacier growth (fig. 35, ch. 21). The models also fail to reproduce notable cooling events such as the Boreal Oscillation, the 5.2 and 2.8 kiloyear events, and only weakly reproduce the Little Ice Age. As a result, climate models are inadequate at reproducing historical climate patterns, making it difficult to have confidence in their ability to accurately project future climates. Surprisingly, the IPCC assessment reports rarely address the clear discrepancy between CO_2 levels and temperature changes over the past 11,000 years, focusing primarily on the Pleistocene concordance.

However, significant discrepancies occur in about half of the interglacial-glacial transitions of the Pleistocene. They show a large cooling phase despite no concomitant changes in CO_2 levels. A notable example is observed at the end of the last interglacial period, about 124,000 years ago. During this period, temperatures gradually decreased over 8,000 years, reaching about half the difference between interglacial and glacial levels, while CO_2 levels remained unchanged (fig. 56).[272]

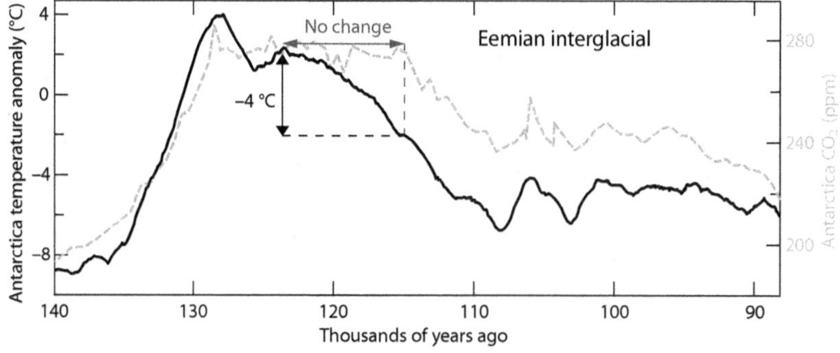

Figure 56. CO_2 does not play a role in the end of an interglacial.

[272] Data from Jouzel, J., et al., 2007. Science, 317 (5839), pp.793–796. doi.org/10.1126/science.1141038 Bereiter, B., et al., 2015. Geophys. Res. Lett. 42 (2), pp.542–549. doi.org/10.1002/2014GL061957

A substantial cooling phase without a concomitant reduction in current CO_2 levels is widely regarded as impossible and is not observed in model simulations under current conditions. In fact, experts in the glacial cycle claim that the onset of a new glacial period is not possible for tens of thousands of years. Nevertheless, there is evidence that glaciations have started several times in the past, driven by orbital changes, without any observable changes in CO_2 levels.

The Enhanced CO_2 Effect hypothesis struggles to explain the 5°C (9 °F) temperature difference between the late Oligocene and the mid-Pliocene. This is particularly difficult because both periods, separated by 20 million years, had similar CO_2 levels below today's levels (fig. B18, ch. 21).

Finally, the Enhanced CO_2 Effect hypothesis and climate models do not provide a satisfactory explanation for the equable climate observed during the early Eocene, characterized by tropical conditions in the polar regions. Surprisingly, this occurred even though CO_2 levels were not much higher than today (ch. 20). Furthermore, if these models cannot explain the climate conditions of the early Eocene, which occurred 50 million years ago, it becomes clear that they are inadequate to explain the transition from these climatic conditions to the glacial conditions of the Pleistocene, which encompasses our current interglacial period. If we remain uncertain about the causes of the early Eocene climate, then it is clear that we lack the knowledge necessary to understand the conditions that had to change for the onset of the ice age cooling.

The current Enhanced CO_2 Effect hypothesis can only explain the warming period since 1976 but cannot explain most of the past climate changes. Therefore, we need a theory that can explain both past and present climate changes. To do this, we need to formulate better hypotheses to replace the current flawed one. Promoting new hypotheses is necessary to advance our understanding of climate.

In summary

We need new climate hypotheses that can explain the many past climate changes that are inconsistent with CO_2 variations. The currently favored hypothesis places excessive sensitivity on climate response to temperature changes while attributing minimal influence to natural causes of climate variability. Consequently, it relies excessively on CO_2 changes and renders any climate change inexplicable without a corresponding CO_2 change. However, there are examples of significant past climate changes that occurred in the absence of significant CO_2 variations. Many of the challenges in explaining such changes stem from the assumption that the Enhanced CO_2 Effect hypothesis is correct. The limited explanatory power of this hypothesis, restricted primarily to the recent warming since 1976, underscores the urgent need for new hypotheses. Unfortunately, the active discouragement of the search for alternative explanations hinders progress in this quest.

CHAPTER 36
INTERNAL VARIABILITY IS NOT WELL UNDERSTOOD

Natural internal variability arises from the redistribution of energy within the coupled atmosphere-ocean system, resulting in oscillations in specific regions and time frames. These oscillations, called modes of variability, are numerous, and scientists are uncertain about their causes or why they change from one phase to another. Climate models reproduce them to some extent as stochastic "random walks," but they do not account for multidecadal trends or their response to natural causes such as the solar cycle. Most climatologists and climate models do not attribute internal variability as a significant factor in climate change, but some scientists disagree. The lack of understanding of the origins of these oscillations and the inability to reproduce their essential characteristics cast doubt on our understanding of internal climate variability and its impact on climate change.

Internal variability and modes of variability

Natural variability refers to changes in climate that occur within the climate system or as a result of external factors that affect the Earth's radiative balance. External factors to the climate system include changes in the Earth's orbit, variations in the amount of solar energy received, or large volcanic eruptions. However, significant orbital changes require long periods of time and show minimal change over a century or two, having little effect on temperature changes during this time. In chapter 38, we will explore how the Enhanced CO_2 Effect hypothesis interprets the effects of solar variations and volcanic eruptions.

Natural internal variability refers to the redistribution of energy within the climate system. According to the IPCC, it is primarily evident in regional rather than global variations in surface temperature. However, this view was challenged in 1994 by a landmark scientific paper that introduced the Atlantic Multidecadal Oscillation and identified it as *"An oscillation in the global climate system of period 65-70 years."*[273] Over the past 30 years, numerous scientists have recognized the global influence of oceanic multidecadal oscillations, which is not reflected in the IPCC's preference to treat them as regional manifestations.

Scientists try to identify recurring patterns within the climate system to reduce the complexity of internal variability. These patterns have inherent spatial characteristics, seasonality, and time scales. They call these recurring patterns modes of variability, but in this book, we refer to them as oscillations. Some of these patterns, such as the El Niño-Southern Oscillation (ch. 18) and the Atlantic and Pacific multidecadal oscillations (ch. 19), have been discussed previ-

---

[273] Schlesinger, M.E. & Ramankutty, N., 1994. Nature, 367 (6465), pp.723–726.
doi.org/10.1038/367723a0

ously. IPCC reports suggest that these patterns result from dynamic properties of atmospheric circulation and interactions between the ocean, atmosphere, and land surfaces, including sea ice. However, scientists are still struggling to explain how these modes of variability with regional spatial patterns appear to affect distant parts of the world. These remote effects are called teleconnections.

Internal variability is not a random walk

Scientists lack a comprehensive understanding of the underlying mechanisms that generate low-frequency climate modes. According to the latest IPCC report, regional climate variability results from a complex interplay of local physical processes, such as thermodynamic and land-atmosphere feedbacks, and broader non-local phenomena.

A notable example of the limitations of climate models is their failure to predict the slowing of global warming in the 2000s. This failure has been attributed to a phase shift in the Pacific Decadal Oscillation and, ultimately, to significant trade wind anomalies that occurred around the year 2000.[274] However, the exact origin of these wind anomalies remains unknown, as does the true nature of the 1997 climate shift discussed in chapter 33.

To illustrate the challenges of understanding the origins of different modes of variability, the Pacific Decadal Oscillation is no longer considered to be a physically coupled mode arising from ocean-atmosphere interactions. Instead, it is now thought to result from three distinct processes: internal atmospheric noise in the North Pacific, midlatitude ocean dynamics mediated by Rossby waves, and tropical forcing transmitted to the midlatitudes via an atmospheric bridge.

Given this complexity, climate models are not expected to accurately reproduce specific changes in these modes of variability. Instead, they focus on capturing the statistical characteristics of these patterns. From this perspective, models have made significant progress in representing internal variability. However, there is a fundamental problem with this view.

The atmosphere-ocean interaction is recognized as the fundamental driver of internal variability on decadal and longer time scales. In this framework, the chaotic atmosphere is a stochastic source of white noise, similar to analog television static noise, with equal power at all frequencies. As the ocean integrates this noise, it produces a spectrum with increasing power at lower frequencies, known as red noise. The result of this overall process is often referred to as a "random walk," which represents a trajectory shaped by a stochastic process where future behavior is independent of past history. The IPCC provides the following description of this phenomenon:

"Another way to picture natural variability and human influence is to think of a person walking a dog. The path of the walker represents the human-induced warming, while their dog represents natural variability."[275]

Figure 57 shows data from the graph accompanying this statement in the IPCC's 6th Assessment Report. The thin gray lines show the effects of natural

---

[274] Farneti, R., 2016. WIREs Clim. Change, 8 (1), p.e441. doi.org/10.1002/wcc.441

[275] Eyring, V., et al., 2021. Climate Change 2021: The Physical Science Basis. 6th AR IPCC. pp. 517–518. doi.org/10.1017/9781009157896.005

variability in three different simulations of the large ensemble of the Community Earth System Model 1 (CESM1), illustrating decadal variations in global surface temperature in the absence of anthropogenic warming (i.e., without the long-term trend). These simulations provide examples of random walks in natural variability, with no discernible pattern or trend in their periods and peak amplitudes. The remaining curves in the figure correspond to different sources.[276]

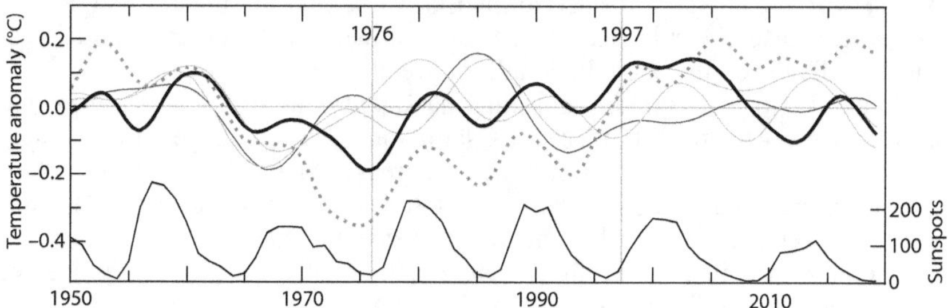

Figure 57. Internal climate variability and solar activity. Gaussian smoothed detrended global surface mean temperature (thick black line) and Atlantic Multidecadal Oscillation (red dotted line), three model simulations of decadal variations in global temperature (thin gray lines) and sunspot number (thin black line). All temperature curves are detrended. Climate shifts in 1976 and 1997 are indicated.

The IPCC acknowledges the existence of multi-decadal and even centennial trends in modes of variability. For example, the North Atlantic Oscillation was dominated by positive values in the early 20th century. However, there was a downward trend between 1920 and 1970, followed by a recovery to consistently high values between 1970 and the early 1990s. Such behavior, characterized by extended periods of positive or negative values, is not replicated by the models and casts doubt on the random walk hypothesis of internal variability.

The dotted gray line represents the Atlantic Multidecadal Oscillation data, smoothed with a Gaussian filter.[277] The behavior of this data is similar to that of the global surface temperature data, indicating that it is an oscillation of the global climate system. This oscillation exhibits abrupt changes, such as those in 1976 and 1997.

The solar cycle is known to have an effect on global temperatures of about ±0.1°C and is shown in the lower part of figure 57. The observed internal variability tends to be in phase with the solar cycle, especially during the most active cycles. This interaction between natural internal and external variability is expected, but climate models are unable to reproduce it. The response of the simulations to the solar cycle is weak, suggesting that the modeled variability comes from some other source.

---

[276] The thick black line is the detrended, Gaussian-filtered HadCRUT4 global surface temperature dataset. The dotted gray line is the Gaussian-smoothed Atlantic Multidecadal Oscillation dataset from NOAA. The thin black line is the number of sunspots from SILSO.

[277] A Gaussian filter is used to reduce noise and smooth the data. Often used to blur images, it has the best combination of suppressing high frequencies while minimizing spatial spread, and is the one most often used by the author.

The long-term trends in internal variability, which change their sign after abrupt shifts, and the response to external natural variability indicate that the IPCC and climate model assumption of internal variability as a random walk is incorrect.

Internal variability is a global phenomenon affecting climate change

Consistent with treating it as a stochastic "random walk," the IPCC reports do not attribute any significance to internal variability in climate change. It is treated as noise occurring at decadal or multidecadal frequencies and is assumed to have no impact on the long-term trend.

Some scientists disagree, suggesting that about one-third of recent warming is due to the Atlantic Multidecadal Oscillation.[278] They suggest that climate models overestimate the warming due to increased GHGs. They offset the excess warming simulated by overestimating the cooling effect of aerosols, leading to an overestimation of the human influence on climate.

Internal variability arises from a combination of local physical processes and large-scale non-local phenomena. As a result, each mode of variability is considered independently, while their interactions are represented by teleconnections. However, as described in box 16 (ch. 19), oscillations in different regions of the Northern Hemisphere appear to change in a coordinated manner in what is called the "stadium-wave hypothesis" (fig. B16, ch. 19).[279]

It appears that oscillations (modes of variability) are a local response to a global phenomenon rather than being locally generated and subsequently interacting through teleconnections. This shift in perspective eliminates the need to consider a wide range of complex teleconnections, thus simplifying the study of internal variability. Instead of multiple causes, we now find that the same global phenomenon causes all modes of variability while their specific characteristics are locally determined. Since we already know that ocean oscillations result from changes in poleward heat transport, we no longer need to search for a different underlying cause for each oscillation.

In summary

The hypothesis of an enhanced CO_2 effect is an obstacle to understanding internal climate variability. Natural climate oscillations are not the result of random stochastic processes. Climatologists and their models lack a fundamental understanding of the causes and mechanisms behind these oscillations or their phase transitions. These oscillations are disregarded as a driving force of climate change and treated as separate phenomena linked by teleconnections. However, multidecadal trends and synchronized phase changes across hemispheres suggest that they are regional manifestations of an underlying global factor. To improve our understanding, it is crucial to develop new hypotheses that explore the causes of internal climate variability and its relationship to climate change.

---

[278] Chylek, P., et al., 2014. Geophys. Res. Lett. 41 (5), pp.1689–1697.
doi.org/10.1002/2014GL059274

[279] Wyatt, M.G., et al., 2012. Clim. Dyn. 38, pp.929–949.
doi.org/10.1007/s00382-011-1071-8

CHAPTER 37
MERIDIONAL TRANSPORT VARIABILITY IS NEGLECTED

Most climatologists believe that changes in meridional transport primarily redistribute heat and do not directly cause climate change. Consequently, these transport changes are overlooked in the IPCC reports. However, it is crucial to recognize that changes in heat transport play a significant role in explaining Arctic warming and in driving multidecadal oceanic oscillations with global implications. They are, therefore, of immense importance for understanding climate dynamics. The potential of changes in heat transport as a causal factor in climate change should be thoroughly investigated rather than prematurely dismissed on the basis of unproven assumptions. A key challenge is the inability of climate models to accurately reproduce variations in transport intensity or their historical patterns. These models assume global offsets between oceanic and atmospheric transport changes despite evidence to the contrary. It is important to recognize that global shifts in heat transport serve as one of the primary determinants of climate behavior over decades.

Heat transport is disregarded in the IPCC reports

In the first chapter of the 6th Assessment Report, the IPCC provides a clear explanation of climate change, defining its causes as follows:

"The natural and anthropogenic factors responsible for climate change are known today as radiative 'drivers' or 'forcers'. The net change in the energy budget at the top of the atmosphere, resulting from a change in one or more such drivers, is termed 'radiative forcing'."[280]

According to the IPCC, heat transport is not considered a radiative forcing and, therefore, not a cause of global climate change. Its effects only contribute to internal or regional variability. This perspective is reflected in the limited attention given to heat transport in the IPCC reports. In the massive 2,391-page 6th Assessment Report, heat transport is only briefly mentioned in a 5-page subsection on ocean heat content.[281]

In this subsection, we learn that climate change is due to heat addition, while changes in ocean circulation cause heat redistribution. The subsection briefly acknowledges the increase in oceanic heat transport toward the Arctic in recent decades, as highlighted in chapter 17 (fig. 27). It then explains that the weakening of the Atlantic Meridional Overturning Circulation reduces the northward flow of oceanic heat in the North Atlantic, and attributes the increased heat transport to stronger wind-driven ocean gyres.

---

[280] Chen, D., et al., 2021. Climate Change 2021: The Physical Science Basis. 6th AR IPCC. p.178. doi.org/10.1017/9781009157896.003

[281] Fox-Kemper, B., et al., 2021. Climate Change 2021: The Physical Science Basis. 6th AR IPCC. pp.1228–1233. doi.org/10.1017/9781009157896.011

Heat transport is a fundamental feature of climate. The weather and climate we experience outside the tropics are determined by differences in insolation and the varying amounts of heat and moisture transported to our location. The atmosphere plays the important role of transporting most of the Earth's heat and moisture, including driving shallow oceanic transport such as the wind-driven gyres. However, the IPCC reports tend to overlook atmospheric transport and do not assign importance to oceanic transport. This is likely due to the limitations of climate models, which struggle to accurately reproduce changes in transport and historical ocean warming patterns.[282]

There is a major contradiction in the IPCC's view of Arctic warming. It assumes that Arctic amplification is solely a consequence of global warming due to various feedbacks. However, the primary cause of Arctic warming since 1997 is the increased transport of heat and moisture to the region by both the atmosphere and the ocean. According to the Bjerknes compensation hypothesis (box 8, ch. 12), changes in atmospheric transport should be balanced by opposite changes in oceanic transport. However, this is not the case in the Arctic, where the atmosphere and the ocean have simultaneously contributed to increased heat transport despite a flattening temperature gradient. Changes in transport, rather than the influence of feedbacks, are the main driver of Arctic warming because the shift in transport and the onset of increased Arctic warming occurred simultaneously after the 1997 climate shift.

The IPCC view, which many climate scientists share, has two major problems. First, it ignores the importance of heat redistribution in climate change. Paradoxically, however, it emphasizes Arctic warming as crucial evidence of human-caused global warming, even though changes in heat redistribution are causing it. To support this claim, global mean surface temperature datasets have been adjusted to include additional warming from the Arctic.[283] This creates a clear contradiction by dismissing redistributed heat as unimportant while simultaneously using its effects as evidence of additional heat in the climate system.

The second problem lies in the IPCC's failure to consider the effects of heat redistribution on a planet where the greenhouse effect is not uniform. While CO_2 may be uniformly mixed, water vapor is not. 75% of the greenhouse effect is due to water vapor when the influence of clouds is included (table 1, ch. 7). However, during the Arctic winter, there is minimal water vapor and few clouds (fig. B4, ch. 7). As a result, the greenhouse effect in the Arctic winter could be up to four times weaker than in the tropics. Consequently, any change in heat transport to the Arctic would lead to changes in radiative fluxes at the top of the atmosphere, emphasizing the role of poleward heat transport as a driving force of climate change.

Internal climate variability reflects changes in heat transport

This book argues extensively that ocean oscillations, also known as modes of variability, reflect changes in the global system responsible for transporting heat to the poles. Chapters 17, 19, 31-34, and 36 provide ample evidence to

---

[282] Bronselaer, B. & Zanna, L., 2020. Nature, 584 (7820), pp.227–233. doi.org/10.1038/s41586-020-2573-5

[283] Morice, C.P., et al., 2021. J. Geophys. Res. Atmos. 126 (3), p.e2019JD032361. doi.org/10.1029/2019JD032361

support this claim. In fact, there is so much evidence that an entire book could be devoted to it.

Let's consider the case of the Pacific subtropical cells, which are shallow meridional cells that connect tropical upwelling zones with subtropical subduction zones in both hemispheres. These cells are important because they have shown that slow variations in the surface temperature of the tropical Pacific Ocean can affect the global climate. The Pacific Decadal Oscillation encompasses the spatial pattern of these cells.

In the mid-1990s, it was discovered that the subtropical cells experienced a significant slowdown after the 1976 climate shift. This slowdown reduced the transport by 13 sverdrups (million cubic meters per second), more than half of the average transport of 21.8 sverdrups (fig. 58, black line).[284] This slowing of the meridional circulation was accompanied by an increase in sea surface temperature of about 0.8°C (1.4 °F) in the central and eastern tropical Pacific (fig. 58, gray line). These changes in ocean temperature and nutrient circulation had corresponding effects on fish populations in the Pacific (fig. 50, ch. 32).[285]

After the 1997 climate shift, there was a notable reversal in transport that led to a subsequent decrease in temperature. It is worth noting that the data presented in the study have been smoothed by the authors, which slightly shifts the timing of the observed changes. However, the authors correctly associate these changes with the climatic shifts identified in numerous studies.

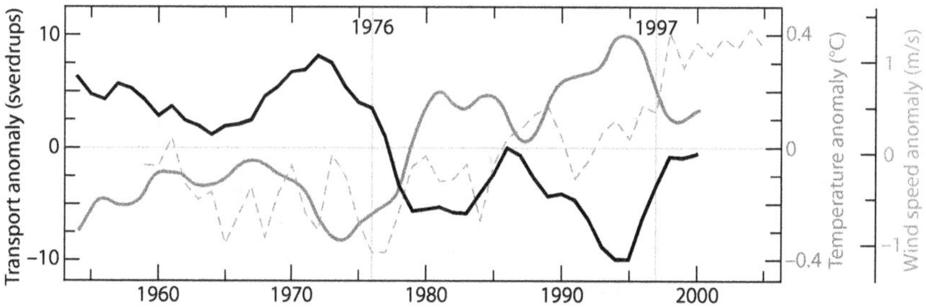

Figure 58. Transport changes in the Pacific subtropical cells. The black line shows changes in the amount of water transported in the tropical region of the central and eastern Pacific. The gray line corresponds to changes in sea surface temperature in the same region. The dashed line represents changes in global ocean wind speed.

Comparing the data from this study with the global ocean wind data presented earlier (fig. B9, ch. 13), it is clear that the changes in the subtropical cell transport can be attributed to variations in the winds that drive it (fig. 58, dashed line).[286] The observed increase in ocean winds is associated with increased atmospheric zonal (east-west) circulation. This enhanced zonal wind circulation has two important effects: first, it reduces the intensity of the meridional (north-south) circulation, which is responsible for transporting heat

[284] Zhang, D. & McPhaden, M.J., 2006. Ocean Model. 15 (3-4), pp.250–273. doi.org/10.1016/j.ocemod.2005.12.005

[285] Chavez, F.P., et al., 2003. Science, 299 (5604), pp.217–221. doi.org/10.1126/science.1075880

[286] Yu, L., 2007. J. Clim. 20 (21), pp.5376–5390. doi.org/10.1175/2007JCLI1714.1

toward the poles, resulting in warming in the midlatitudes; second, it causes the Earth to rotate faster, thereby shortening the length of the day.

Interestingly, the observation that the Earth's rotational speed had accelerated in the early 1970s was the only evidence-based prediction that the climate would soon begin to warm.[287] This mechanism of climate change, driven by changes in heat transport, has nothing to do with GHGs.

Knowledge to be incorporated into theory and modeling

The Pacific Decadal Oscillation, along with other modes of variability, is the expression of meridional heat transport regimes forced by the atmosphere over the ocean. This simple and comprehensive explanation is supported by a range of evidence, including studies of atmospheric and oceanic circulations, marine ecosystems, and even the Earth's rotation.

The climate change hypothesis based on changes in CO_2 is insufficient to explain the significant climate shifts we observe, such as the remarkable period of intense warming from 1976 to 1997, which can be attributed, at least in part, to variations in meridional heat transport. Unfortunately, climate models cannot accurately reproduce the variability in heat transport because they are based on the flawed premise that atmospheric and oceanic transport changes should cancel each other out. It is clear that we need a new hypothesis that can effectively account for the crucial impact of changes in heat transport on climate.

In summary

Transport changes in subtropical cells that are part of the shallow meridional overturning circulation in the Pacific indicate that global changes in poleward heat transport are the main driver of multidecadal internal climate variability. These changes result from changes in the global atmospheric circulation. When the east-west (zonal) component of the circulation strengthens, it leads to an acceleration of the Earth's rotation, a reduction in poleward heat transport through the atmosphere and ocean, and surface warming outside the polar regions. The opposite occurs when the global atmospheric circulation strengthens in the north-south (meridional) direction.

These changes in transport have a direct impact on climate trends. However, climate models fail to accurately reproduce any of these observations, indicating that the prevailing hypothesis of an enhanced CO_2 effect does not adequately account for a fundamental property of climate — the poleward transport of heat.

---

[287] Lambeck, K. & Cazenave, A., 1976. Geophys. J. Int. 46 (3), pp.555–573. doi.org/10.1111/j.1365-246X.1976.tb01248.x

CHAPTER 38
UNSOLVED SOLAR QUESTION

According to IPCC reports, the impact of changes in solar activity on climate is thought to be minimal. However, this assessment is based primarily on small variations in total solar irradiance. It does not take into account a substantial body of evidence on the indirect effects of solar activity, which suggests a significant solar influence on the winter atmospheric circulation. In addition, paleoclimatic evidence strongly implicates solar variability as a primary driver of climate variability on centennial scales. In particular, past periods of remarkably low solar activity coincide with some of the most abrupt climatic events. These compelling findings have led many paleoclimatologists to argue that solar variability plays a critical role in shaping climate, a perspective at odds with the stance of the IPCC reports. If we consider that changes in solar activity have only a minor impact, it becomes difficult to explain the occurrence of the Little Ice Age. This is because there were no concurrent changes in CO_2 levels during the initial 400-year cooling period or notable volcanic eruptions during most of the Little Ice Age.

The effect of solar activity on climate, according to IPCC reports

According to the IPCC, the prevailing scientific consensus is that variations in solar activity, whether short-term or long-term, have a negligible effect on the Earth's climate. In contrast, the warming effect of increased levels of anthropogenic greenhouse gases is thought to be much stronger than any influence from recent variations in solar activity.

Satellites have been observing the Sun's energy output for more than 40 years and have found variations of less than 0.1%. Based on this data, the IPCC estimates that the warming effect of GHGs emitted by human activities since 1750 is more than 270 times greater than the small additional warming contributed by the Sun itself over the same period.

In its 6th Assessment Report, the IPCC acknowledges that solar activity in the second half of the 20th century exceeded that of 90% of the past 9,000 years. However, the increase in radiative forcing from the Maunder Minimum (1645-1715) to the second half of the 20th century is estimated to be only 0.09-0.35 W/m^2.[288]

The effect of solar variations on climate is considered so insignificant that, despite very high solar activity compared to the Holocene average, climate models produce cooling in the second half of the 20th century in the absence of anthropogenic warming (fig. 55, ch. 35). This is due to the assumption that the combined influence of three moderate-intensity volcanic eruptions (Agung in 1963, El Chichón in 1982, and Pinatubo in 1991) outweighed the effects of the 20th-century solar grand maximum.

---

[288] Forster, P., et al., 2021. Climate Change 2021: The Physical Science Basis. 6th AR IPCC. pp.957–958. doi.org/10.1017/9781009157896.009

Indirect solar effects are ignored

The IPCC reports and climate models consider the effects of small variations in total solar irradiance, as discussed in chapter 2. However, they ignore compelling evidence that the Sun influences climate through indirect mechanisms involving dynamic changes in atmospheric circulation. Unfortunately, these indirect effects have not been adequately incorporated into climate models, resulting in their limited representation.

Here are some notable pieces of evidence for the existence of an important indirect solar effect on climate that is unlikely to result from small changes in surface insolation:

* The solar cycle signal on global temperature amounts to 0.1°C. This signal is four times larger than expected from energy changes alone (ch. 28).
* The temperature changes induced by the solar cycle show a highly irregular pattern at the Earth's surface, with certain regions near 60°N experiencing temperature increases of more than 1°C, while other regions experience cooling (fig. 44a, ch. 28).
* The heat budget of the tropical oceans shows temperature fluctuations related to the solar cycle that are almost ten times larger than those caused by changes in total solar irradiance alone (box 14, ch. 17).
* Significant dynamical effects associated with the solar cycle have been observed in the atmosphere, influencing various phenomena such as the Hadley cell, subtropical jets, the polar vortex, the jet stream, winter blocking events, and the pressure gradient between the Icelandic Low and the Azores High, among others (ch. 28).
* The El Niño-Southern Oscillation, an important climate oscillation, shows a response to the solar cycle, particularly in the timing and frequency of La Niña events (figs. 28 & 29, ch. 18).
* The polar stratosphere experiences temperature variations of up to 10°C (18 °F) as a result of the solar cycle. This effect is particularly pronounced in winter when there is no direct solar radiation in this region (fig. B22, ch. 29).
* The solar cycle affects the Earth's rotation rate, which cannot be attributed solely to small changes in total solar irradiance (fig. 47, ch. 30).

A large body of evidence documented in numerous scientific papers supports the conclusion that the solar influence on climate is primarily atmospheric in nature. This effect cannot be attributed solely to small changes in surface irradiance. Instead, the evidence strongly suggests that the stratosphere-troposphere coupling is the underlying source of this influence. Changes in solar UV radiation and its effect on the ozone layer are thought to initiate the complex signaling pathway discussed in detail in chapter 29.

Climate models generally fail to reproduce these effects properly, in part because of their inadequate representation of stratospheric and ozone-related effects. Unfortunately, the scientific evidence for these indirect effects is ignored in the IPCC reports. The IPCC's conclusion that variations in solar activity play a minimal role in the Earth's climate is influenced by a biased selection of evidence. A more neutral scientific assessment would recognize the wealth of evidence that the Sun influences climate through indirect pathways that remain poorly characterized but could have important effects.

An important consideration is that if we accept that the Sun has a greater influence on climate, this would require a reduction in the influence of GHGs and aerosols. This implies that the hypothesis of an enhanced CO_2 effect on climate, as currently proposed, is flawed and should be revised accordingly.

Paleoclimate evidence identifies solar variability as a major driver of climate change

The IPCC reports rely on paleoclimate proxy evidence to assert that the current climate change is highly unusual and that current temperatures are most likely the highest for a considerable period of time. However, when examining the paleoclimatic consequences of past variations in solar activity, the IPCC reports consider the proxy evidence inconclusive.

The evidence is anything but inconclusive. Over the past 11,700 years, there have been four distinct 200-year periods of lowest solar activity. They are known as Spörer-type solar grand minimums. Although these periods represent less than 7% of the Holocene, they coincide with four of its most notable abrupt climatic events. These events were characterized by substantial cooling and major shifts in precipitation patterns (ch. 23). If the reduced solar activity during these periods played a role in the extraordinary climatic conditions that occurred, we could not exclude the possibility that the high solar activity observed during the second half of the 20th century has had an influence on the observed warming trend since 1976.

Paleoclimate scientists openly acknowledge a significant solar influence on climate, a viewpoint not adequately reflected in the IPCC reports. Let us quote some of their perspectives to further emphasize this point.

"In view of these findings, we call for an in-depth multi-disciplinary assessment of the potential for solar modulation of climate on centennial scales."[289]

"On a centennial scale, the successive climatic events which punctuated the entire Holocene in the central Mediterranean coincided with cooling events associated with deglacial outbursts in the North Atlantic area and decreases in solar activity during the interval 11700–7000 cal BP, and to a possible combination of NAO-type circulation and solar forcing since ca. 7000 cal BP onwards."[290]

"Our results imply that small variations in solar irradiance induced pronounced cyclic changes in northern high-latitude environments. They also provide evidence that centennial-scale shifts in the Holocene climate were similar between the subpolar regions of the North Atlantic and North Pacific, possibly because of Sun-ocean-climate linkages."[291]

The belief that the paleoclimate evidence strongly supports the significant influence of solar variations on climate change on centennial time scales is widely held in the field. The three influential papers cited above incorporate

---

[289] Rohling, E.J., et al., 2002. Clim. Dynam. 18 (7), pp.587–593.
   doi.org/10.1007/s00382-001-0194-8
[290] Magny, M., et al., 2013. Clim. Past, 9 (5), pp.2043–2071.
   doi.org/10.5194/cp-9-2043-2013
[291] Hu, F.S., et al., 2003. Science, 301 (5641), pp.1890–1893.
   doi.org/10.1126/science.1088568

the expertise of 50 authors who are highly respected in the field of paleoclimatology. Of the 28 papers presenting proxy evidence for the 2,500-year climate cycle discussed in chapter 23, 16 explicitly attribute changes in solar forcing as the likely cause, while only one study rejects this possibility.[292]

The IPCC practices anti-solar biased evidence selection by not including this paleoclimate evidence and expert opinion in its reports.

A large solar effect is needed to explain the Little Ice Age

The climate patterns observed over the past 2000 years are consistent with a millennial cycle of solar activity (fig. B21, ch. 23). The onset of the Little Ice Age cannot be attributed to changes in GHG levels since CO_2 remained unchanged between 1100 and 1500 AD when most of the cooling occurred. Nor can volcanic eruptions explain the Little Ice Age since there were no notable volcanic eruptions for three hundred years, from 1460 to 1765. Without acknowledging the major impact of low solar activity, the causes of the Little Ice Age remain unexplained.

The evidence that highlights this problem of explainability comes from the application of causal identification techniques within systems theory. These techniques allow a comparison between forced identification, using forcings specified by the IPCC, and free identification, where no specific forcings are assumed. This analysis finds that solar activity contributes significantly to the explanation of both the Medieval Warm Period and the Little Ice Age. Consequently, the IPCC hypothesis of low climate sensitivity to solar activity is refuted.[293]

Four reasons to look for a better climate change hypothesis

In the last four chapters, we have thoroughly examined the main flaws of the Enhanced CO_2 Effect hypothesis of climate change, which ultimately make it an unsatisfactory explanation.

1. The hypothesis overemphasizes changes in CO_2 as the driving force behind climate change and thus fails to account for numerous instances of past climate variations that occurred independently of CO_2 fluctuations. This limitation severely reduces its explanatory power and limits it to effectively explaining only the most recent warming trend observed since 1976.

2. The Enhanced CO_2 Effect hypothesis also contributes to the confusion regarding internal climate variability by neglecting its importance and presenting it as a random-walk stochastic process. However, this approach fails to acknowledge the presence of multi-decadal trends, known as climate regimes, and the synchronized phase changes across hemispheres. These observations suggest that regional modes of variability are rooted in an underlying global driver. Furthermore, the hypothesis conveniently ignores the 1997 climate shift, which contradicts its premises while selectively attributing different explanations to some of its effects.

---

[292] Vinós, J., 2022. Climate of the Past, Present and Future: A scientific debate. Critical Science Press. pp.67–88.

[293] de Larminat, P., 2016. Annu. Rev. Control, 42, pp.114–125. doi.org/10.1016/j.arcontrol.2016.09.018

3. Another notable shortcoming of the Enhanced CO_2 Effect hypothesis is that it ignores the critical role of changes in poleward heat transport in explaining Arctic warming and influencing global-scale multidecadal oceanic oscillations. These changes in heat transport have altered climate trends and contributed to a pronounced warming trend in the Arctic, similar to what occurred in the 1920s. While the hypothesis attributes Arctic warming solely to Arctic amplification driven by positive feedback factors, compelling evidence indicates that the substantial increase in heat transport to the Arctic since 1997, and not the feedback factors alone, is the primary driver of this warming phenomenon.

4. The hypothesis blatantly overlooks the substantial evidence that solar variations exert a significant influence on climate, far beyond what would be expected from the energy involved alone. This influence is not because the climate is overly sensitive to such changes but because solar variability operates through an indirect and nonlinear mechanism within the atmosphere. Although the details of this mechanism are not yet fully understood, extensive evidence, including changes in the Earth's rotation rate, strongly supports its existence.

Regrettably, the IPCC reports clearly favor the Enhanced CO_2 Effect hypothesis of climate change, selectively omitting evidence that contradicts it while inappropriately downplaying doubts and uncertainties about its key claims.

In summary

The assessment of the solar influence on climate falls short in the IPCC reports because important indirect effects are inadequately considered and important paleoclimatic evidence for a strong solar influence is ignored. Unfortunately, climate models are of little help in understanding these indirect mechanisms due to limited knowledge of the associated pathways and deficiencies in the representation of the stratosphere. Furthermore, the Enhanced CO_2 Effect hypothesis struggles to account for the multitude of climate changes that have occurred independently of substantial CO_2 variations. It fails to acknowledge the existence of distinct climate regimes and shifts that influence temperature trends, which are closely linked to changes in poleward heat transport, and define the phases of internal modes of climate variability. In particular, the hypothesis misrepresents recent Arctic warming as solely due to feedback-driven Arctic amplification rather than recognizing it as a transport-related phenomenon, as supported by substantial evidence.

SECTION 10 KEY ISSUES

By making the climate overly sensitive to warming or cooling from any cause through a large feedback response, the Enhanced CO_2 Effect hypothesis succeeds in making changes in CO_2 the main driver of climate change despite its small direct effect without feedback. This requires that natural causes have a small effect on temperatures; otherwise, the feedback response to them would also be very large. As a result, this hypothesis cannot explain many past climate changes that occurred without significant changes in CO2 levels.

The Enhanced CO_2 Effect hypothesis treats internal climate variability as the result of random stochastic processes, failing to acknowledge the existence of multidecadal climate regimes with synchronized phase changes across hemispheres. It ignores the 1997 climate shift, which contradicts its premises, and misattributes its effects to other explanations.

This hypothesis ignores the critical role of changes in global atmospheric circulation and poleward heat transport in explaining Arctic warming and influencing global-scale multidecadal oceanic oscillations. It dismisses the role of changes in heat transport as a cause of global climate change on the basis of unproven assumptions.

This hypothesis blatantly overlooks the substantial evidence that solar variations exert a significant influence on climate, far beyond what would be expected from the energy involved alone. This influence is not because the climate is overly sensitive to such changes but because solar variability operates through an indirect and non-linear mechanism within the atmosphere.

The prevailing hypothesis does not adequately explain a fundamental property of climate, poleward heat transport. There is a need for new hypotheses that do.

SECTION 11. THE WINTER GATEKEEPER HYPOTHESIS

CHAPTER 39
A WIND-WALLED ICE REALM

Each winter, the polar vortex forms in the atmosphere around the polar regions, creating an extremely cold region where heat must be transported from lower latitudes. Inside the vortex, the atmosphere becomes highly transparent to infrared radiation. The winds that form the polar vortex are exceptionally strong and act as a barrier that severely restricts heat transport. The strength of the vortex determines the amount of heat transported to the Arctic during winter and, thus, the amount of heat lost from the climate system through outgoing longwave radiation. Scientific evidence shows that between 1976 and 1997, the polar vortex was stronger, characterized by faster surrounding winds, resulting in winter cooling in the Arctic. Since 1997, however, the vortex has weakened, leading to accelerated winter warming in the Arctic relative to the rest of the planet.

The Arctic in winter

As winter approaches in the Arctic, daylight hours decrease rapidly, causing temperatures to drop. The colder air causes the moisture in the atmosphere to condense. Due to the Earth's rotation, winds near the poles tend to circulate in a cyclonic pattern (counterclockwise in the Northern Hemisphere). This circulation is driven by the transport of angular momentum through the atmosphere (box 6, ch. 9). As a result, winds blow in the same direction as the Earth's rotation when they are closer to the pole. In late fall, the temperature contrast between the tropics and the Arctic intensifies, accelerating the westerly winds (from west to east). This process leads to the formation of the polar vortex.

Polar vortices are a natural phenomenon on all rotating planets with an atmosphere. On Earth, two polar vortices occur in winter, one in the troposphere and one in the stratosphere, as explained in box 7 (ch. 11). The tropospheric vortex is influenced by changes in the stratospheric vortex, indicating their coupling.

During the transition from fall to winter, the Arctic experiences a progressive cooling. As a result, the surface freezes, releasing latent heat stored in the water since the previous spring thaw. However, this heat is now lost to space through a process called radiative cooling. Infrared emissions continue in the polar darkness, aided by the dryness of the atmosphere.

Due to the lack of water vapor in the air, the polar night sky remains mostly clear throughout the winter. Clouds and snowfall typically occur when storms from lower latitudes reach the polar regions. However, clear skies are quickly restored by two key factors: thermal inversions, which cause the surface to cool more than the air above it, and strong radiative cooling from the cloud tops.

In winter, the polar regions resemble a realm from another planet. The atmosphere is uniquely transparent to infrared radiation, unlike any other place on Earth in the past 540 million years. Understanding the extraordinary conditions in these regions is critical to understanding climate change on our planet. In winter, the Arctic receives no sunlight and the atmosphere contains minimal amounts of water vapor, resulting in a greatly diminished greenhouse effect. As a result, there is a pronounced radiative cooling that causes heat to dissipate. The main source of heat is at lower latitudes, but to be transported, it must cross the polar vortex wind wall.

A wind barrier limits heat transport

Figure 59, taken from a recent study of the correlation between solar activity and atmospheric circulation via the stratospheric polar vortex, highlights two related aspects of this phenomenon.[294] Figure 59a shows the robust winds that form the walls of the vortex. These winds can reach sustained speeds of 160 km/h (100 mph), creating a formidable barrier that restricts the poleward circulation responsible for transporting heat and moisture. Negative speeds indicate easterly winds.

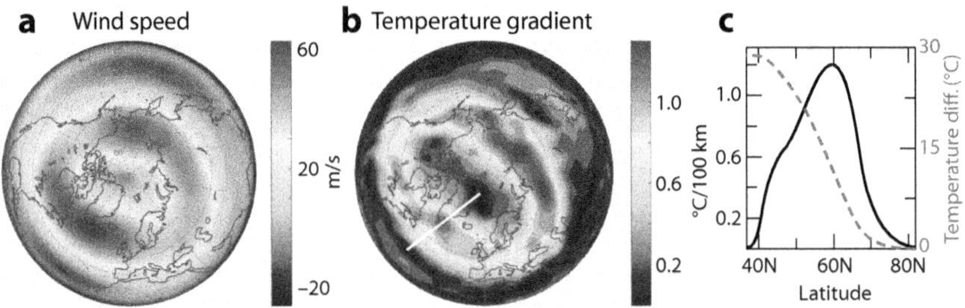

Figure 59. The polar vortex. a) Stratospheric wind speed in January 2005. Positive wind speeds correspond to westerly winds (west to east). b) Temperature gradient in the stratosphere in January 2005. c) Temperature gradient (black line) and temperature difference (dashed gray line) corresponding to the white transect in b).

The barrier effect is evident in Figure 59b, where the temperature gradient is higher than anywhere else in the hemisphere. This gradient abruptly separates the extremely cold air inside the vortex from the much warmer air outside. The wind wall of the vortex acts as a formidable barrier, blocking the entry of warmer air into the polar regions.

Figure 59c graphs the values corresponding to the white transect shown in figure 59b. The graph shows the prominent temperature gradient (black line) that peaks at 60°N. This temperature gradient maintains a 30°C (50 °F) temperature difference between the two sides of the vortex (dashed gray line).

The importance of the polar vortex becomes clear when we consider the consequences of its weakening. In such cases, warm air from lower latitudes enters the Arctic region and rises above the cold polar air. This displacement results in the expulsion of substantial masses of cold air from the Arctic into the midlatitudes of the Northern Hemisphere. As a result, these regions experience unusually cold winter conditions. This phenomenon has become increasingly common during the first two decades of the 21st century.

But to fully understand climate change, we should also look at the polar vortex from the opposite perspective – how its changes affect the conditions inside the vortex. During winter, the Arctic lacks a heat source and relies on heat transported through the atmosphere from outside the polar vortex. This is

---

[294] Reproduced from Veretenenko, S., 2022. Atmosphere, 13 (7), p.1132. doi.org/10.3390/atmos13071132

primarily due to the limited contribution of the ocean to the Arctic's winter energy budget (ch. 10 & 16). Once most of the ocean is covered by sea ice, its ability to transfer heat to the atmosphere is greatly reduced. Because the Arctic receives more heat from the midlatitudes than the Antarctic, it becomes the region on our planet with the highest net energy loss. Throughout this book, we have referred to it as the largest heat sink to space.

The Arctic warms more in winters with a weakened polar vortex due to increased heat inflow. As a result, it emits more infrared energy to space, leading to a greater loss of energy from the climate system. This effect is unavoidable and cannot be compensated for elsewhere because the greenhouse effect in the Arctic is considerably weaker in winter than in other regions. Since the weaker greenhouse effect increases the effectiveness of infrared radiative cooling, the transport of more heat into the Arctic in winter inevitably leads to a greater reduction in the energy content of the climate system.

The transport of heat into the Arctic during winter plays a crucial role in the dynamics of climate change, with the polar vortex acting as a formidable barrier to this heat transfer. It's important to note, however, that the strength of this wind wall varies from winter to winter, much like a gate that can be more open or closed.

Stratospheric and tropospheric heat transport

During winter, heat is transported to the Arctic through the stratosphere and the troposphere. Chapters 13-16 provide a comprehensive overview of the poleward transport in both layers. In particular, the stratosphere contributes 20% of the atmospheric poleward heat transport above 70°N during winter.[295] However, most of this heat is lost as outgoing longwave radiation. Similarly, a substantial fraction of the heat transported by the troposphere is lost by the same mechanism due to the negative energy balance of the Arctic in winter.

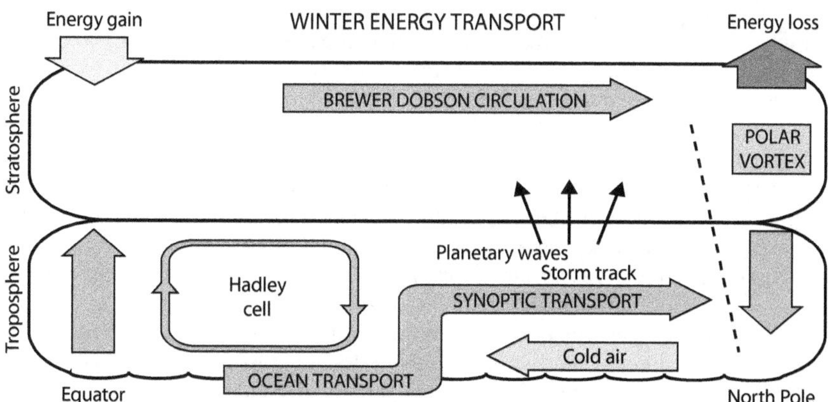

Figure 60. Schematic of meridional transport in winter (I). The largest net energy gain of the system occurs in the tropics, while the largest net energy loss occurs in the polar regions in winter. The medium gray arrows represent the heat transport between these two zones.

---

[295] Cardinale, C.J., et al., 2021. J. Clim. 34 (11), pp.4261–4278.
    doi.org/10.1175/JCLI-D-20-0722.1

Figure 60 shows how heat is transported to the Arctic in winter through the two layers of the atmosphere. While these transports in the stratosphere and troposphere are influenced differently by various factors, they are not completely independent. Upward coupling occurs at the tropical tropopause, where most of the stratospheric air comes from the upward motion of the Brewer-Dobson circulation. In addition, planetary waves (box 10, ch. 14) play a role in driving the Brewer-Dobson circulation, weakening the polar vortex and contributing to additional upward coupling. On the other hand, changes in the polar vortex lead to downward coupling.

Much of the heat transported by the tropical oceans is transferred to the atmosphere beyond the location of the Hadley cell. From this latitude, large-scale (synoptic) meteorological systems play a crucial role in poleward heat transport through storm tracks.

Meridional heat transport is linked to vortex strength

Understanding the complex process of how heat moves from the tropics to the poles is a challenging task. It involves many variables, which explains why our scientific understanding of this vital aspect of the climate system is so poor. One thing is clear, however: for heat to reach the Arctic in winter, most of it must cross the barrier created by the vortex.

The strength of the vortex varies from winter to winter. In some winters, the zonal winds responsible for maintaining the vortex weaken, causing the jet stream to meander. This meandering allows a greater influx of warm air into the Arctic, resulting in the release of more cold air. The result is harsh winter conditions across the northern midlatitude continents. However, like other aspects of heat transport, vortex strength shows long-term trends spanning decades. These trends in vortex strength play a crucial role in shaping the evolution of Arctic winter temperatures over time.

Figure 61a provides insight into the interannual winter vortex strength. The graph shows a clear biannual pattern change, as pointed out by the author of the study.[296] Notably, it also shows a period between 1976 and 1997 when the vortex had a higher average strength, indicated by the dotted lines on the graph. Consequently, winters with weak vortices became less frequent during this period. Since 1997, however, there has been an average weakening of the vortex, resulting in a resurgence of weak vortex winters. This phenomenon sheds light on the increase in the frequency of cold winters, which has puzzled climate scientists because climate models predict the opposite.[297]

The polar vortex was stronger during the 1976-1997 climate regime (fig. 61, gray area). This was attributed to faster zonal winds, indicating less atmospheric wave activity. In particular, between 1976 and 1997, the zonal wind speed remained consistently above average for most of the winters. This trend is illustrated by the cumulative value of the zonal wind speed anomaly, which continuously increased during these years. However, a significant shift oc-

---

[296] Data from NCEP reanalysis in Christiansen, B., 2010. J. Clim. 23 (14), pp.3953–3966. doi.org/10.1175/2010JCLI3495.1

[297] Cohen, J., et al., 2020. Nat. Clim. Change, 10 (1), pp.20–29. doi.org/10.1038/s41558-019-0662-y

curred after 1997, when the zonal wind speed was predominantly below average for most of the winters (fig. 61b).[298]

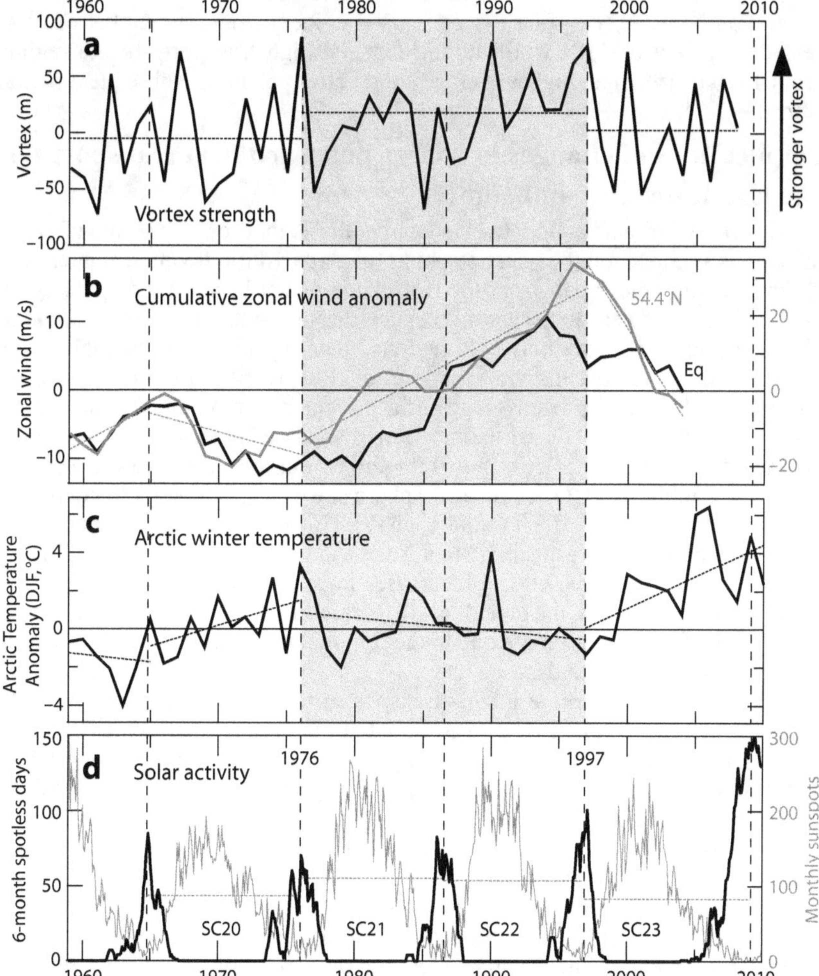

Figure 61. Polar vortex, zonal wind, Arctic temperature, and solar cycle. The vertical dashed lines correspond to solar cycle minimums. The gray area corresponds to the 1976-1997 low transport climate regime. The dotted lines correspond to the linear trends, or mean values if horizontal, of the different periods considered.

When the polar vortex is stronger due to faster winds, it leads to a cooling of the Arctic in winter. This occurs because the intensified vortex creates a greater barrier for heat to penetrate the Arctic region. As a result, the surface temperature anomaly in the 80-90°N region decreased during the 1976-1997 period, which was characterized by the strongest vortex. However, a significant shift

---

[298] Lu, H., et al., 2008. J. Geophys. Res. Atmos. 113, D10114.
    doi.org/10.1029/2007JD009647

occurred after 1997 when the vortex weakened, resulting in a sharp increase in the winter surface temperature anomaly (fig. 61c).[299]

The years 1976 and 1997 showed minimal solar activity, which served as a separation between two solar cycles. This is evident from the number of spotless days (fig. 61d). It's worth noting that the high transport climate regime from 1976 to 1997 had higher average solar activity than the subsequent period after 1997.[300]

Implications of changes in winter poleward heat transport for the Earth's energy imbalance

This book presents a climate change hypothesis that focuses on the effect of variations in the amount of heat reaching the Arctic during winter. Winter conditions set the poles apart from the rest of the planet because they provide an easier means for heat to escape into space through outgoing longwave radiation. Despite being extremely cold regions, they experience substantial energy loss because infrared emission is proportional to temperature on the absolute (Kelvin) scale. The Earth's average surface temperature on the Kelvin scale is 287.65 K (14.5°C, 58 °F), while the Arctic has an average winter temperature of 243.15 K (–30°C, –22 °F), only 15% lower. Some scientists are concerned about the doubling of CO_2 in the atmosphere since CO_2 is thought to contribute 19% to the greenhouse effect (table 1, ch. 7). However, the direct effect of this increase alone would result in a small increase in the total greenhouse effect, since the effect of increased CO_2 decreases logarithmically with its concentration. On the other hand, the greenhouse effect in the Arctic during winter is less than half of the global average due to the minimal presence of water vapor and, consequently, fewer clouds.

The increase or decrease in heat transported to the Arctic during winter directly affects the amount of energy radiated outward by the climate system. When more heat is transported, more energy is lost. Conversely, when less heat is transported, more energy is conserved. This phenomenon forms the basis of this new hypothesis, known as the "Winter Gatekeeper hypothesis." The hypothesis takes into account the variability of heat transport to the polar regions. Most of this transport must pass through a barrier, the polar vortex. Several "gatekeepers" influence the amount of heat that passes through this barrier, often working in opposite directions. The hypothesis was initially developed from evidence that changes in the amount of heat transported to the Arctic were involved in some unexplained features of climate change. It soon became clear that this pathway was shared by several causal factors, most notably changes in solar activity.

Because the transport of heat from the tropics to the poles is one of the most complex and least understood aspects of Earth's climate, it will take several chapters to explain the hypothesis in detail. The scientific information needed to understand its complexity has been provided in the previous chapters.

A major shortcoming of current climate change theories is that they overlook the important heterogeneity of the greenhouse effect on our planet. This

---

[299] Data from the Danish Meteorological Institute.
    ocean.dmi.dk/arctic/meant80n_anomaly.uk.php
[300] Data from SILSO. www.sidc.be/SILSO/home

variability results from changes in the concentration of the primary greenhouse gas, water vapor, which ranges from zero to three percent in the lower troposphere between the winter polar regions and the tropics. Because of this substantial variability in the greenhouse effect, the transport of varying amounts of heat from the tropics to the poles does not maintain climate neutrality. Instead, it serves as an overlooked driver of climate change by altering the radiative flux at the top of the atmosphere. Consequently, changes in heat transport lead to changes in the Earth's energy imbalance, ultimately leading to climate change.

How important is this forcing effect? The paleoclimatic evidence reviewed in chapters 44 and 45 suggests that it is very important and probably the main driver of climate change. To use a simple analogy, increasing CO_2 levels could be like installing double-glazed windows in a house, while changing heat transport to the polar regions could be like having a window open or closed all the time in winter. Before replacing the gas heater in the house with an electric heat pump, it would be a good idea to check that window.

In summary

During the winter, the polar regions exhibit unique characteristics, particularly with respect to their weak greenhouse effect. Heat is transported from lower latitudes to these regions, where it is efficiently radiated into space. However, the formation of the polar vortex acts as a significant barrier, limiting the heat loss from the climate system. As a result, the polar vortex plays a critical role in regulating climate change. Scientific evidence suggests that different climate regimes, such as the period from 1976 to 1997, are characterized by different strengths of the polar vortex, corresponding to the level of heat transport to the poles that defines each regime.

The Winter Gatekeeper hypothesis of climate change is based on the understanding that the variable winter transport of heat to the polar regions leads to changes in radiative flux at the top of the atmosphere. In support of this hypothesis, paleoclimatic evidence indicates that it is the primary driver of climate change across all time scales.

CHAPTER 40
MULTIPLE GATEKEEPERS

Several factors influence the winter heat transport to the Arctic through the polar vortex. These include volcanic eruptions, the Quasi-Biennial Oscillation, the El Niño-Southern Oscillation, multi-decadal ocean oscillations, and solar activity. These factors act as "gatekeepers" that regulate the passage of heat through the polar vortex. Changes in Arctic winter transport affect the amount of energy the Earth emits to space. Therefore, these gatekeepers are climate forcings that can change the climate. However, they interact with each other, affecting transport and modifying their effects, sometimes reinforcing or opposing each other and heat transport. It is important to note that the climatic impact of each gatekeeper varies over different time scales. Volcanic eruptions, the Quasi-Biennial Oscillation, and the El Niño-Southern Oscillation have short-term effects. In contrast, multidecadal ocean oscillations establish climate regimes that last for decades, and changes in solar activity can last for decades or even centuries.

Multiple modulations of poleward heat transport

The global circulation of the ocean and atmosphere is highly complex, with multiple modes of variability, oscillations, teleconnections, and modulations. At its core, however, this complexity can be traced back to a simple underlying cause — the variable transport of energy from its point of entry to its point of exit. This fundamental climate variable is critical in determining the latitudinal temperature gradient and, ultimately, whether the planet is experiencing a glacial, interglacial, or hothouse period.

To understand the impact of changes in heat transport to the Arctic on global climate, it is essential to study the causes of meridional transport variability. The atmosphere plays a key role in this process, transporting heat through two interconnected pathways: the stratosphere and the troposphere. Over ocean basins, both the atmosphere and the ocean contribute to this transport. These two atmospheric pathways are closely coupled, with the strongest coupling occurring in winter. This is when temperature contrasts and atmospheric wave generation in the troposphere are most intense, leading to the development of the polar vortex and pronounced temperature gradients in the stratosphere.

Heat transport in the stratosphere is influenced by several factors that affect temperature gradients across latitudes and altitudes. These factors include ozone, solar activity, volcanic aerosols, and the Quasi-Biennial Oscillation. They all affect the strength of the zonal wind circulation and play a role in determining the degree of transmission of the planetary waves that drive heat transport in the stratosphere.

The El Niño-Southern Oscillation plays a role in tropospheric transport and is influenced by its conditions. However, it also acts as a regulator of stratospheric transport by influencing the strength of the Brewer-Dobson circulation.[301] This involvement in stratospheric dynamics contributes to the coupling between stratospheric and tropospheric transport. In addition, scientists have observed interactions among solar activity, the Quasi-Biennial Oscillation, and the El Niño-Southern Oscillation, which collectively influence the stratospheric conditions that determine meridional transport.[302]

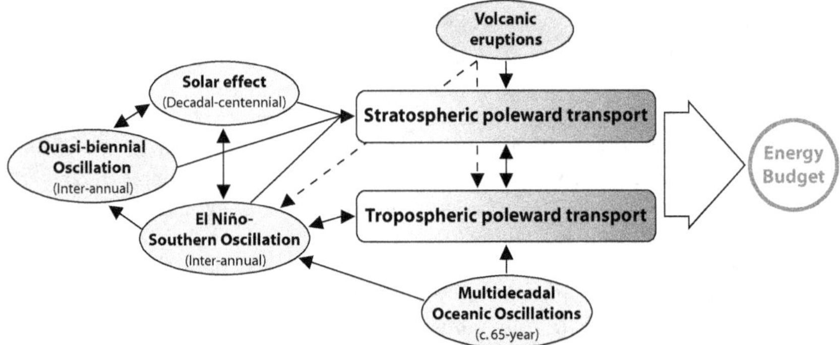

Figure 62. Poleward heat transport diagram. The ovals are the gatekeeper factors that modulate heat transport and determine the amount of heat that reaches the Arctic in winter, affecting the planetary energy budget. Solid arrows indicate known interactions and dashed arrows indicate indirect interactions.

As discussed in chapter 25, volcanic eruptions have several dynamic effects, primarily by directly affecting poleward heat transport in the stratosphere and indirectly in the troposphere. Thus, strong tropical volcanic eruptions can lead to winter warming in the Northern Hemisphere. This occurs by strengthening the polar vortex while also inducing El Niño conditions in the Pacific.

Multidecadal ocean oscillations are the primary modes of internal variability. These oscillations undergo coordinated phase changes (box 16, ch. 19), establishing distinct climate regimes (ch. 31-34). These regimes are characterized by specific transport intensities and mean vortex strengths, as discussed in the previous chapter (fig. 61, ch. 39). Their impact on climate is substantial, as poleward tropospheric heat transport through the coupled atmosphere-ocean system accounts for most of the winter heat transport to the Arctic. Multidecadal ocean oscillations were responsible for warming in the early 20th century and cooling in the mid-20th century. They have also contributed to warming in the late 20th century and continue to influence the climate regime in the early 21st century.

---

[301] Domeisen, D.I., et al., 2019. Rev. Geophys. 57 (1), pp.5–47. doi.org/10.1029/2018RG000596

[302] Labitzke, K., 1987. Geophys. Res. Lett. 14 (5), pp.535–537. doi.org/10.1029/GL014i005p00535 Calvo, N. & Marsh, D.R., 2011. J. Geophys. Res. Atmos. 116, D23112. doi.org/10.1029/2010JD015226 Salby, M. & Callaghan, P., 2000. J. Clim. 13 (2), pp.328–338. doi.org/10.1175/1520-0442(2000)013<0328:CBTSCA>2.0.CO;2 Taguchi, M., 2010. J. Geophys. Res. Atmos. 115, D18120. doi.org/10.1029/2010JD014325

Interaction of Arctic transport gatekeepers

Heat transport through the polar vortex is a complex process influenced by many factors, as shown in figure 62. These factors can be viewed as the gatekeepers of the polar vortex, influencing it by regulating the generation and propagation of planetary waves. The propagation of these waves depends on the zonal wind speed in the stratosphere, which is strongly influenced by latitudinal and vertical temperature gradients, as well as the phase of the Quasi-Biennial Oscillation.

The five vortex gatekeepers are interconnected and influence each other. However, in a given winter season, they can have opposite effects on heat transport. The spatial and temporal integration of their signals is highly complex and will require considerable effort by scientists to unravel. To illustrate their effects across latitudes, we can include them in the transport diagram presented in the previous chapter (fig. 60).

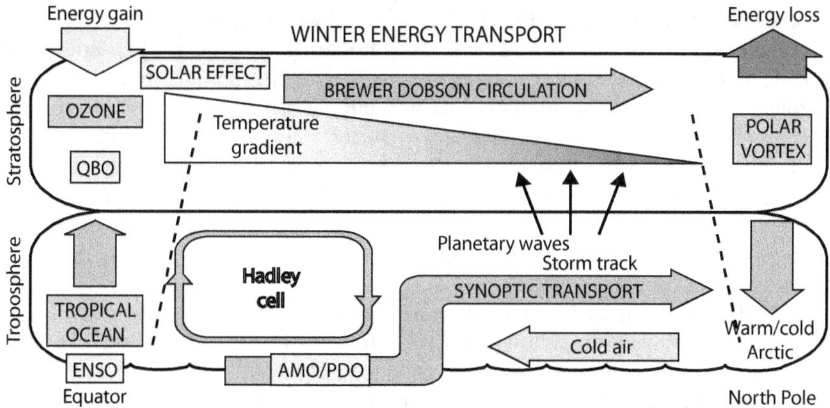

Figure 63. Schematic of meridional transport in winter (II). The ozone layer and the tropical ocean define the main energy entry points in terms of heat transport, and the region bounded by the polar vortex is the main net exit point. The solar effect resulting from the ozone-UV interaction, the Quasi-Biennial Oscillation (QBO), the El Niño-Southern Oscillation (ENSO), and the multidecadal ocean oscillations, represented by the Atlantic Multidecadal Oscillation (AMO) and the Pacific Decadal Oscillation (PDO), are the main modulators of heat transport that determine the amount of heat reaching the Arctic and act as gatekeepers of the polar vortex.

Climate energetics highlights three critical locations for heat transport. First, the ozone layer of the tropical stratosphere receives most of the UV energy. Second, the tropical ocean receives most of the incoming solar energy. Finally, in winter, the region inside the polar vortex serves as the primary net energy sink, as discussed in the previous chapter.

Heat from the tropical ocean is transported to the poles in three ways. First, convection transports some of the heat into the stratosphere, forming the upward branch of the Brewer-Dobson circulation. Second, the Hadley circulation transports another portion through the troposphere. Finally, the ocean itself contributes to the transport. The state of the El Niño-Southern Oscillation modulates the distribution of this energy. During La Niña, oceanic transport is favored, while neutral conditions increase atmospheric transport. In contrast, El

Niño directs a significant amount of oceanic heat into both the stratosphere and troposphere. Scientists have extensively studied the gatekeeping effect of El Niño on the polar vortex, which is evident in the polar stratospheric temperatures shown in figure B23 (ch. 29).[303]

Although the ozone layer absorbs slightly more than 1% of the total solar energy (the UV part), it accounts for 5% of the atmospheric energy absorption. Notably, its absorption varies thirty times more with solar activity than in the visible spectrum. This variability contributes to significant radiative and dynamical changes in the stratosphere over the solar cycle. Since the stratosphere has an average density of 25 times less than the troposphere, the effect of absorbed solar energy on stratospheric temperature is enormous. Without ozone, the stratosphere would be 50°C (90 °F) colder and the tropopause would not exist. Interestingly, the ozone layer is a unique characteristic of the Earth, as no other known planet has such a feature.

The presence of ozone in the stratosphere allows the absorption of solar energy, which in turn leads to the establishment of a temperature gradient. This gradient depends on factors such as the amount of UV energy, the amount of ozone, and its distribution. Changes in solar activity trigger a response from the ozone, which affects this temperature gradient and gives rise to a solar effect in the stratosphere — the solar gatekeeper. During winter, the zonal wind circulation, characterized by westerly winds, responds to variations in this temperature gradient caused by changes in solar activity. This affects the propagation of planetary waves into the stratosphere (fig. 67, ch. 42). These planetary waves play a key role in driving the Brewer-Dobson circulation and exerting a weakening effect on the vortex.

The Quasi-Biennial Oscillation exerts an important phase-dependent influence on the polar vortex, known as the Holton-Tan effect (box 11, ch. 14). During climate regimes characterized by enhanced poleward heat transport, such as the period between 1945 and 1975 and since 1998, the easterly phase of the Quasi-Biennial Oscillation weakens the polar vortex. However, in climate regimes characterized by reduced poleward heat transport, such as the period between 1976 and 1997, the polar vortex remains strong in most winters, even during the easterly phase of the oscillation (fig. 64; see also fig. 61a, ch. 39).

Tropical volcanic eruptions that release large amounts of sulfate into the stratosphere cause significant stratospheric warming and ozone depletion. These changes contribute to a stronger polar vortex in the following winter. In addition, they indirectly affect tropospheric transport because volcanic aerosols reduce solar radiation more in the tropics than at mid- and high latitudes. This change in the latitudinal temperature gradient reduces transport. The result is a warmer winter in the midlatitudes of the Northern Hemisphere after the eruption, despite reduced solar radiation.

---

[303] Garfinkel, C.I. & Hartmann, D.L., 2007. J. Geophys. Res. Atmos. 112, D19112. doi.org/10.1029/2007JD008481

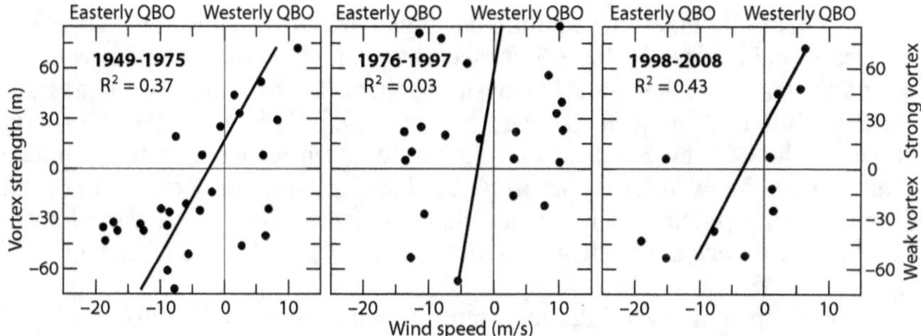

Figure 64. Effect of the Quasi-Biennial Oscillation on vortex strength for different climate regimes. The polar vortex is weakest during the easterly phase of the Quasi-Biennial Oscillation, except during a low transport regime (1975-1997, middle panel), when the vortex remains strong during most winters.[304]

Integrating these complexities into an explanation of climate change

The many gatekeepers that influence the strength of the polar vortex and the heat transported to the Arctic add complexity to our understanding of climate change. It is important to recognize that Arctic warming is not solely a consequence of human-induced climate change, but rather a mechanism by which the planet increases its energy loss. Examining the influence of different gatekeepers over different timescales is necessary to formulate a comprehensive climate change hypothesis that encompasses this complexity.

Volcanic eruptions with significant climatic effects are rare. There were only three such eruptions in the 20th century, and none have been observed so far in the 21st century if we exclude the unusual Hunga Tonga submarine eruption of 2022. Our understanding from these past eruptions is that their noticeable effects tend to fade after the first few years. Historical evidence, including the 1815 eruption of Mt. Tambora, supports this interpretation (ch. 24). Thus, while volcanic forcing can be intense, it should not be overemphasized because of its transient nature.

The phase of the Quasi-Biennial Oscillation plays a crucial role in determining the strength of the vortex during a given winter. An analysis of the data for the period 1949-2008 revealed that 75% of the winters with weak vortices occurred when the oscillation was in its easterly phase (fig. 64). This finding highlights the significant influence of the Quasi-Biennial Oscillation on the global stratospheric circulation, taking into account the various factors that influence vortex strength. However, the effect is reversed when the phase changes one to two years later, with 72% of the winters during the westerly phase of the oscillation showing a strong vortex (fig. 64). Thus, the Quasi-Biennial Oscillation alone does not have a direct effect on climate. However, it does increase the sensitivity of the vortex to other gatekeepers, such as solar activity, during its easterly phase (fig. B23, ch. 29).

---

[304] QBO data from NOAA. Vortex strength data from NCEP reanalysis in Christiansen, B., 2010. J. Clim. 23 (14), pp.3953–3966. doi.org/10.1175/2010JCLI3495.1

The El Niño-Southern Oscillation has a global climate impact, occurs every 2-7 years, and can persist for 2-3 consecutive winters. During El Niño winters, the polar vortex weakens and polar stratospheric temperatures rise (fig. B23, ch. 29), while La Niña winters have the opposite effect.[305] While the short-term effects of the El Niño-Southern Oscillation do not influence long-term climate change, El Niño events have an irregular distribution, with periods of several decades during which their frequency changes significantly. These variations in El Niño occurrence are associated with the phase of the Pacific Decadal Oscillation.

Over the span of several decades, multidecadal ocean oscillations (modes of variability) play a crucial role in establishing different climate regimes and triggering shifts between them. These oscillations primarily affect the troposphere, which is responsible for transporting most of the heat to the poles. As a result, the climate regimes determined by these globally coordinated oscillations dominate over other factors.

During the period from 1976 to 1997, a low transport regime occurred, leading to a weakening of the Holton-Tan effect. As a result, the influence of the Quasi-Biennial Oscillation on the polar vortex decreased (fig. 64, middle panel). In addition, the low-transport conditions led to increased heat accumulation in the subsurface of the tropical ocean, making it easier for the El Niño-Southern Oscillation to intensify, leading to an increase in the frequency of El Niño events.

In the early 20th century, a climate regime characterized by low transport explains the warming trend of the planet since the 1910s. This warming occurred despite low solar activity. Later, when solar activity increased in the 1930s, it further contributed to the pronounced warming of that period. Between 1945 and 1975, a climate regime characterized by high transport took place, which explains the cooling period in the middle of the 20th century. This cooling period occurred despite the presence of high solar activity. The cooling would probably have been even more pronounced if solar activity had been low during this period.

Although not recognized as a natural climate forcing, multidecadal ocean oscillations play an important role in shaping our climate. These oscillations occur over extended periods, and when they lead to increased heat transport toward the poles, they result in greater energy loss from the climate system. As a result, the planet experiences either cooling or reduced warming. Conversely, periods of low transport have the opposite effect, causing increased warming.

The impact of these multidecadal ocean oscillations far exceeds that of recognized climate forcings such as volcanic eruptions. Although the cause of these oscillations remains unknown, it is important not to assume that the alternating phases imply the absence of long-term effects. Several studies have shown that the Atlantic Multidecadal Oscillation, for example, is not stationary. Since 1850, it has shown a strong amplification that coincides with the warming trend observed during the industrial era.[306]

---

[305] Garfinkel, C.I. & Hartmann, D.L., 2007. J. Geophys. Res. Atmos. 112, D19112. doi.org/10.1029/2007JD008481

[306] Moore, G.W.K., et al., 2017. Sci. Rep. 7 (1), p.40861. doi.org/10.1038/srep40861

In the next chapter, we discuss the role of solar variability as a climate forcing within the Winter Gatekeeper hypothesis.

In summary

Variations in winter heat transport to the Arctic lead to climate change. Changes in Arctic heat transport are linked to changes in the strength of the polar vortex. During any given winter, several factors can contribute to a weaker vortex, resulting in increased Arctic warming and more severe winter conditions for the midlatitude continents of the Northern Hemisphere. These factors include the easterly phase of the Quasi-Biennial Oscillation, an El Niño event, or low solar activity. Conversely, the opposite conditions or a tropical volcanic eruption can strengthen the vortex and result in milder winters. However, observing noticeable climate changes requires a sustained effect over several winters. This can only be achieved by climate regimes induced by multidecadal ocean oscillations or sustained changes in average solar activity.

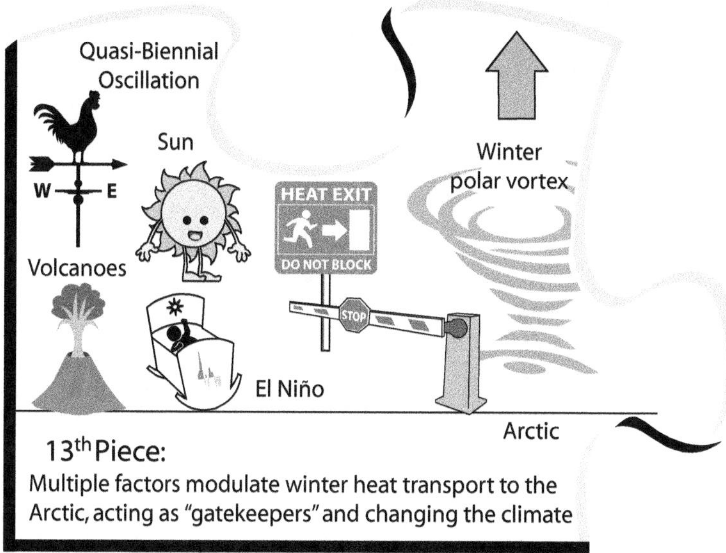

CHAPTER 41
THE SUN AS A CENTENNIAL GATEKEEPER

Instead of making assumptions about how solar activity affects climate, we should rely on decades of research. Scientists have long known that solar activity affects the strength of the vortex, planetary rotation, and winter atmospheric circulation. The mechanism behind these effects was proposed 50 years ago and has been studied extensively by many scientists. Some models have even reproduced it, as recognized by the IPCC.

The Winter Gatekeeper hypothesis links this mechanism to the ability to modify climate by changing the radiative flux at the top of the atmosphere. High solar activity leads to a reduction in poleward heat transport, which in turn reduces winter energy loss in the Arctic. Conversely, low solar activity has the opposite effect. It's important to note that solar activity is not the only factor influencing this pathway. Its effect on climate becomes important on longer timescales, spanning several decades to centuries, when other factors diminish their influence.

The Modern Solar Maximum, the longest period of above-average solar activity in at least 600 years, has contributed to the increase in energy content within the climate system during 20th-century global warming through this mechanism. This contribution is independent of temperature trends and solar activity levels.

It is a mistake to assume how solar activity should affect climate

The IPCC reports, which focus only on small variations in total solar irradiance, suggest that changes in solar activity have a negligible effect on climate. However, these reports overlook a large body of evidence on the indirect effects of solar activity, including the findings discussed in chapters 28-30. This evidence supports the view that solar activity has a substantial influence on the state of the winter atmospheric circulation, as demonstrated by its effect on the Earth's rotation rate.

The effect of solar activity on the polar vortex is not a recent discovery. It was first documented in 1959 and has been the subject of scientific research ever since.[307] However, most climate scientists and the IPCC have largely ignored this research, similar to the case of solar effects on the Earth's rotation. The temperature of the polar stratosphere is affected by solar activity, but the true significance of this gatekeeper effect, as discovered by Karin Labitzke,

---

[307] Palmer, C.E., 1959. J. Geophys. Res. 64 (7), pp.749–764. doi.org/10.1029/JZ064i007p00749 Labitzke, K., 1987. Geophys. Res. Lett. 14 (5), pp.535–537. doi.org/10.1029/GL014i005p00535 Veretenenko, S., 2022. Atmosphere, 13 (7), p.1132. doi.org/10.3390/atmos13071132

becomes clear when the influence of other gatekeepers is considered (fig. B23, ch. 29)).

This is an important explanation for the lack of understanding of the solar influence on climate over the past two centuries. The Sun provides 99.9% of the energy that drives the climate system, leading many to assume that if solar variations affect climate, it would be clearly reflected in temperature patterns. Even NASA agrees with this assumption, as evidenced by their statement:

"One of the 'smoking guns' that tells us the Sun is not causing global warming comes from looking at the amount of solar energy that hits the top of the atmosphere. Since 1978, scientists have been tracking this using sensors on satellites, which tell us that there has been no upward trend in the amount of solar energy reaching our planet."[308]

This statement conceals a widely held but unsupported assumption that if trends in solar activity and temperature don't coincide, then the Sun cannot have a significant effect on climate. It ignores the possibility that decadal temperature trends may be influenced by other factors that mask a significant effect of multidecadal shifts in solar activity, which can substantially alter the magnitude of warming and cooling trends. For example, during periods of sustained above-average solar activity, such as 1935-2000, cooling trends are attenuated while warming trends are enhanced, resulting in long-term warming patterns similar to those observed.

The Sun as a winter gatekeeper

The physical phenomenon responsible for the solar effect on climate was first proposed in 1974 and attributed to changes in the propagation of planetary waves in the atmosphere.[309] Since then, scientists have worked to elucidate the mechanism responsible, which is explained in chapter 29 (fig. 46). This well-established top-down mechanism originates in the stratosphere and is supported by data assimilation reanalysis. The IPCC 5th Assessment Report acknowledges this amplification mechanism, noting that it has been simulated in certain models and can account for small local and seasonal climate anomalies associated with the 11-year solar cycle.[310] However, the report concludes with high confidence that changes in total solar irradiance did not contribute to global warming between 1986 and 2008. However, the justification for this confidence is questionable for the reasons discussed above.

The top-down amplification mechanism is not considered to be a driver of global climate change because it is not thought to significantly alter the energy content of the climate system. In general, a change in the energy imbalance at the top of the atmosphere is required to induce changes in the global climate, with few exceptions.

For the first time, the Winter Gatekeeper hypothesis links the recognized top-down amplification mechanism, the known solar effect on the polar vortex, and climate change. The hypothesis proposes that the dynamic influence of solar activity on the winter atmospheric circulation and the polar vortex leads

---

[308] climate.nasa.gov/faq/14/is-the-sun-causing-global-warming/

[309] Hines, C.O., 1974. J. Atmos. Sci. 31 (2), pp.589–591.
doi.org/10.1175/1520-0469(1974)031<0589:APMFTP>2.0.CO;2

[310] Bindoff, N.L., et al., 2013. Climate Change 2013: The Physical Science Basis. 5th AR IPCC. pp.885–886.

to important changes in heat transport toward the winter pole. During the extended polar night, much of this heat is emitted from the planet as outgoing infrared radiation, inducing the necessary change in the energy imbalance at the top of the atmosphere to affect the global climate.

Figure 65 shows the gatekeeper role of solar activity as proposed by the hypothesis. It does not show changes in zonal wind speed and planetary wave propagation, which are essential mediators in this process. It is important to keep in mind that the outcome of each winter is influenced not only by solar activity but also by other gatekeepers. Moreover, these gatekeepers do not act in isolation; they interact and influence each other.

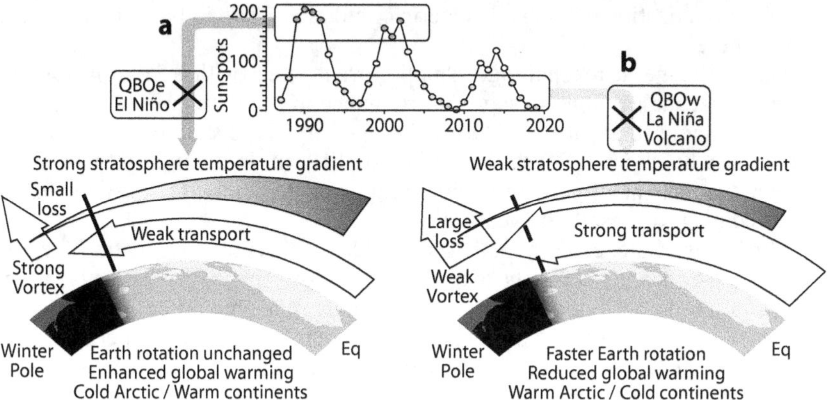

Figure 65. The Sun as winter gatekeeper. Role of solar activity in meridional heat transport and winter energy loss in the Arctic.

During winters of high solar activity (fig. 65a), increased ozone and enhanced ozone heating due to increased UV radiation contribute to a strong temperature gradient in the stratosphere. This enhanced gradient, in turn, strengthens the zonal winds that impede the propagation of planetary waves. As a result, the polar vortex remains strong throughout the winter, reducing heat transport to the pole and decreasing heat loss in this region. However, the easterly phase of the Quasi-Biennial Oscillation and El Niño conditions can counteract this effect of high solar activity. The climatic result of high solar activity is increased global warming and a pattern of cold Arctic and warm continents in winter.

During winters of low solar activity (fig. 65b), the reduced UV radiation leads to a reduced temperature gradient in the stratosphere, resulting in a weakened polar vortex. This, in turn, increases meridional heat transport and intensifies heat loss at the winter pole. However, factors such as the westerly phase of the Quasi-Biennial Oscillation, La Niña conditions, and volcanic aerosol forcing can counteract this effect. The increased meridional transport also contributes to an acceleration of the Earth's rotational speed as zonal winds decrease, leading to a decrease in atmospheric angular momentum. As a climatic consequence, low solar activity is associated with reduced global warming and a winter pattern characterized by warm Arctic and cold continents.

This hypothesis sheds light on several long-standing questions about the potential influence of solar activity on climate.

A duration-dependent solar effect

A major argument against attributing a more important role to the Sun in climate change is the apparent lack of substantial influence of the 11-year solar cycle. Modern climate analyses based on satellite data since 1979, covering nearly four complete solar cycles, have shown that the observed changes, while significant, are relatively modest (fig. 44a, ch. 28).

In striking contrast, paleoclimatic studies provide substantial evidence for the dramatic climate consequences associated with extended periods of low solar activity, as discussed in chapter 27. These studies show that during a prolonged solar minimum of 200 years, there is compelling evidence for a profound reorganization of the atmospheric circulation, especially in the Northern Hemisphere.

The challenge in reconciling a small current effect with a large historical effect is that the assumed changes in irradiance in both scenarios are believed to be extremely small (0.1%). During the 11-year solar cycle, total solar irradiance varies by about 1.1 W/m^2 over the satellite era. In addition, according to the 6th Assessment Report, millennial-scale changes of only 1.5 W/m^2 have occurred over the past 9,000 years.[311]

The Winter Gatekeeper hypothesis provides a plausible explanation for this apparent paradox. It suggests that changes in solar activity indirectly affect global or hemispheric temperature by influencing the amount of energy lost in the Arctic during winter. In winters with low solar activity, more energy is lost, while in winters with high solar activity, the energy loss is reduced. However, it's important to note that all the gatekeepers participate in determining the outcome in any given winter, so the result can be different from what the solar effect induces. During a solar cycle, some years have high solar activity while others have low activity (fig. 65). As a result, the cumulative effect within a single solar cycle is almost imperceptible, as any changes in one direction are largely offset within a few years.

However, when solar activity remains low for long periods of time, over several decades or even centuries, the frequency of Arctic winters with enhanced energy loss increases greatly. While the average effect of other gatekeepers tends to decrease over longer periods, the solar effect accumulates. The planet lacks the means to compensate for the additional energy loss, resulting in a growing cumulative deficit over time. As a result, the overall temperature of the planet decreases.

Regions along the primary energy transport pathways to the poles, particularly the North Atlantic region encompassing Europe and North America, experience the earliest, longest, and most pronounced cooling. Nevertheless, the energy drain affects the entire planet. Although the Arctic region initially warms due to the increased energy influx facilitated by increased transport, it eventually cools along with the rest of the planet. A significant increase in the frequency of cold winters accompanies this cooling trend. In addition, enhanced poleward transport leads to increased moisture transport, resulting in increased snowfall and glacier expansion. These conditions are consistent with observations during the Little Ice Age.

---

[311] Gulev, S.K., et al., 2021. Climate change 2021: The physical science basis. 6th AR IPCC. p. 297. doi.org/10.1017/9781009157896.004

Only when solar activity ceases to be low can the planet stop the continuous energy loss and begin the recovery process. However, it would take a long period of above-average solar activity for the planet to replenish the energy lost during a solar grand minimum.

The impact of the Modern Solar Maximum

The effect of solar activity on climate is directly related to the duration of the unusual solar activity. The IPCC's 6th Assessment Report acknowledges that solar activity in the second half of the 20th century was in the top 10% of solar activity observed over the past 9,000 years. This extended period of above-average solar activity, spanning seven decades, constitutes a solar grand maximum and has been termed the Modern Solar Maximum (fig. 66).[312]

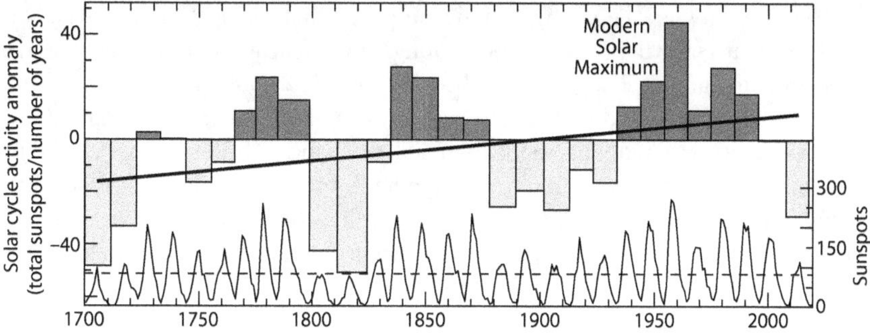

Figure 66. Deviation of the activity of each solar cycle from the mean. The thick line is the linear trend. The horizontal dashed line is the average sunspot number from 1700 to 2020.

This graph of sunspot data provides a clear representation of the activity level of each solar cycle, taking into account its duration. It highlights the remarkable shift in solar activity, showing the transition from the lower 10% during the Maunder Minimum (1645-1715) to the upper 10% between 1935 and 2000.

Throughout the Modern Solar Maximum, there has been a marked increase in the number of years characterized by high solar activity. As a result, the amount of energy lost in the Arctic during winter decreased. This increased energy storage by the planet played an important role in the observed global warming trend during this period. It is absurd to argue that the simultaneous occurrence of the longest period of high solar activity in at least 600 years, along with the period of the most pronounced warming in at least 600 years, is purely coincidental, as the IPCC reports do when they claim a lack of solar contribution to recent warming.

The Winter Gatekeeper hypothesis provides a comprehensive explanation for the role of the Modern Solar Maximum in contributing to recent global warming. It explains why factors such as small energy changes and discrepancies between temperature and solar activity trends are inconsequential. The critical factor is the persistence of above-average solar activity over several

---

[312] Damon, P.E., et al., 1997. Radiocarbon, 40 (1), pp.343–350.
  doi.org/10.1017/S003382220001821X

decades until recently. The effect is most pronounced in the Arctic during winter, leading to an increase in the overall energy content of the climate system. Surface temperature trends over the rest of the planet are influenced by changes in solar activity but also by other factors. As a result, relying solely on surface temperature trends is not a reliable way to measure the solar effect on climate. Instead, a better indicator of the solar influence on climate is the frequency of cold winters in the Northern Hemisphere or the warming observed in the Arctic during winter.

Box 26. A definition of the Winter Gatekeeper hypothesis

The Winter Gatekeeper hypothesis proposes that changes in the heat and moisture that reach the polar regions during winter, particularly the Arctic, play a major role in climate change. This is because the winter polar regions have a very low greenhouse effect due to the lack of atmospheric water vapor, the Earth's main GHG. Combined with the lack of solar radiation, this leads to an effective loss of energy to space through outgoing infrared radiation. Consequently, changes in the transport of heat to the polar regions during winter have an impact on the energy budget of the planet. The Arctic is particularly important in this hypothesis because its weaker polar vortex allows for greater variations in heat transport.

Any factor that affects the atmospheric zonal circulation, the generation and propagation of planetary waves, or the strength of the polar vortex acts as a vortex gatekeeper capable of regulating the transport of heat to the Arctic during winter. These gatekeepers include the Quasi-Biennial Oscillation, El Niño-Southern Oscillation, volcanic eruptions, multi-decadal ocean oscillations (modes of variability), and solar activity.

The effect is small and easily reversible from year to year, affecting short-term weather patterns. Over several decades, however, the climate effect is primarily driven by multidecadal ocean oscillations. These oscillations cause significant changes in temperature trends, leading to climate regimes that undergo abrupt shifts after a few decades. Over periods of a century, solar activity emerges as the dominant force, especially during periods of prolonged and unusual activity lasting several decades. Solar activity has played a major role in some of the most notable climatic events of the Holocene and has contributed significantly to the warming observed in the 20th century since 1935. On a millennial time scale, solar activity plays a critical role in shaping major climate periods that span several centuries, as its long cycles give rise to clusters of large solar minimums, such as those experienced during the Little Ice Age.

Long-term changes in the transport of heat and moisture to the poles, the mechanism behind the Winter Gatekeeper hypothesis, are also the basis for the drastic climate changes the planet has experienced over tens of thousands to millions of years.

Over tens of thousands of years, this meridional transport mechanism contributes to the Milankovitch glacial cycle. It translates changes in the Earth's obliquity into changes in heat transport distributed over different latitudes, facilitating the necessary adjustments in heat and moisture transport required to form or melt ice sheets.

On a timescale of millions of years, the meridional transport mechanism was instrumental in the transition from the equable climate of the early Eocene to the ice age of the late Cenozoic. This change was driven by changes in continental arrangements, oceanic passageways, and mountain ranges that increased the efficiency of poleward heat transport, leading to planetary cooling and an ice age.

The Earth's climate functions as a thermodynamic heat engine, similar to the internal combustion engine in a car (fig. B26). Heat is generated primarily in the engine block, which corresponds to the tropics. While some heat is radiated directly from this engine, two cooling systems, analogous to radiators, are located in the polar regions of both hemispheres. The cooling process is facilitated by air driven by the atmosphere (acting as a cooling fan) and by a cooling fluid in the ocean. To keep the engine temperature within acceptable limits and prevent overheating, a thermostat regulates the cooling system by adjusting the amount of heat directed to the radiators.

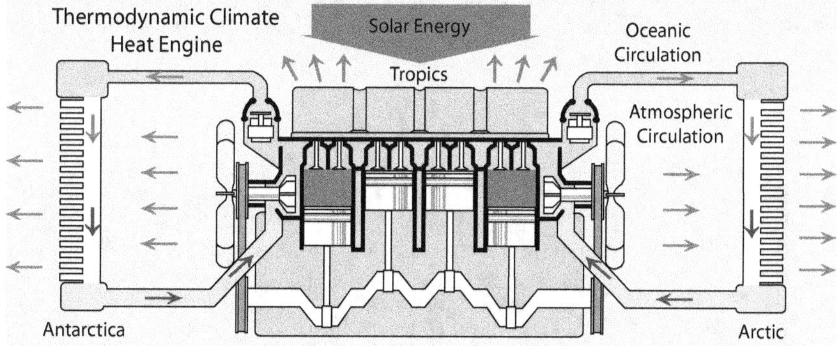

Figure B26. The thermodynamic climatic heat engine has two coolers. The Arctic radiator is more efficient because it receives more heat.

Similarly, the Earth has two cooling mechanisms resulting from its axial tilt. Directing more heat to these mechanisms cools the planet while directing less heat to them warms it. The Arctic cooling mechanism is particularly effective because of the Earth's hemispheric asymmetry. Over the past 50 million years, its increased efficiency has led to the onset of the current Ice Age. However, since the end of the Little Ice Age, and especially between 1976 and 1997, less heat has been directed to the Arctic, leading to further warming. While human emissions of CO_2 have contributed to this warming, the influence of natural winter gatekeepers on Arctic temperature and warming trends remains prominent. Therefore, much of the global warming observed in recent decades can be attributed to natural causes.

In summary

After 50 years of research, scientists have successfully explained how variations in solar activity affect the winter atmospheric circulation and the polar vortex. However, understanding how these effects can cause global climate change is the missing piece of the puzzle. The Winter Gatekeeper hypothesis links changes in poleward heat transport to energy loss in the Arctic during winter. During periods of high solar activity, strong vortex formation occurs, reducing poleward transport and heat loss at the winter pole. Conversely, low solar activity favors the opposite effect. The winter gatekeeper effect of changes in solar activity rises above the background noise of other gatekeepers when solar activity deviates substantially from average levels for several decades. Its impact on climate becomes profound when this unusual activity persists for 50 to 200 years.

According to this hypothesis, the Modern Solar Maximum, which occurred between 1935 and 2000, resulted in reduced energy loss in the Arctic. As a result, cooling trends were reduced and warming trends increased, contributing significantly to the observed long-term warming during the 20th century. The Winter Gatekeeper hypothesis suggests that climate responds to changes in the amount of heat directed to the poles on all time scales, helping to explain phenomena such as the glacial cycle or the transition from a warm world to an ice age during the Cenozoic.

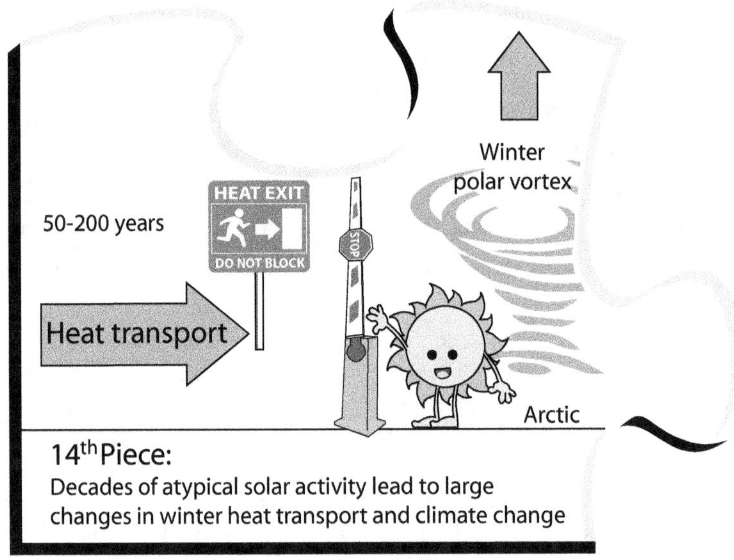

Section 11 Key Issues

In winter, more heat is transported to the polar regions, where it is efficiently radiated to space. The formation of the polar vortex limits this loss and regulates climate change. Different climate regimes have different levels of heat transport, corresponding to different strengths of the polar vortex.

Each winter, several factors contribute to the strength or weakness of the polar vortex: the Quasi-Biennial Oscillation, the El Niño-Southern Oscillation, solar activity, multidecadal oceanic oscillations, and volcanic eruptions. However, a sustained effect over several winters is required to observe noticeable climate changes. This can only be achieved by climate regimes induced by multidecadal ocean oscillations or sustained changes in the mean decadal solar activity.

Scientists have explained how variations in solar activity affect the winter atmospheric circulation and the polar vortex. The Winter Gatekeeper hypothesis explains how they change the climate by altering the amount of energy lost in the Arctic due to changes in poleward heat transport. The effect is cumulative and time-dependent, taking decades to become noticeable. Thus, the 70-year Modern Solar Maximum contributed significantly to 20th-century global warming.

SECTION 12. THE EVIDENCE

CHAPTER 42
SOLAR GATEKEEPING EVIDENCE

Solar activity has a direct effect on the polar vortex, which in turn affects heat transport to the Arctic during winter. Several lines of evidence support this connection. Solar activity plays a role in determining the temperature of the polar stratosphere and in regulating the atmospheric circulation during winter, which influences the rotation of the planet. This regulation occurs by modifying the propagation of planetary waves, which affect the strength of the polar vortices. When solar activity is lower, these important waves have a greater amplitude, resulting in a weaker vortex. The importance of this solar modulation mechanism is evident in the inverse relationship between solar activity and Arctic temperatures. This relationship has been observed in climate proxies for over 4,000 years and has contributed to the cooling of Greenland between the 1970s and 1990s, as well as much of the Arctic warming observed since 1997. In addition, solar activity affects the frequency of extremely cold winters in the midlatitudes of the Northern Hemisphere.

The Sun in the Winter Gatekeeper hypothesis

The Winter Gatekeeper hypothesis introduces several new ideas:

- Climate change is primarily driven by persistent changes in the amount of heat and moisture transported to the Arctic during winter.
- This heat easily escapes from the planet in the Arctic region, affecting the Earth's energy budget. This is due to the low greenhouse effect caused by the lack of water vapor.
- Variations in the generation and propagation of planetary waves affect the strength of the polar vortex and, thus, the poleward transport of heat.
- Numerous factors, known as gatekeepers, contribute to this mechanism and consequently influence climate change.
- Among the various natural causes of climate change on centennial time scales, solar variability emerges as the most important factor through this particular mechanism.

This hypothesis is supported by evidence that changes in the transport of heat to the poles have an important effect on the overall energy budget of the planet. In addition, there is compelling evidence that the Sun plays a critical role in regulating the strength of the polar vortex and heat transport. In this chapter, we will review the evidence for the role of solar activity in this context.

Evidence for the mechanism by which solar activity affects climate

Throughout this book, we have presented compelling evidence that solar activity acts as a gatekeeper, regulating the amount of heat transported to the Arctic during winter by modulating the strength of the polar vortex. One of the earliest pieces of evidence comes from the 1987 groundbreaking research of Karin Labitzke, discussed in box 23 (fig. B23, ch. 29). Labitzke's work demonstrated the influence of solar activity on the winter temperature of the polar stratosphere. Considering that this temperature is directly affected by the heat transport through the polar vortex (ch. 39), the connection between solar activity and the strength of the polar vortex becomes clear.

In chapters 11 and 16, we discussed the increased need to transport heat to the poles during winter, which intensifies atmospheric circulation and affects the Earth's rate of rotation. As discussed in chapter 30, research dating back to the 1970s has established a clear relationship between solar activity and the Earth's rotation rate during the winter season (fig. 47, ch. 30). In addition, it was observed as early as 1988 that solar activity has a substantial effect on the winter tropospheric circulation in the Northern Hemisphere, especially when the influence of the Quasi-Biennial Oscillation, another gatekeeper, is taken into account.[313] By affecting the Earth's rotation, solar activity provides further evidence of its influence on the winter atmospheric circulation and the poleward heat transport facilitated by this circulation.

In 1974, Colin Hines proposed a compelling idea for how solar activity could affect climate by influencing the strength of the vortex, the winter atmospheric circulation, and the rate of rotation of the planet.[314] Hines proposed that solar variations would likely affect climate through changes in the propagation of planetary waves, emphasizing their importance in mid- and high latitudes during winter. Although the study of planetary waves is challenging, there is evidence that solar activity affects the propagation of these waves in the lower stratosphere between 55°N and 75°N (fig. 67).[315] While the Quasi-Biennial Oscillation plays a noticeable role in the periodic fluctuations of planetary wave flux every 2-3 years, the influence of solar activity remains clear. The authors of the study conclude that the amplitude of planetary waves is related to the 11-year solar cycle. The amplitude decreases during periods of high solar activity and increases during periods of low solar activity. This implies that the solar effect may account for about 25% of the variability in wave amplitude, which is quite significant considering the relatively small change in solar energy.

The available evidence strongly supports every aspect of the hypothesis about how solar activity affects climate. When solar activity is high, the amplitude of planetary waves decreases, resulting in a stronger vortex, no significant strengthening of the winter atmospheric circulation, no effect on the Earth's rotation, and less heat reaching the Arctic. Conversely, when solar activity is low, the amplitude of the planetary waves increases, resulting in a weaker vortex, an intensification of the winter atmospheric circulation, an acceleration of the Earth's rotation, and a greater amount of heat reaching the Arctic. It's important to note that this effect does not manifest itself every winter due to the influence of other gatekeepers, notably the Quasi-Biennial Oscillation and the El Niño-Southern Oscillation.

---

[313] van Loon, H. & Labitzke, K., 1988. J. Clim. 1 (9), pp.905–920. doi.org/10.1175/1520-0442(1988)001<0905:ABTYSC>2.0.CO;2

[314] Hines, C.O., 1974. J. Atmos. Sci. 31 (2), pp.589–591. doi.org/10.1175/1520-0469(1974)031<0589:APMFTP>2.0.CO;2

[315] Powell Jr, A.M. & Jianjun, X., 2011. J. Atmos. Sol. Terr. Phys. 73 (7-8), pp.825–838. doi.org/10.1016/j.jastp.2011.02.001

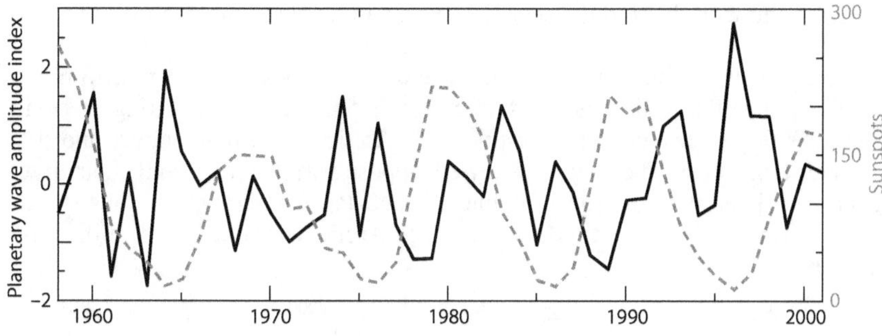

Figure 67. Solar cycle and planetary wave amplitude.

With this understanding, our focus must now shift to examining the climatic consequences of changes in solar activity through this mechanism.

Evidence for the climatic effect of solar activity (I). Arctic winter temperature

Scientists have been looking for evidence of a solar effect on climate for over a hundred years, although not in the best place to find it. The primary influence of solar variability lies in its ability to alter the transport of heat into the Arctic, with the most pronounced effect occurring during winter. Consequently, this altered heat transport leads to warmer winters in the Arctic and colder winters in the midlatitudes of the Northern Hemisphere due to reduced solar activity. Scientists have not looked for a solar effect on the Arctic in winter because there is no sunlight then, but we should focus our search for evidence on Arctic winter temperatures and midlatitude cold winters.

It is important to note that an exact correspondence between solar activity and winter weather at mid and high latitudes should not be expected, as poleward heat transport is also influenced by factors such as the Quasi-Biennial Oscillation, El Niño-Southern Oscillation, volcanic eruptions, and multidecadal ocean oscillations. Nevertheless, long-term trends and changes in winter climate should be consistent with a significant solar influence.

One of the most interesting scientific climate debates of the 21st century is the relationship between Arctic amplification and the occurrence of severe winter weather in the midlatitudes. Since 1997, Arctic winter temperatures have experienced an increasingly rapid warming trend relative to the global average (fig. 68a). During the same period, winter land temperatures in eastern North America and eastern Eurasia have experienced minimal warming, coinciding with an increase in the occurrence of severe winter weather events. This divergence between Arctic and midlatitude temperature trends has puzzled scientists because it was not predicted by climate models. As the authors of one study point out, *"This is the strongest observational evidence that some unaccounted-for mechanism has been offsetting greenhouse-gas-forced warming over the Northern Hemisphere midlatitudes."*[316] That unaccounted-for mechanism is indirect solar forcing, the Cinderella of climate studies. It does

---

[316] Cohen, J., et al., 2020. Nat. Clim. Change, 10 (1), pp.20–29.
   doi.org/10.1038/s41558-019-0662-y

much of the work behind the scenes, remains invisible to most, and receives no credit.

To uncover the hidden Cinderella of climate studies, we can turn to one of its revealing glass slippers — Arctic winter temperature. Looking at trends since 1960, we see a clear contrast between Arctic winter temperature and solar activity (fig. 68b). They have followed opposite trajectories, and their divergence has been particularly pronounced since 1997. Solar activity has significantly decreased during this period, while Arctic warming has significantly increased.

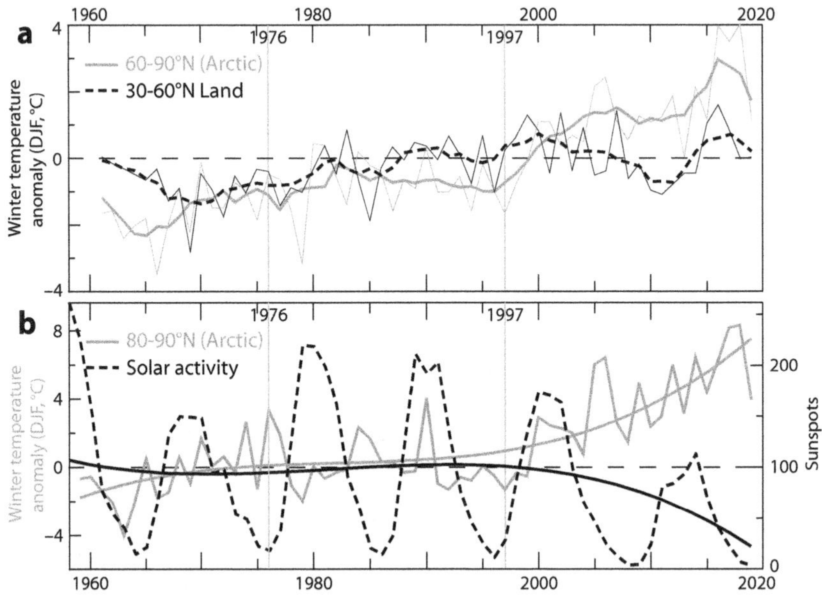

Figure 68. Recent Arctic warming and solar activity. a) Divergent trends in Arctic winter temperatures (solid gray line) and midlatitude land winter temperatures (dashed black line) since 1997. b) Opposing trends in Arctic winter temperatures (solid gray line) and solar activity (dashed black line).[317]

Furthermore, the previous period of strong Arctic warming in the 1920s (fig. 53, ch. 34) coincided with a period of low solar activity, as the modern solar maximum did not begin until about 1935.

Paleoclimate data provide compelling evidence that the inverse relationship between solar activity and Arctic climate is not a coincidence limited to two isolated periods in the last hundred years. The results from two core samples are particularly revealing. One, located near the Gulf of Alaska in the North Pacific storm track, provides a proxy for temperature, while the other, located in the Chukchi Sea, provides a proxy for sea ice cover.[318] Both cores are sensi-

---

[317] Arctic temperature data from the Danish Meteorological Institute. ocean.dmi.dk/arctic/meant80n_anomaly.uk.php Solar data from SILSO. www.sidc.be/SILSO/home

[318] Porter, S.E., et al., 2019. J. Geophys. Res. Atmos. 124 (20), pp.10784–10801. doi.org/10.1029/2019JD031023

tive to poleward heat transport to the Arctic through the Bering gateway. Figure 69 compares data from these proxies with solar activity from a reconstruction covering the past 800 years.[319]

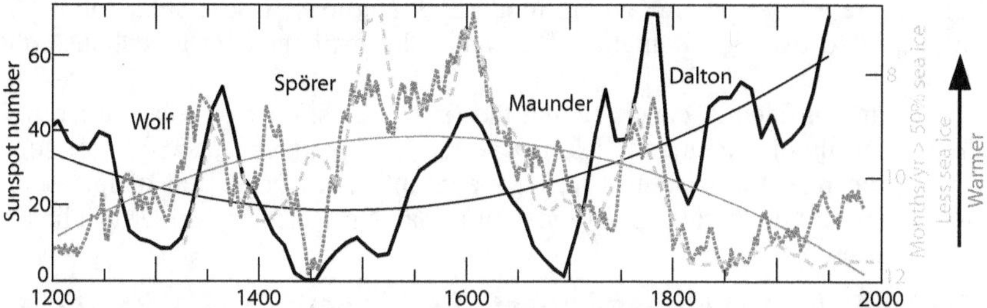

Figure 69 Solar-climate relationship along the Bering gateway during the Little Ice Age. Solar activity data are represented by a solid black line with quadratic regression (thin black line). The Alaska ice core temperature proxy data are represented by a dotted medium gray line with quadratic regression (thin gray line). The dashed light gray line represents the inverted Chukchi Sea ice cover data. Names correspond to solar minimums.

The relationship between the two climate proxies and solar activity is clear, but the effect is the opposite of what might be expected. As solar activity decreased leading up to the Spörer Minimum, there was a significant warming trend and a substantial decrease in sea ice cover in the Alaska region. Proxy records indicate that around 1500, the Alaskan region experienced much warmer temperatures and the Chukchi Sea had much less sea ice than it does today. However, after 1700, as solar activity began to increase, there was a significant cooling effect and an increase in sea ice cover.

The simplest explanation for the warming of this region during the Little Ice Age, when most of the planet was cooling, is that a substantial amount of heat from the Pacific Ocean was transported through this region to the Arctic. The high-latitude region became unusually warm due to a large influx of heat from the south.

Another study provides compelling evidence that the anomalously low temperatures in Greenland from the 1970s through the early 1990s were caused by the Modern Solar Maximum.[320] Leading researchers in paleoclimatology reconstructed Greenland's temperature patterns over the past millenniums and compared them with temperature reconstructions from the Northern Hemisphere. The results support previous conclusions that solar variability over the past 4,000 years correlates with significant opposite temperature anomalies in Greenland. In other words, when solar activity decreased (increased), Greenland experienced warming (cooling).

Surprisingly, the authors failed to make an obvious connection between the recent warming of Greenland since 1997 and the reduced solar activity observed in the past two decades. Since the title of the paper states that Green-

---

[319] Wu, C.J., et al., 2018. Astron. Astrophys. 615, p.A93.
    doi.org/10.1051/0004-6361/201731892
[320] Kobashi, T., et al., 2015. Geophys. Res. Lett. 42 (14), pp.5992–5999.
    doi.org/10.1002/2015GL064764

land's cooling in the late 20th century was forced by the Modern Solar Maximum, it logically follows that the end of the solar maximum contributed to some of the warming observed there in the 21st century. This oversight highlights a biased perspective within modern climatology, where solar forcing is only used to explain anomalies that cannot be explained by greenhouse gas forcing.

We can confidently conclude that the first glass slipper of solar variability, the Cinderella of climate studies, fits perfectly. The gatekeeper effect of solar activity on heat transport has had a major influence on Arctic temperatures over the past 4,000 years. Even today, it remains the primary driver of climate in the region.

Evidence for the climatic effect of solar activity (II). Extreme winter weather in the midlatitudes

Climate Cinderella's second glass slipper resolves the aforementioned scientific debate: the divergence in winter temperature trends between the Arctic and midlatitudes that climate models failed to predict. Understanding why this is a problem is not easy. When the Arctic experiences winter warming, it indicates an influx of warm air. Consequently, the cold air displaced by this warm air must move toward the midlatitudes, leading to colder winters in those regions. So why are scientists puzzled by this phenomenon?

Their confusion stems from an over-reliance on climate models that lack essential properties related to heat transport, preventing them from providing a straightforward explanation. These models fail to interpret Arctic amplification as a heat transport phenomenon. As a result, scientists look for answers in the unforeseen consequences of sea ice loss or possible stratospheric linkages rather than recognizing the more obvious implications of heat redistribution.

According to the Winter Gatekeeper hypothesis, a significant decrease in solar activity should weaken the polar vortex, leading to winter warming in the Arctic and winter cooling in the midlatitudes. Remarkably, this is exactly what researchers have observed. The Arctic has warmed, while the midlatitudes of the Northern Hemisphere have experienced an increased frequency of extreme winter cold since 1997, signaling the Sun's influence on our climate. The changing frequency of cold winters in the midlatitudes is the second glass slipper in our climate Cinderella story.

Figure 70 shows the percentage of land area between 20°N and 50°N that experienced winter months colder than one standard deviation below the 1951-1980 mean (black line).[321] With the onset of global warming in 1976, there was a marked decrease in the frequency of cold winters. This decrease is consistent with both the Enhanced CO_2 Effect hypothesis and the low transport regime of the Winter Gatekeeper hypothesis between 1976 and 1997. Both are likely to have played a role in this phenomenon.

---

[321] Cohen, J., et al., 2014. Nat. Geosci. 7 (9), pp.627–637. doi.org/10.1038/NGEO2234

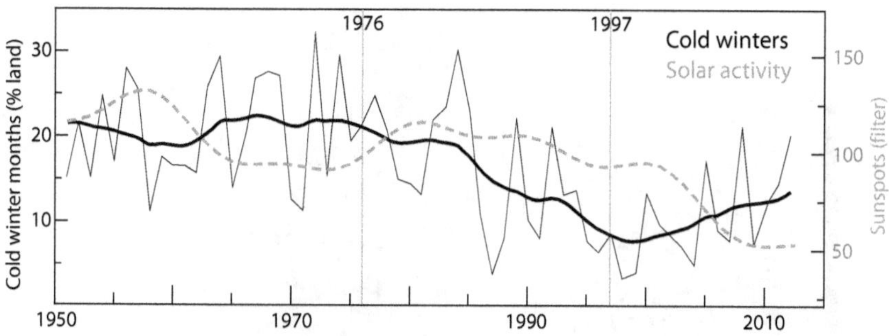

Figure 70. Solar activity and frequency of cold winters. The solid black line shows the percentage of land with a cold winter in the midlatitudes of the Northern Hemisphere. The dashed gray line is a Gaussian smoothing of sunspot numbers.

Since 1998, however, the frequency of cold winters has increased and only the Winter Gatekeeper hypothesis provides a plausible explanation for this phenomenon. Various factors can influence the poleward transport of heat, but there is a clear correlation when we compare the frequency of cold winters in the midlatitudes with solar activity. Figure 70 shows the interdecadal variability of solar activity by smoothing the sunspot number (shown by the dashed gray line). It is noticeable that when solar activity increases or decreases rapidly on interdecadal timescales, the frequency of cold winters shows an opposite trend. This striking correlation causes the two prominent curves in figure 70 to appear as distorted reflections of each other. The inverse correlation is not more accurate because other factors contribute to the climate trend.

In conclusion, we can affirm that the second glass slipper of solar variability, the Cinderella of climate, also fits perfectly. The gatekeeper effect of solar activity on heat transport has indeed influenced the frequency of cold winters in the midlatitudes of the Northern Hemisphere, and this influence continues.

With her two glass slippers, this Cinderella can dance. The solar effect on climate is not what we imagined, but it is very relevant to climate change.

In summary

Solar activity is responsible for altering the stratosphere, thereby affecting the propagation of planetary waves. These waves carry a large amount of energy and momentum that, when they reach the polar vortex, weaken it and allow more heat to reach the Arctic during winter. This phenomenon is enhanced when solar activity is low, as has been the case in recent decades. Consequently, the increased heat input leads to a warming of the Arctic and an expulsion of cold polar air, resulting in a higher frequency of extremely cold winter episodes in the midlatitudes of the Northern Hemisphere. Although these changes have been observed recently, they have not been attributed to the decrease in solar activity. The unexpected increase in the frequency of extreme winter cold has surprised scientists, as it was not predicted by climate models. However, the inverse relationship between solar activity and Arctic temperatures has persisted for at least 4,000 years, as indicated by climate proxies. It can only be explained by the variable heat transport mechanism that responds to solar activity.

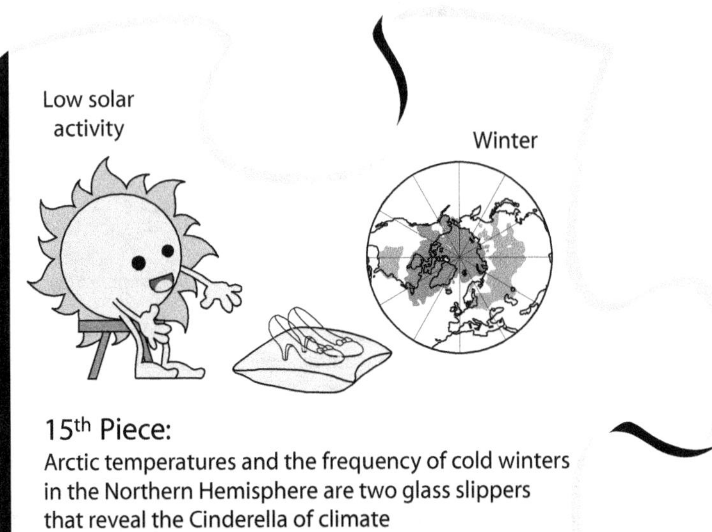

15th Piece:
Arctic temperatures and the frequency of cold winters in the Northern Hemisphere are two glass slippers that reveal the Cinderella of climate

CHAPTER 43
HEAT TRANSPORT MODIFIES THE ENERGY BUDGET

The Winter Gatekeeper hypothesis proposes that changes in the amount of heat transported into the Arctic during winter play an important role in driving climate change. Evidence for this hypothesis has been found for its three main aspects. First, changes in heat transport affect energy distribution, as indicated by different temperature trends across latitudes due to changes in vortex strength. Second, these changes affect the pattern of infrared emissions, as observed in the analysis of outgoing longwave radiation in the Arctic. Finally, the altered emission pattern leads to changes in the planetary energy budget, as evidenced by changes in the Earth's energy imbalance and the rate of ocean warming. Consequently, the Winter Gatekeeper hypothesis is a viable explanation for climate change.

Viability of the Winter Gatekeeper hypothesis

The first part of the book consists of 16 chapters that examine the processes by which the planet acquires its energy, circulates it through the climate system, and releases it back into space. While this topic may seem unexciting to many, it is a fundamental basis for understanding climate change. Understanding energy dynamics is crucial because any hypothesis that attempts to explain climate change must account for the necessary energy changes. In general, the global climate cannot undergo substantial changes without corresponding changes in the energy it contains. Thus, the energy changes must be consistent with the predictions of the hypothesis.

That's why no credible alternative to the Enhanced CO_2 Effect hypothesis has emerged since the 1960s. Milankovitch's orbital theory of glaciation explains the energy changes that drive the glacial cycle, but the gradual orbital changes are insufficient to account for substantial climate changes that occur within a few centuries. Some proposed alternatives fall short of the required energy changes, while others are not yet supported by the available evidence.

The Winter Gatekeeper is a thermodynamic hypothesis focusing on heat transport within the climate system. It is supported by evidence that changes in heat transport directly affect the energy content of the planet. I have reviewed thousands of scientific papers looking for evidence that would render the hypothesis unworkable or inconsistent with our knowledge of how climate has changed in the past or is changing in the present. However, these efforts were unsuccessful, underscoring the robustness of the hypothesis and its ability to comprehensively explain past climate change.

Changes in transport alter the energy distribution

Throughout most of the book, we have reviewed numerous lines of evidence pointing to the role of multidecadal oceanic oscillations in shaping climate regimes and triggering abrupt shifts. These oscillations reflect changes in heat transport conditions, as supported by the results discussed in chapters 13 (fig.

B9), 17 (figs. 26 & 27), 19 (figs. 30 & 31), 31-33 (figs. 48, 50 & 51), 37 (fig. 58), 39 (fig. 61), 40 (fig. 64), and 42 (figs. 68 & 70).

Additional evidence for this notion is found in the notable disparity in warming rates at different latitudes in the Northern Hemisphere. A strong contrast is observed between the low transport regime from 1976 to 1997 and the subsequent high transport regime since 1997. This disparity is illustrated in figure 71a, which shows the different surface and lower tropospheric temperature trends across latitudes for a low transport period (1985-1997) and a high transport period (2002-2013). The authors of the study referred to these periods as the pre-hiatus and hiatus regimes, respectively, with hiatus being the scientific name for the pause.[322]

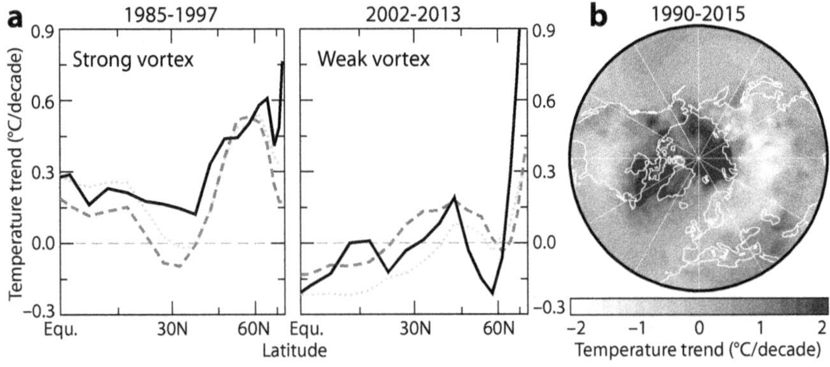

Figure 71. Changes in heat transport modify its distribution. a) Surface and lower tropospheric temperature trends versus latitude for two periods. The black solid line corresponds to the surface temperature data, while the discontinuous gray lines represent two lower troposphere satellite data.[323] The horizontal axis is proportional to the sine of latitude to reflect the area/latitude ratio. b) Winter surface temperature trends for a period when most years had a weak vortex configuration.

Looking at the consistent amount of solar energy received at a given latitude from year to year, it becomes clear that the remarkable temperature variations observed are primarily due to differences in heat transport. During periods of strong polar vortex and low transport, less heat is transported north of 60°N, resulting in increased warming in the mid and low latitudes south of the polar vortex. Conversely, during periods of weak polar vortex and high transport, more heat is transported toward the poles, leading to enhanced warming in the Arctic. However, this increased transport also leads to a noticeable cooling effect in the midlatitudes due to the exchange of air masses with the higher latitudes.

Another study provides valuable insights into the influence of the polar vortex on temperature trends and the resulting occurrence of cold extremes in the

[322] Gleisner, H., et al., 2015. Geophys. Res. Lett. 42 (2), pp.510–517. doi.org/10.1002/2014GL062596

[323] The black line is from HadCRUT4 data. The dashed medium gray line is from UAH data. The dotted light gray line is from RSS data.

midlatitudes.[324] Figure 71b from this study provides a visual representation highlighting the regions where different winter temperature trends manifest during a weak polar vortex and high transport period. In particular, the results reveal an intriguing phenomenon where cold Arctic air is driven into Eurasia and eastern North America. This event can be attributed to the influx of warm, moisture-laden air into the Arctic from the ocean basins and western regions of North America.

Based on the observations, it can be concluded that changes in poleward heat transport significantly affect the distribution of heat within the climate system, leading to distinct temperature trends. These differences in heat distribution are particularly evident during winter and are closely related to variations in vortex strength.

The change in heat distribution modifies outgoing radiation emissions

As the Arctic began to warm in 1997, it logically experienced an increase in outgoing infrared radiation. Figure 72 shows the data on outgoing longwave radiation in the 70-90°N zone at the top of the atmosphere, and its analysis is very revealing.[325]

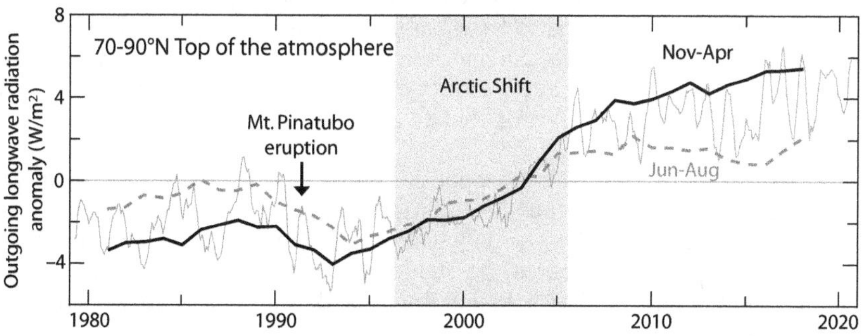

Figure 72. Change in outgoing longwave radiation in the Arctic region. The thin line is a 7-month moving average. The black line is a 5-year mean of the cold season values. The dashed gray line is a 5-year mean of the summer values.

The eruption of Mt. Pinatubo in 1991 caused a temporary decrease in emissions. This is consistent with the strengthening effect of volcanic eruptions on the northern polar vortex, leading to a warm winter in the northern midlatitudes and a cold winter in the Arctic after the eruption, as discussed in chapter 25.

Since 1997, however, a notable trend has emerged as Arctic infrared emissions have increased. While emissions are highest in the summer, when the Arctic receives more sunlight, the increase in cold-season emissions has been more pronounced. The result has been a reduction in seasonality. In this book, this period of abrupt increase in heat transport, especially during the cold season, is referred to as the Arctic Shift. The phenomenon is also evident in ocean heat transport toward the Arctic region (fig. 27, ch. 17).

---

[324] Kretschmer, M., et al., 2018. Bull. Amer. Meteor. Soc. 99 (1), pp.49–60. doi.org/10.1175/BAMS-D-16-0259.1

[325] Data from KNMI explorer climexp.knmi.nl/select.cgi?field=noaa_olr

The Arctic Shift has had a significant impact, resulting in a remarkable increase of 8 W/m^2 in infrared emissions to space during the cold season and about half that amount during the summer. Figure 71b shows that the affected area covers about 10% of the Northern Hemisphere, underscoring the substantial redistribution in the source of outgoing emissions caused by the Arctic Shift.

The change in emissions alters the global energy budget

As we have noted throughout this book, the polar regions have a greatly reduced wintertime greenhouse effect compared to the rest of the planet. This is due to the extremely low levels of water vapor in the cold polar atmosphere since water vapor and clouds account for about 75% of the greenhouse effect.[326] Even for a given average global surface temperature, the origin of the emissions is critical. If more emissions originate in the Arctic during winter, the total energy emitted by the planet increases. This is because these emissions come from lower altitudes, and the polar atmosphere is less opaque to the passage of infrared radiation.

Increased heat transport to the Arctic during the winter has a similar effect to an abrupt reduction in atmospheric CO_2, as it facilitates the escape of infrared radiation to space. However, the effect is much greater because the greenhouse effect in the Arctic in winter is less than half that in the tropics, while halving (or doubling) CO_2 levels would change the greenhouse effect by only a small percentage. This change directly affects the energy budget of the planet and serves as an overlooked driver of climate change, exerting an influence that has not been considered.

Between 1976 and 1997, the opposite trend occurred. The Arctic experienced a decrease in wintertime heat transport, resulting in reduced energy loss through the large infrared emission window provided by the Arctic atmosphere during this season. Consequently, a significant portion of the observed warming in recent decades is due to natural factors and cannot be attributed solely to human emissions and aerosols.

There is compelling evidence that the 1997 Arctic Shift has altered the global heat budget. A recent study shows a decrease in the Earth's energy imbalance since 2000 (fig. 73, black line).[327] Although the imbalance remains positive, indicating continued warming, the rate has slowed over time. The authors were surprised by this decrease in the energy imbalance, given the continued emission of GHGs. To validate their findings, they further examined changes in ocean warming rates over time, using variations in ocean heat content as an alternative measure of energy imbalance (ch 6). The analysis revealed a transition in ocean behavior during the Arctic Shift, from a faster warming trend to a slower one (fig. 73, dashed gray line). The convergence of these independent lines of evidence increased the authors' confidence in their results and provides robust support for the Winter Gatekeeper hypothesis.

---

[326] Schmidt, G.A., et al., 2010. J. Geophys. Res. Atmos. 115 (D20). doi.org/10.1029/2010JD014287

[327] Dewitte, S., et al., 2019. Remote Sens. 11 (6), p.663. doi.org/10.3390/rs11060663

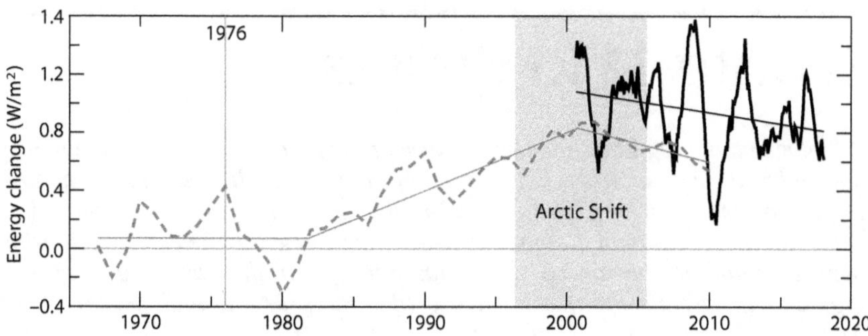

Figure 73. Changes in the Earth's energy imbalance and ocean warming rate over time. The energy imbalance data (black line) is a 12-month moving average, and the ocean warming data (dashed gray line) is a 10-year average.

The Winter Gatekeeper hypothesis offers an explanation for two longstanding questions in climatology: how solar activity influences climate and the extent of natural factors contributing to recent warming. In doing so, it emerges as a more comprehensive hypothesis than the enhanced CO_2 effect, demonstrating a superior ability to explain past and present climate variations.

In summary

Persistent changes in the amount of heat transported to the Arctic during winter have far-reaching implications for heat distribution, latitudinal temperature trends, Arctic infrared emissions, and the overall energy budget of the planet. These changes in heat transport reveal an overlooked and underappreciated natural driver of climate change. During winter, the Arctic atmosphere is a more transparent window for infrared emissions, allowing heat to escape from the planet more easily. The decrease in heat transport to the Arctic between 1976 and 1997 led to global warming as the Earth retained more heat. Conversely, the increase in heat transport to the Arctic since 1997 has led to warming in the region and a slowing of global warming. These findings underscore the significant natural component of climate change that has not been adequately addressed in IPCC reports or by most climate scientists.

SECTION 12 KEY ISSUES

Solar activity regulates the temperature of the polar stratosphere, the winter atmospheric circulation, and the rotation of the planet. It does this by modifying the propagation of planetary waves that affect the strength of the polar vortices. This solar modulation mechanism has caused an inverse relationship between solar activity and Arctic temperatures over the past 4,000 years. It regulates the frequency of extremely cold winters in the midlatitudes of the Northern Hemisphere.

As proposed by the Winter Gatekeeper hypothesis, changes in heat transport affect energy distribution, as indicated by different temperature trends across latitudes due to changes in vortex strength. These changes affect the pattern of infrared emissions, as observed in the analysis of outgoing longwave radiation in the Arctic. The altered emission pattern leads to changes in the planetary energy budget, as indicated by changes in the Earth's energy imbalance and the rate of ocean warming.

PART IV. A BETTER HYPOTHESIS

SECTION 13. EXPLAINING PAST CLIMATE CHANGE

CHAPTER 44
SOLVING CLIMATE PUZZLES OF THE DISTANT PAST

The Winter Gatekeeper hypothesis has considerable explanatory power for past climatic puzzles. The Oligocene and Miocene climates are particularly challenging. Most of the CO_2 decline of the past 50 million years occurred during the Oligocene when levels dropped from 800 to 300 ppm. Despite this remarkable decline in CO_2 levels, the Oligocene ended with an extended warming period of 2.5 million years in a world significantly warmer than today. Coinciding with this warming trend and the decline in CO_2 was the gradual emergence of the Antarctic Circumpolar Current, which reduced heat and moisture transport toward the South Pole. The Winter Gatekeeper hypothesis suggests that this reduction caused Antarctica to cool while the rest of the world warmed. By creating extremely cold Antarctic Bottom Water, the Antarctic Circumpolar Current also sequestered CO_2.

Earth experienced its warmest phase in 34 million years during the Mid-Miocene Climate Optimum, despite CO_2 levels comparable to today's. This period can be attributed to the culmination of warming effects resulting from reduced heat loss in the southern polar region. However, it ended when geographic and orographic changes increased heat loss in the northern polar region. This shift initiated a long-term global cooling trend that lasted until the end of the Last Glacial Maximum, about 20,000 years ago.

The mark of a good hypothesis

In chapter 35, we learned about scientific hypotheses. They are tentative proposals based on observations and supported by some of the available evidence. In cases where experimentation is impractical, hypotheses are tested against previously unexamined or newly acquired evidence. The strength of a hypothesis lies in its explanatory power, which can be measured by several factors. A hypothesis has high explanatory power when it explains a large number of facts, illuminates puzzling observations, has strong predictive power, relies less on authority and more on empirical observations, makes minimal assumptions, and is easily falsifiable.

The Winter Gatekeeper hypothesis outperforms the Enhanced CO_2 Effect hypothesis by this criterion. This new hypothesis emerges as a thermodynamically sound explanation for the evidence supporting a role for observed changes in heat transport in climate change. Surprisingly, it also reconciles the large paleoclimatic effect of changes in solar activity (as indicated by proxies) with the comparatively small effect observed by modern instrumentation. The principle of uniformitarianism states that processes occur in the same way and with the same intensity in the past and in the present. The proposed mechanism by which solar activity affects climate operates through UV-induced changes in ozone, modulation of planetary waves, and the strength of polar vortices, thereby altering the meridional transport of heat. Recognizing that meridional heat transport and vortex strength are fundamental climate features influenced by multiple factors, the hypothesis was extended to include all contributing factors, referred to as "gatekeepers."

Suddenly, the hypothesis gained impressive explanatory power, offering compelling explanations for various climate phenomena. It shed light on events such as the Little Ice Age (ch. 27) and the increase in cold winters in the Northern Hemisphere since 1997. Surprising results, such as the simultaneous occurrence of the global warming pause from 1998 to 2014 and the Arctic amplification, become clearer and more understandable through the lens of this hypothesis. In addition, numerous climate puzzles that had not been considered in the development of the hypothesis were easily resolved with the insights it provided. This strengthened my confidence in the fundamental accuracy of the hypothesis. In the next three chapters, we will examine how the hypothesis explains several instances of climate change that challenge alternative explanations.

The Late Oligocene Warm Period

During the early Eocene, about 50 million years ago, the Earth experienced a hothouse climate. Around this time, however, global temperatures began a long downward trend that culminated in the Late Cenozoic Ice Age. The exact causes of this temperature decline remain uncertain. One plausible hypothesis, however, is that it was due to the gradual emergence of a passageway between the Atlantic and the Arctic.[328] This interpretation is consistent with the principles of the Winter Gatekeeper hypothesis and is discussed in detail in chapter 20.

As the planet cooled, the polar regions experienced a more pronounced temperature drop, reducing their greenhouse effect during winter. This created a positive feedback loop that led to increased energy loss from the planet and further cooling. At the time, Antarctica was in close proximity to South America and Australia, with only shallow water separating them. This proximity allowed warm currents to flow toward Antarctica, bringing heat and promoting energy loss (fig. 74a).

As the cooling progressed, Antarctica experienced the formation of ice sheets at high elevations and developed a stronger response to orbital forcing. At the same time, the continent was physically separated from other land masses by the opening of the Drake Passage and the Tasman Gateway. This geographic shift paved the way for the development of the Antarctic Circumpolar Current, which is driven by the Coriolis effect acting on wind and water.

As the Antarctic Circumpolar Current became stronger, it gradually blocked the influx of heat from the tropics, increasing the cooling of the southern polar region. About 34 million years ago, Antarctica reached a tipping point, causing ice sheets to cover the continent in less than a million years. This marked the beginning of the Oligocene.

---

[328] Vahlenkamp, M., et al., 2018. Earth Planet. Sci. Let. 498, pp.185–195. doi.org/10.1016/j.epsl.2018.06.031

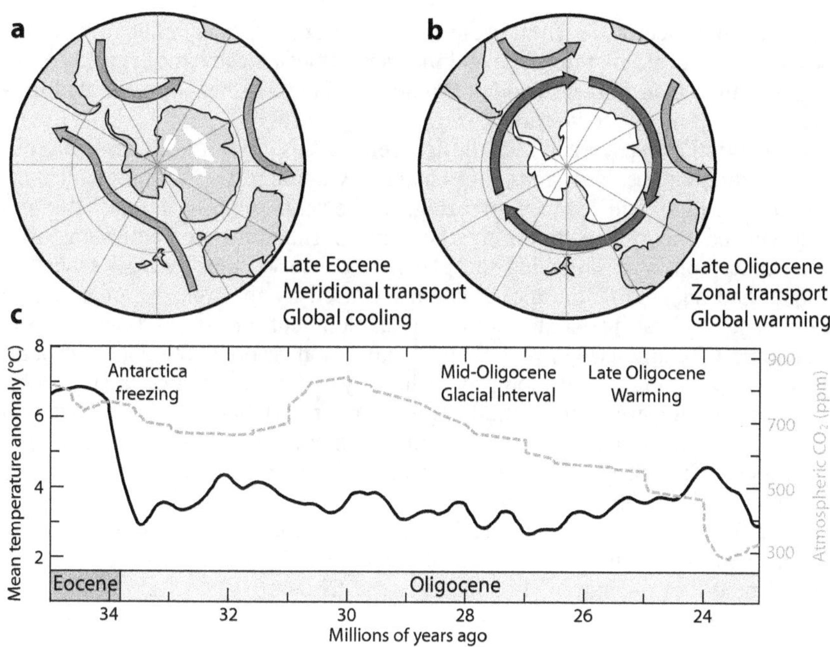

Figure 74. Explanation of global warming during the Oligocene. a) During the late Eocene, warm water currents brought heat and moisture to Antarctica. b) During the late Oligocene, a well-developed Antarctic Circumpolar Current reduced heat input to Antarctica. c) The development of the Antarctic Circumpolar Current during the Oligocene greatly reduced CO_2 levels while promoting global warming during the late Oligocene.[329]

The climate during this 11-million-year epoch has been defined as an enigma.[330] The available evidence suggests that the Eocene showed no clear trend in CO_2 levels (fig. B18, ch. 21). In contrast, the Oligocene witnessed a large decrease in CO_2, accounting for most of the decline observed throughout the Cenozoic. Levels fell from about 800 to 300 ppm, more than halving in concentration. Remarkably, despite these lower CO_2 levels, temperatures during this period were several degrees warmer than today (fig. 74c). To complicate the interpretation, after a cold period known as the Middle Oligocene Glacial Interval, which lasted from 28 to 26.5 million years ago, the temperature began an irregular rise that lasted about 10 million years, eventually leading to the Mid-Miocene Climate Optimum.

Therefore, about 26.5 million years ago, the climate transitioned into what is known as the Late Oligocene Warming. This period spanned about 2.5 million years and witnessed a remarkable global temperature increase of about 2°C (3.6 °F), even as CO_2 levels halved from 600 to 300 ppm. Climatologists have struggled to understand this warming phase because existing climate models

---

[329] Data for the figure from Westerhold, T., et al., 2020. Science, 369 (6509), pp.1383–1387. doi.org/10.1126/science.aba6853 CO_2 data kindly provided by T. Westerhold.
[330] O'Brien, C.L., et al., 2020. PNAS. 117 (41), pp.25302–25309. doi.org/10.1073/pnas.2003914117

cannot reproduce it. To complicate matters, despite the apparent midlatitude warming indicated by terrestrial and marine proxies and several degrees higher temperatures than observed today, Antarctica remained heavily glaciated during the Late Oligocene Warming.[331]

The Winter Gatekeeper hypothesis explains this mystery. Antarctica's climatic isolation resulted from the Antarctic Circumpolar Current, which reduced heat input and led to its freezing. At the same time, isolation and freezing reduced energy loss by limiting heat transport and infrared emissions from the colder continent, allowing the planet to conserve more energy. Once the planet had adapted to the global cooling caused by the freezing of an entire continent, it began to warm due to the development and strengthening of the Antarctic Circumpolar Current. This phenomenon resolves the apparent paradox of a warming world alongside a heavily glaciated Antarctica (fig. 74b). Despite the increase in the latitudinal temperature gradient, heat transport was mitigated by the circumpolar current, the Southern Annular Mode, and the polar vortex. The decrease in poleward heat transport to the Antarctic pole likely triggered the late Oligocene warming.

The formation of the Antarctic Bottom Water, the densest body of water on Earth with an average temperature of 1.5°C (35 °F), can be attributed to the development of the extremely cold Antarctic Circumpolar Current. This water mass occupies the deepest parts of the oceans connected with the Southern Ocean, below 4,000 m (13,000 ft) depth. The formation of this water mass likely played a significant role in the decline of CO_2 levels during the Oligocene epoch. Cold water has a greater capacity to dissolve CO_2, resulting in its sequestration in the deep ocean. This understanding challenges the prevailing view that CO_2 was a major contributor to climate change throughout the Cenozoic. As a result, it explains how the late Oligocene experienced warming despite the decrease in CO_2 absorbed by the ocean.

The Mid-Miocene Climatic Optimum

The warming trend that began in the late Oligocene, about 26.5 million years ago, peaked ten million years later during the Mid-Miocene Climatic Optimum, between 16.9 and 14.7 million years ago. By this time, about two-thirds of the cooling that occurred during the Eocene-Oligocene transition, marked by the Antarctic glaciation, had been reversed. During this remarkable climate period, the planet experienced temperatures 5 to 8 °C (9-14 °F) warmer than today, with similar CO_2 levels of about 400 ppm. For this reason, climatologists consider it an intriguing period.[332]

Climate models cannot reproduce the shallow latitudinal temperature gradient observed during the mid-Miocene, characterized by warm conditions in the tropics and midlatitudes. The models require a minimum of 800 ppm CO_2 to achieve this representation, indicating that they are missing about half of the forcing necessary to explain this climate optimum. The missing forcing is likely to be even greater since the CO_2 forcing itself tends to be overestimated. This overestimation occurs because the models do not include the effects of

---

[331] Hauptvogel, D.W., et al., 2017. Paleoceanography, 32 (4), pp.384–396. doi.org/10.1002/2016PA002972

[332] Goldner, A., et al., 2014. Clim. Past, 10 (2), pp.523–536. doi.org/10.5194/cp-10-523-2014

indirect solar forcing, a powerful driver of climate change. The fact that more than half of the required forcing is missing from the models suggests that meridional heat transport, rather than CO_2, is a more important climate driver.

The Winter Gatekeeper hypothesis offers a possible explanation for the mid-Miocene paradox. Over a period of 10 million years, the planet experienced a gradual warming due to the increasing climatic isolation of Antarctica, which facilitated energy conservation. At the same time, the tectonic changes discussed in chapter 20 gradually changed the planet's circulation pattern from predominantly zonal to predominantly meridional (fig. 33, ch. 20). This change favored the loss of energy at the opposite pole. After the mid-Miocene climate optimum, the escalating Arctic energy loss reached a critical threshold, setting the planet on a trajectory toward a much colder bipolar ice age.

Surprisingly, many scientists view the Miocene epoch as a potential analog for our future climate.[333] This perspective stems from the fact that during the Miocene, CO_2 levels were similar to current levels and temperatures were 5-8°C (9-14 °F) higher than today. These projections are consistent with the expected future warming if emissions continue unabated for about a century. However, these scientists do not adequately address the persistent mismatch between CO_2 and temperature trends throughout the Cenozoic (fig. B18, ch. 21), especially during the Oligocene (fig. 74). Figure B18b (ch. 21) clearly illustrates the lack of correlation between CO_2 and temperature changes. These observations suggest that the tectonic changes during this period were the primary drivers of climate cooling and CO_2 reduction. Even existing models do not support the notion that we are heading toward a Miocene-like climate within a few centuries, as they struggle to explain the complexities of the Miocene climate itself. According to the Winter Gatekeeper hypothesis, the next glacial period is due in a few thousand years.

Helping Milankovitch

Milankovitch's orbital theory of glaciation provides a solid explanation of the underlying cause of the glacial cycle. However, climatologists are still trying to understand how subtle variations in incoming solar radiation at the top of the atmosphere translate into massive changes in ice volume at the Earth's surface. Some authors, including myself, argue for the central role of changes in the Earth's axial tilt, known as obliquity, in producing interglacial periods.[334] Extensive evidence supports the idea that changes in obliquity produce a much more pronounced global climatic response (fig. 75) than changes in precession, the axial wobbles that gradually alter the orientation of the axis and affect seasonal variations.

Precession doesn't change the total energy received at each latitude each year. Rather, it affects the seasonal distribution of that energy, resulting in milder or more extreme summers and winters, albeit with contrasting effects in each hemisphere. On the other hand, obliquity plays a different role by altering the total annual energy received at each latitude, affecting both hemispheres in

---

[333] Steinthorsdottir, M., et al., 2021. Paleoceanogr. Paleoclimatol. 36 (4), p.e2020PA004037. doi.org/10.1029/2020PA004037

[334] Vinós, J., 2022. Climate of the Past, Present and Future: A scientific debate. Critical Science Press. pp.5–25.

the same way. This property makes obliquity particularly well suited to driving a glacial cycle that occurs at the same time in both hemispheres.

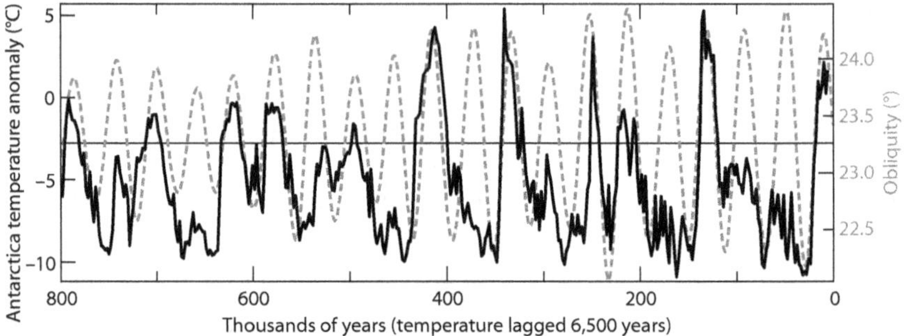

Figure 75. Temperature changes due to changes in axial tilt. The thick black line is the Antarctic temperature anomaly with a lag of 6,500 years, and the dashed gray line is the axial tilt. The thin horizontal line represents the level above which an interglacial is considered to have occurred.[335]

One challenge in understanding the importance of obliquity is that the resulting energy changes are more pronounced near the poles while small in tropical and midlatitude regions. Nevertheless, scientists find it intriguing that numerous paleoclimate records from tropical and subtropical regions show a strong obliquity signal. In particular, the behavior of the monsoons and the Intertropical Convergence Zone show a substantial response to obliquity, suggesting that its variations have a discernible impact on global atmospheric circulation.

The key to solving this puzzle is to understand how obliquity affects the summer insolation gradient. The tilt of the Earth's axis directly affects the amount of incoming solar radiation that high latitudes receive during the summer months but not during the winter, when high latitudes receive no sunlight. The importance of summer conditions in driving the glacial cycle was recognized as early as 1869, half a century before Milankovitch's theory. The dependence of the summer insolation gradient on obliquity is shown in figure B19 (ch. 21). Consequently, variations in obliquity also affect the summer latitudinal temperature gradient and the transport of heat and moisture to the poles.

During the Miocene, a notable influence of obliquity (axial tilt) on Antarctic ice sheet evolution emerged, consistent with the growing influence of meridional transport on climate evolution throughout the Cenozoic. According to a recent study, this link is due to obliquity-induced changes in the meridional temperature gradient. These changes directly affect the position and intensity of the Antarctic Circumpolar Current, which in turn alters heat transport across the Antarctic continental margin.[336]

Enhanced moisture transport plays a critical role in the formation of the massive ice sheets that define glacial periods, a concept recognized as early as

[335] Data for the figure from Jouzel, J., et al., 2007. Science, 317 (5839), pp.793–796. doi.org/10.1126/science.1141038 and from Laskar, J., et al., 2004. Astron. Astrophys. 428 (1), pp.261–285. doi.org/10.1051/0004-6361:20041335

[336] Levy, R.H., et al., 2019. Nat. Geosci. 12 (2), pp.132–137. doi.org/10.1038/s41561-018-0284-4

the 19th century. In 1872, John Tyndall argued: *"So natural was the association of ice and cold that even celebrated men assumed that all that is needed to produce a great extension of our glaciers is a diminution of the Sun's temperature. Had they gone through the foregoing reflections and calculations, they would probably have demanded more heat instead of less for the production of a 'glacial epoch'."*[337]

As obliquity decreases, the gradient of summer insolation between latitudes becomes more pronounced. As a result, atmospheric and oceanic circulation intensifies, facilitating the transport of greater amounts of heat and moisture toward the poles. This shift also leads to a strong change in the seasonality of precipitation, with the summer contribution becoming dominant and the origin of moisture shifting toward the equator, toward warmer oceanic sources.[338]

The Winter Gatekeeper hypothesis revolves around changes in heat and moisture transport to the poles. This hypothesis emphasizes the importance of winter-related changes in this transport, which are very important for sub-Milankovitch-scale climate variations, as discussed in detail throughout the book. However, it's important to note that the planet also responds to changes in the redistribution of heat and moisture during the summer. The same principle that currently influences climate patterns is responsible for translating orbital variations in insolation into climatic effects during the glacial cycle. The magnitude of this transport plays a critical role: the greater the transport, the colder the planet, while the smaller the transport, the warmer the planet.

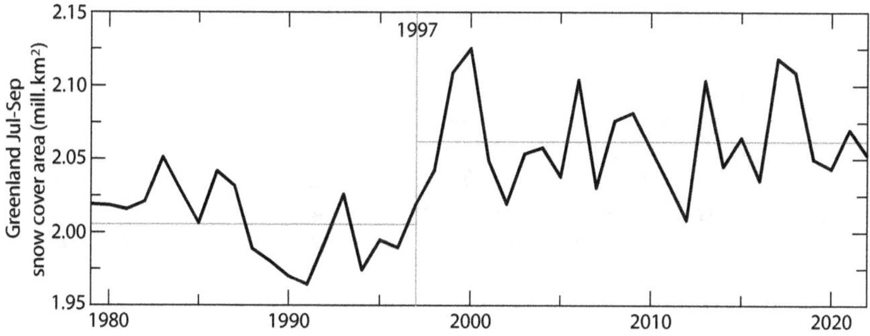

Figure 76. Summer snow cover in Greenland. The gray horizontal lines are period means.

Figure 76 provides compelling evidence that variations in transport, as postulated by the Winter Gatekeeper hypothesis, have an impact on summer snowfall at high latitudes. Specifically, the figure shows changes in the snow-covered area of Greenland from July to September. Despite the overall warming trend observed in the Arctic region, the intensification of transport following the 1997 climate shift has resulted in a notable expansion of the snow-

---

[337] Kukla, G. & Gavin, J., 2004. Glob. Planet. Change, 40 (1-2), pp.27–48. doi.org/10.1016/S0921-8181(03)00096-1

[338] Masson-Delmotte, V., et al., 2005. Science, 309 (5731), pp.118–121. doi.org/10.1126/science.1108575

covered area during the summer. The increase amounts to more than 50,000 km^2 (20,000 square miles).[339]

The Winter Gatekeeper hypothesis provides valuable insights into the mechanism by which small variations in solar radiation at the top of the atmosphere lead to massive changes in ice volume at the ground. On the other hand, the Enhanced CO$_2$ Effect hypothesis falls short of providing a comprehensive explanation of the glacial cycle, particularly with respect to cooling and ice accumulation at the end of interglacial periods despite elevated CO$_2$ levels (fig. 56, ch. 35).

The Earth's tilt is slowly decreasing, reducing summer insolation at the poles and increasing the summer latitudinal temperature gradient by cooling the polar regions. The change is so slow that it is imperceptible, often obscured by temporary warming periods like the current one. But over the next few thousand years, the tilt change will gradually increase heat and moisture transport, leading to more summer snowfall at high latitudes, similar to what is shown in figure 76. The planet will be colder than it is now, and the greater winter transport will increase energy loss and cooling. The onset of these irreversible trends is called glacial inception. Reaching full glacial conditions is a very long process, usually taking about 15,000 years (longer than the Holocene), during which the long-term cooling is sometimes interrupted by periods of warming. This process has always occurred over the past 2.5 million years, regardless of CO$_2$ levels. The belief of many scientists that this time will be different is not based on evidence.

In summary

The Winter Gatekeeper hypothesis provides a compelling framework for understanding climate changes in the distant past that have long puzzled researchers and challenged alternative hypotheses. It effectively explains the Late Oligocene Warm Period and the Mid-Miocene Climate Optimum, providing logical explanations based on known changes in heat transport during these times. In addition, it provides insights into how relatively small changes in the Milankovitch forcing at the top of the atmosphere translate into massive changes in the volume of ice sheets at the Earth's surface. Recent observations of changing summer snow cover in Greenland further support this understanding. If the Winter Gatekeeper hypothesis is correct, it would mean that CO$_2$ variations were not the driving force behind the profound climate changes that occurred during the transition from the early Eocene hothouse to the late Pleistocene icehouse.

---

[339] Data from Rutgers University. climate.rutgers.edu/snowcover/

CHAPTER 45
HOLOCENE CLIMATE PUZZLES

Estimates of climate forcing based on the Enhanced CO₂ Effect hypothesis fail to reproduce key features of the Holocene climate. The primary long-term driver of this climate period is orbital forcing, with global temperature changes driven primarily by summer insolation in the Northern Hemisphere. In contrast, CO₂ levels during the Holocene did not reflect global temperature variations but rather responded to Southern Ocean conditions, which in turn were influenced by changes in Southern Hemisphere summer insolation of the opposite sign. Climate models struggle to simulate the temperature changes observed during the Holocene, suggesting an inadequate response to orbital forcing and an exaggerated response to CO₂ variations. In addition, the official hypothesis cannot explain the frequent abrupt climate events of the Holocene. Some of these events have a clear solar origin, indicating an important misunderstanding of solar forcing. The Winter Gatekeeper hypothesis provides a plausible explanation for these Holocene climate puzzles.

Drivers of the Holocene climate

The Holocene resulted from two key orbital events: a maximum of obliquity (axial tilt) about 9,500 years ago and a maximum of summer insolation at 65°N (climatic precession due to axial orientation) about 11,000 years ago. These orbital parameters increased the energy received during summer in the high latitudes of the Northern Hemisphere. Combined with the strong feedback effect of sea level rise, reduced ice reflectivity (albedo), smaller latitudinal temperature gradients, and higher GHG concentrations, this energy change caused the melting of a large volume of ice that had accumulated outside the polar regions during 100,000 years of glaciation.

About 9,500 years ago, the Holocene reached its climatic optimum. While the improvement in insolation conditions stopped, the orbital forcing is slow-acting and tends to show inertial effects for thousands of years. As a result, the remnants of the ice sheets continued to melt for about 2,000 years, maintaining optimal climatic conditions until about 6,000 years ago. At that point, reduced summer insolation in the Northern Hemisphere and the high latitudes at both poles gradually eroded the climatic optimum, giving way to the Neoglaciaciation.

The main long-term driver of the Holocene climate is orbital forcing. Figure 77a plots mean summer insolation at 60° latitude for both the Northern and Southern Hemispheres. Global temperatures correlate with Northern Hemisphere insolation but with a lag of about 2000 years (fig. 77b, black line).[340] Similarly, temperatures in the high latitudes of the Southern Hemisphere corre-

---

[340] After Marcott, S.A., et al., 2013. Science, 339 (6124), pp.1198–1201. doi.org/10.1126/science.1228026 Data reprocessed, see ch. 21.

late with Southern Hemisphere insolation, also with a comparable lag (fig. 77b, dotted light gray line).[341]

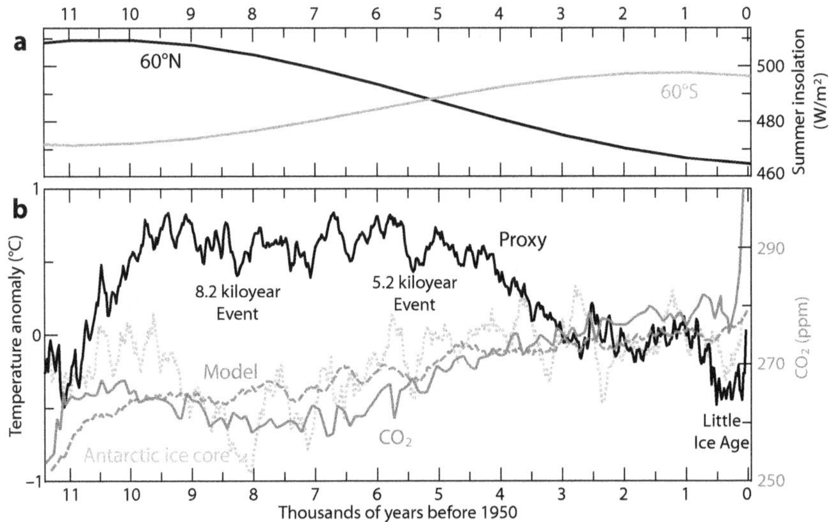

Figure 77. Holocene climate drivers. a) Summer insolation at 60°N (black line) and 60°S (gray line).[342] b) A global temperature reconstruction from proxies (black line) is compared with an Antarctic temperature reconstruction (dotted light gray line) and an atmospheric CO_2 reconstruction, both from ice cores. The dashed grey line shows a multi-model simulation of Holocene temperatures. The three major Holocene climate events are indicated by their names.

CO_2 levels have played a minor role in determining the Holocene climate. As highlighted in chapters 21 and 22, CO_2 levels fluctuated within a range of 20 ppm for most of the Holocene, whereas now they are increasing by 20 ppm in just eight years. Changes in CO_2 during the Holocene were relatively small and unlikely to have had a significant impact on climate. In addition, variations in CO_2 correspond more closely to Southern Hemisphere temperatures than to global temperatures (fig. 77b, gray line).[343] This pattern is consistent with the analysis of CO_2 levels during the Oligocene, as discussed in the previous chapter (fig. 74, ch. 44). It suggests that the Southern Ocean and the Antarctic Circumpolar Current played a key role in regulating CO_2 levels before human emissions became the dominant factor.

Model simulations of Holocene climate tend to produce a shape more similar to Southern Hemisphere temperatures than to global temperatures, indicating too little response to Northern Hemisphere insolation and too much re-

[341] Data from Jouzel, J., et al., 2007. Science, 317 (5839), pp.793–796. doi.org/10.1126/science.1141038

[342] Data from Laskar, J., et al., 2004. Astron. Astrophys. 428 (1), pp.261–285. doi.org/10.1051/0004-6361:20041335

[343] Data from Monnin, E., et al., 2004. Earth Planet. Sci. Lett. 224 (1–2), pp.45–54. doi.org/10.1016/j.epsl.2004.05.007

sponse to CO_2 changes (fig. 77b, dashed gray curve).[344] Thus, climate theory and the forcings derived from the Enhanced CO_2 Effect hypothesis fail to reproduce the main features of the Holocene climate. The Winter Gatekeeper hypothesis, on the other hand, has no problem explaining the Holocene climate.

In the previous chapter, we observed that about 30 million years ago, when the Antarctic Circumpolar Current emerged and Antarctica became climatically isolated, the global climate began to depend primarily on the variable amount of heat transported to the Arctic. This dependence is one of the reasons why the global climate is more sensitive to boreal summer insolation. During the early Holocene, high summer insolation in the high latitudes of the Northern Hemisphere led to a reduction in the latitudinal temperature gradient and less energy loss in the Arctic due to reduced poleward heat transport. As a result, the Earth experienced the warm temperatures of the Holocene Climatic Optimum.

The combination of decreasing axial tilt and reduced summer insolation in the Northern Hemisphere accentuated the latitudinal temperature gradient, gradually increasing the heat transported to the Arctic. As a result, the Earth experienced a cooling trend during the Neoglaciation. Antarctica has experienced the opposite trend because the decrease in obliquity, which led to a decrease in solar radiation, was offset by a simultaneous increase in summer insolation due to precession changes.

In the coming millenniums, the Southern Hemisphere is expected to experience a cooling trend due to a decrease in summer insolation resulting from changes in obliquity and precession. The Northern Hemisphere will experience opposite trends in these two orbital parameters, but overall, there will be a net increase in summer insolation at most latitudes. Nevertheless, both hemispheres will experience an increasing latitudinal insolation gradient (fig. B19, ch. 21), signaling the approaching end of the Holocene.

Drivers of Holocene abrupt climate events

As discussed in chapters 22 and 23, paleoclimate studies have identified more than 20 abrupt climate events during the Holocene. A comprehensive analysis of these events and their potential causes has been published.[345] Among the most important events in terms of their climatic impact are the Little Ice Age, the 8.2, 5.2, 4.2, and 2.8 kiloyear events, and the Boreal Oscillation. Numerous studies have been conducted on all of these events. The Little Ice Age has already been discussed in chapters 23 and 26. To emphasize the limitations of current climate theory in explaining the well-documented abrupt climate changes of the Holocene and to highlight the merits of the Winter Gatekeeper hypothesis, we will review the 2.8 kiloyear event.

This event was first recognized as an abrupt climate change by peat stratigraphic studies in 1912, even before the Little Ice Age was discovered.[346] Swedish botanist Rutger Sernander linked it to the Fimbulvintern, or the legen-

---

[344] Liu, Z., et al., 2014. PNAS, 111 (34), pp.E3501–E3505.
doi.org/10.1073/pnas.1407229111

[345] Vinós, J., 2022. Climate of the Past, Present and Future: A scientific debate. pp.45–65. Critical Science Press.

[346] Fries, M., 1956. "Fimbulvintern" ur vegetations-historisk synpunkt. Fornvännen 51, pp.5–10.

dary Great Winter, described in Nordic Bronze Age sagas as lasting many years. This mythical long winter has been depicted in modern fantasy works such as the Game of Thrones books and television series. The event marked a significant change in Scandinavia, ending the warm climate that had previously allowed for the cultivation of grapes and ushering in a wetter and colder climate.

The 2.8 kiloyear event was a globally synchronized abrupt climate change that left clear traces in many proxies, leaving no doubt about its occurrence and climatic consequences.[347] Figure 78 shows key proxies that help us understand what happened 2,800 years ago and the likely cause.

Reconstructions of solar activity reveal the presence of two solar grand minimums that occurred 2,990 and 2,790 years ago, separated by a span of 200 years. The second minimum is of the Spörer type (fig. 78a, black line).[348] This pattern mirrors the occurrence of the Wolf and Spörer solar minimums during the Little Ice Age in 1280 and 1480. In addition, a reconstruction of the Northern Hemisphere temperature shows a strong correlation with solar activity (fig. 78a, dashed gray line).[349] It shows a substantial decrease of 0.7°C within a century and an overall cooling of 1°C over the entire event.

According to the reconstructions, Iceland's summer sea surface temperature reflects a remarkable 1°C decrease during the first solar grand minimum, followed by an additional 1°C decrease during the second one (fig. 78b, black line).[350] Meridional transport analysis shows that the polar circulation brings increased amounts of non-sea salt potassium to Greenland during periods of intensification, which are normally associated with winter conditions. This increase indicates a strengthening of the Siberian high-pressure system, which facilitates enhanced poleward heat transport. In the context of the 2.8 kiloyear event, non-sea-salt potassium levels reach their highest point in thousands of years, indicating a large increase in poleward heat transport (fig. 78b, dashed gray line).[351]

The El Niño-Southern Oscillation is a critical component of the heat transport system, and a proxy analysis (reviewed in box 15, ch. 18) shows that the 2.8 kiloyear event was flanked by enhanced El Niño activity. However, during the solar grand minimum, this activity decreased to low levels (fig. 78c, black line).[352] This suggests that during periods of increased poleward heat transport, typically during solar grand minimums, there is insufficient accumulation of excess subsurface heat in the equatorial Pacific to trigger numerous Niño events.

[347] Chambers, F.M., et al., 2007. Earth Planet. Sci. Lett. 253 (3–4), pp.439–444. doi.org/10.1016/j.epsl.2006.11.007

[348] Wu, C.J., et al., 2018. Astron. Astrophys. 615, p.A93. doi.org/10.1051/0004-6361/201731892

[349] Kobashi, T., et al., 2013. Clim. Past, 9 (5), pp.2299–2317. doi.org/10.5194/cp-9-2299-2013

[350] Jiang, H., et al., 2015. Geology, 43 (3), pp.203–206. doi.org/10.1130/G36377.1

[351] Data from Mayewski, P.A., et al., 2004. Quat. Res. 62 (3), pp.243–255. doi.org/10.1016/j.yqres.2004.07.001 with Gaussian smoothing.

[352] Moy, C.M., et al., 2002. Nature, 420 (6912), pp.162–165. doi.org/10.1038/nature01194

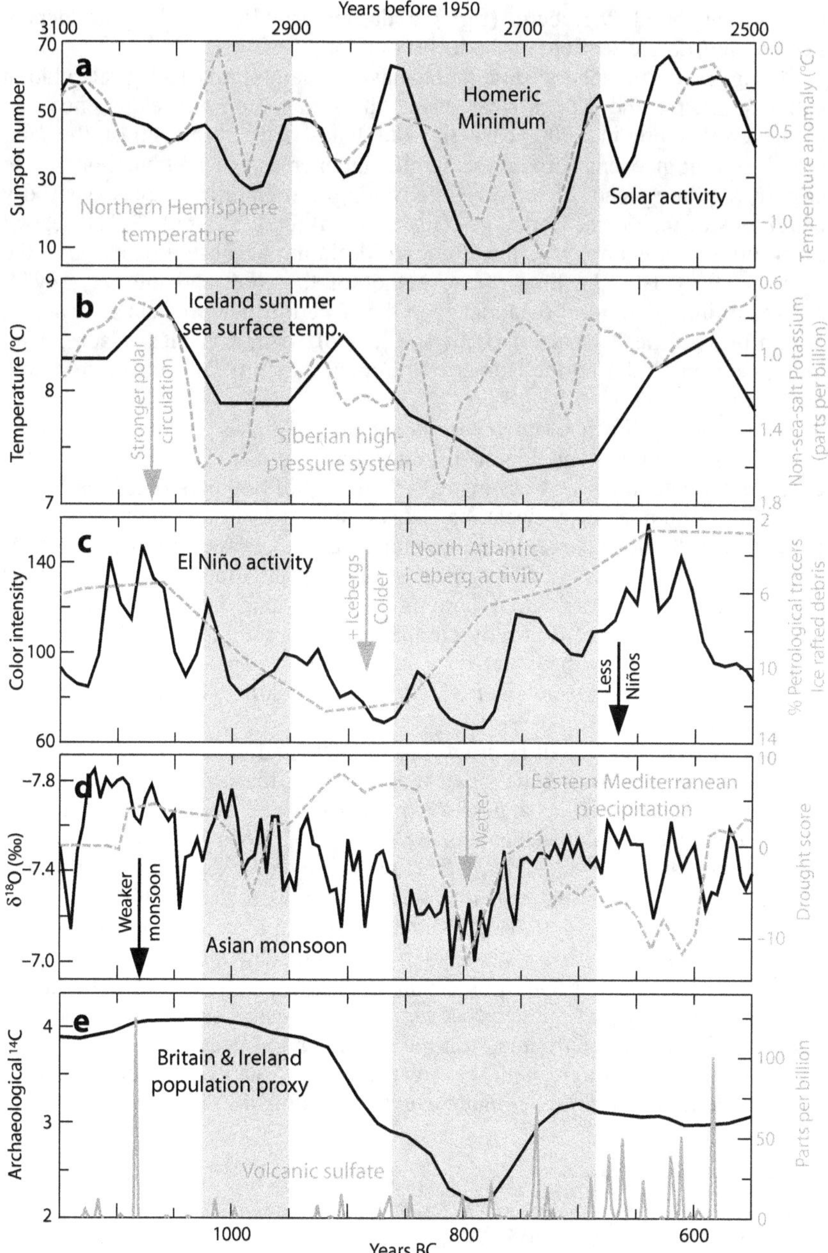

Figure 78. The 2.8 kiloyear event. The gray areas correspond to the two solar grand
minimums that showed a strong climate response.

The temperature decrease indicated by the proxies coincides with the largest
increase in iceberg activity observed in the North Atlantic in 4,000 years, according to a proxy that measures the deposition of high-latitude petrologic

tracers transported by icebergs (fig. 78c, dashed gray line).[353] They are deposited on the Atlantic seafloor as the icebergs melt.

Changes in atmospheric circulation have a major impact on precipitation patterns. During the 2.8 kiloyear event, the Asian monsoon experienced its greatest weakening in 2,000 years (fig. 78d, black line).[354] In contrast, the solar grand minimum events brought a significant increase in precipitation to the eastern Mediterranean, effectively interrupting and ending the previously dry conditions of the Greek Dark Ages (fig. 78d, dashed gray line).[355] The second grand minimum is known as the Homeric Minimum because it occurred in the eighth century BC, the time of the author of the Iliad and the Odyssey. It marked a time of cultural renaissance in the eastern Mediterranean after a dark period of centuries characterized by the loss of writing in most places. The increased rainfall during the Homeric Minimum may have played a role in fostering this renaissance.

Over much of Europe, however, the impact of these climatic changes on human societies was predominantly negative. In the British Isles, semi-quantitative data allow us to estimate the severity of population decline by examining the number of radiocarbon dates obtained from archaeological sites. During the 2.8 kiloyear event, the probability distribution of radiocarbon dates drops to half its value, indicating a catastrophic population decline – the most significant since the arrival of farmers in the British Isles (fig. 78e, black line).[356] The authors of the study emphasize the correlation of this and other population declines with Spörer-type solar minimums and suggest that food supply crises resulting from solar activity-induced climate changes were likely responsible.

Scientists are making little progress in understanding the causes of abrupt climate events because of the small magnitude of the accepted solar forcing and the inability of CO_2 changes to explain Holocene events. Some attempts have been made to link these events to volcanic eruptions because their effects are overestimated in models (ch. 24). However, a closer examination of figure 78e (gray line) shows that volcanic activity remained remarkably low for a period of 330 years after 1080 BC, precisely during a period of great climatic change, coinciding with the first solar minimum and most of the Homeric Minimum.[357] Only after that did volcanic activity increase, coinciding with the period of warming and climate recovery. This indicates that volcanic activity had a minimal effect on climate during and after the event. It becomes clear that if solar activity alone could have caused the 2.8 ka event, it could also be responsible for other abrupt climatic events, including the Little Ice Age.

---

[353] Bond, G., et al., 2001. Science, 294 (5549), pp.2130–2136.
   doi.org/10.1126/science.1065680
[354] Wang, Y., et al., 2005. Science, 308 (5723), pp.854–857.
   doi.org/10.1126/science.1106296
[355] Kaniewski, D., et al., 2013. PloS one, 8 (8), p.e71004.
   doi.org/10.1371/journal.pone.0071004
[356] Bevan, A., et al., 2017. PNAS, 114 (49), pp.E10524–E10531.
   doi.org/10.1073/pnas.1709190114
[357] Zielinski, G.A., et al., 1996. Quat. Res. 45 (2), pp.109–118.
   doi.org/10.1006/qres.1996.0013

The Winter Gatekeeper hypothesis provides a superior explanation for abrupt climate events associated with solar grand minimums. As discussed in chapter 41, the proposed mechanism results in increased poleward heat transport. As a result, we would expect an increase in colder winters in the midlatitudes of the Northern Hemisphere, leading to a gradual cooling of the planet due to increased energy loss in the Arctic. Such a change would induce an atmospheric reorganization that would shift atmospheric jets, storm tracks, and monsoons toward the equator. As a result, some regions would become drier and others wetter.

While most Holocene solar grand minimums coincide with abrupt climatic events, not all do. The problem arises because the proxy method, the ^{14}C production rate, doesn't directly indicate solar activity but instead records the arrival of cosmic rays on Earth. Typically, cosmic ray variations of up to a few centuries are caused primarily by variations in the Sun's magnetic field, which correspond to changes in its activity. However, there are some exceptions.

About 9,600 years ago, there was a 2.8% increase in ^{14}C production, exceeding levels observed even during a Spörer-type large solar minimum. Remarkably, this increase persisted for 400 years, twice as long as the Spörer minimum. Despite this substantial increase in ^{14}C production, the largest of the Holocene, no significant climate change has been detected for most of this period. This lack of a climate signature led me to explore alternative factors that could have influenced cosmic rays during this period.

Interestingly, during the early Holocene, a massive star exploded in the constellation Vela, only 800 light-years away, making it the closest known supernova to us. The peculiar shape, intensity, and duration of the ^{14}C production spike 9,600 years ago make it the most plausible candidate for the Vela supernova imprint on the ^{14}C record. Despite such a substantial increase in ^{14}C production, the lack of associated climatic effects poses a significant challenge to any hypothesis linking cosmic rays to climate change.[358]

It is important to recognize that while the proxy record captures all solar grand minimums, not every significant increase in ^{14}C production corresponds to a solar grand minimum. This aspect is often overlooked in various studies and reconstructions of past solar activity. The most reliable indicator that an increase in ^{14}C production corresponds to a solar grand minimum is the synchronous detection of its effect in climate proxies.

In summary

The Holocene is the result of changes in orbital forcing and strong feedbacks. In the early Holocene, the Earth's axis was more tilted and at the same time oriented so that the Northern Hemisphere received more energy during the summer. This led to the warm climate of the Holocene Optimum. However, as these two factors diminished over time, the planet gradually cooled, leading to the Neoglaciation. In contrast, CO_2 levels responded to conditions in the Southern Ocean, which were influenced by increasing summer solar energy in the Southern Hemisphere, as the two hemispheres have opposite trends in summer insolation. Climate models fail to accurately reproduce Holocene cli-

---

[358] Svensmark, H., 1998. Phys. Rev. Lett. 81 (22), 5027.
  doi.org/10.1103/PhysRevLett.81.5027

matic conditions, suggesting an inaccurate assessment of climate forcings. The Winter Gatekeeper hypothesis emphasizes the critical role of the Northern Hemisphere latitudinal temperature gradient in the Earth's energy balance. Accordingly, it suggests that the observed cooling is primarily caused by the increased poleward heat transport resulting from orbital changes and the subsequent accentuation of this gradient.

The 2.8 kiloyear event is one of several abrupt climate events within the Holocene. This global phenomenon caused major changes in temperature, atmospheric circulation, precipitation patterns, El Niño activity, and even the human population. Notably, it is unrelated to CO_2 fluctuations or volcanic eruptions but correlates with two solar grand minimums that occurred 2,990 and 2,790 years ago. The Enhanced CO_2 Effect hypothesis does not explain this and other abrupt climate phenomena observed during the Holocene. However, the Winter Gatekeeper hypothesis provides a plausible explanation for the synchronicity between variations in solar activity and climatic effects during the 2.8 kiloyear event. It proposes that changes in atmospheric circulation alter heat transport to the Arctic, ultimately affecting the global energy budget.

CHAPTER 46
EXPLAINING RECENT CLIMATE CHANGE

According to direct and proxy climate records, the Little Ice Age ended in the mid-1840s, followed by a significant warming trend. However, this warming has not been consistent over time. Instead, it has occurred in alternating 30-year warming and cooling periods within the broader long-term warming trend. The IPCC does not consider this climate pattern to be significant and attributes substantial long-term climate change only to human-induced greenhouse gas emissions and aerosols. This position contrasts sharply with observations of substantial warming and glacial retreat between 1845 and 1940, which account for almost half of the total change, even though only 10% of total human CO_2 emissions occurred during this period. While climate models reproduce the long-term warming trend, they often struggle to accurately reproduce the timing of specific changes, especially during the early 20th-century warming and mid-20th-century cooling. The Winter Gatekeeper hypothesis offers an explanation for the timing of temperature changes, attributing most of the long-term warming during the 20th century to the Modern Solar Maximum.

Global warming started after the Little Ice Age

The Little Ice Age was identified by studying evidence from glaciers in the western United States. It was found that these glaciers were not remnants of the Pleistocene but had formed and expanded during the late Holocene, reaching their maximum size between the 16th and 18th centuries. Using the same criteria, it can be determined that the Little Ice Age ended around 1845 when glaciers began to retreat worldwide. This retreat is supported by old photographs of the Rhône Glacier in the Alps (fig. 79a). The global trend of glacier retreat is well documented (fig. 79b, thick black line) and continues today. Several proxies (fig. 79b, thin lines) confirm that modern warming began after a particularly cold period from 1809 to 1843. This interval coincides with four major volcanic eruptions (fig. 79b, gray bars), including the 1815 eruption of Mt. Tambora discussed in chapter 24.

A few years after an extraordinary cluster of powerful eruptions not seen since 1300, the Little Ice Age ended, marking the beginning of modern global warming. The warming soon after the eruptions supports the idea that volcanic eruptions have an acute but short-term effect on the climate. The exact reasons for the onset of global warming in the late 1840s remain unknown to scientists, although it is clear that natural factors triggered it. Analysis of climate proxies (fig. 79b) indicates that much of the warming and glacial retreat occurred before 1900, well before human emissions became noticeable. Furthermore, most of these changes occurred before 1960, when human emissions accelerated significantly. Therefore, most of the observed global warming over the past 175 years is due to natural causes, with human emissions becoming important only in the past 60 years. However, the specific natural causes responsible and their mechanisms remain unclear.

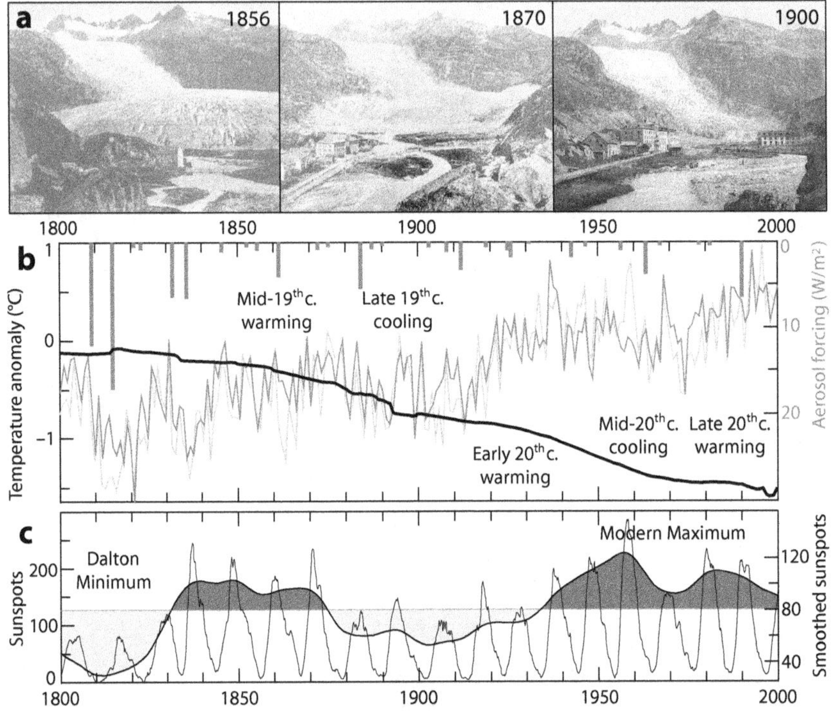

Figure 79. Two centuries of warming. a) Old photographs of the Rhône Glacier, Switzerland. b) The thick black line represents the global mean glacier retreat (unscaled, with a range of 2.4 km, 1.5 miles). The thin medium gray line records tree growth anomalies and the thin light gray line is a multi-proxy reconstruction of Northern Hemisphere temperature variations. The dark gray bars represent a reconstruction of the global volcanic aerosol forcing. c) The number of sunspots represents solar activity. The thick line shows a smoothing of the data, with above-average areas in dark gray and below-average areas in light gray.[359]

The Winter Gatekeeper hypothesis offers an explanation for the end of the Little Ice Age. After a period of low solar activity known as the Dalton Minimum, which lasted from 1795 to 1835, solar activity recovered and became high between 1835 and 1875 (fig. 79c). This period coincides with the warming observed in the mid-19ᵗʰ century. While changes in solar activity have a minimal direct influence on surface temperatures, the increased solar activity likely increased global warming indirectly by reducing poleward heat transport. The resulting reduction in Arctic energy loss contributed to the overall increase in global temperatures. At the same time, the associated reduction in

---

[359] Data for glacier length from Oerlemans, J., 2005. Science, 308 (5722), pp.675–677. doi.org/10.1126/science.1107046 Data for climate proxies and volcanic eruptions from Sigl, M., et al., 2015. Nature, 523 (7562), pp.543–549. doi.org/10.1038/nature14565 Sunspot data from SILSO www.sidc.be/SILSO/home

moisture transport led to reduced snowfall, resulting in a double impact on glaciers from rising temperatures and reduced precipitation.

The hypothesis that the recent climate change is due to human emissions

It is interesting to observe that the temperature changes over the last two centuries show a pattern characterized by alternating periods of warming and cooling, each lasting about 30 years (fig. 79b). This distribution shows that the 19th century had two cooling periods and one warming period, while the 20th century had the opposite pattern. Based on this alone, one would expect a higher rate of warming in the 20th century. Furthermore, the long-term warming trend results from the cooling periods, causing a smaller decrease in temperature compared to the increase during the warming periods.

This climate structure reflects the influence of multidecadal oceanic oscillations, also known as climate modes of variability, as it oscillates in sync with the Atlantic Multidecadal Oscillation. As discussed previously (ch. 19 & 36), these multidecadal oscillations lead to changes in poleward heat transport and act as the primary drivers of climate change on the multidecadal time scale. Significantly, they modify the Earth's energy budget by altering the outgoing longwave radiation, but their role as a major climate forcing remains unrecognized.

Scientists working with the IPCC believe that modes of variability simply redistribute energy within the climate system. According to the IPCC's most recent report, these modes are predominantly observed as regional surface temperature variations rather than global ones. However, some climate scientists disagree and argue that these modes have a global influence.[360] This influence is evident in the global surface temperature record (fig. 31, ch. 19).

According to the IPCC's 6th Assessment Report FAQ 3.1, historical temperature changes can be explained by only three categories of climate forcing.[361] Figure 80 shows the average values of these forcings as proposed by climate models. The contribution to observed temperatures is calculated on the basis of the Enhanced CO_2 Effect hypothesis, taking into account human GHGs, human aerosols, and natural forcings (solar + volcanic). However, there appears to be an error in the original IPCC figure, as it suggests that human aerosols have a warming effect and human GHGs have a cooling effect during the first two decades, which seems unlikely.

The three categories shown in figure 80, represented by thin lines, are derived from available data and estimated using models to determine their effect on the radiative flux at the top of the atmosphere. This radiative forcing is then used as input to climate models. In the case of the IPCC figure, shown in figure 80, the radiative forcing has been converted into the surface temperature change caused by each forcing category and compared to the global temperature from observations (thick black line). The challenge for climate modelers is to reproduce past observed temperatures using only these three types of forc-

---

[360] Schlesinger, M.E. & Ramankutty, N., 1994. Nature, 367 (6465), pp.723–726. doi.org/10.1038/367723a0

[361] Eyring, V., et al., 2021. Climate Change 2021: The Physical Science Basis. 6th AR IPCC. p.516. doi.org/10.1017/9781009157896.005

ing. Because the success of models is measured by their ability to simulate past climates, climate modelers may also adjust numerous parameters within the models for factors that cannot be directly calculated until this is achieved.

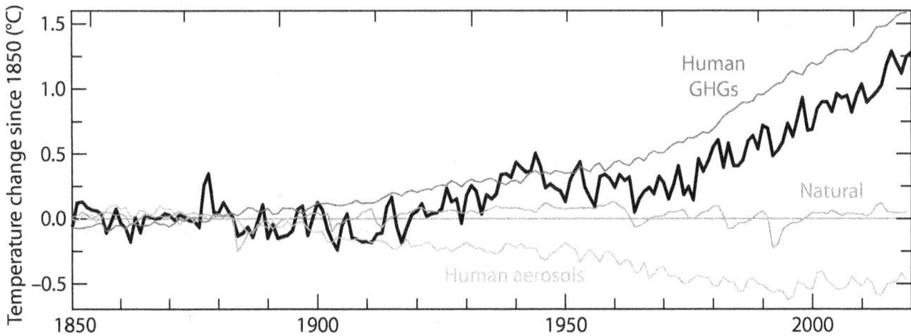

Figure 80. Drivers of climate change since 1850. Global surface temperature changes in observations compared to average multi-model simulations of the response to human GHGs (dark gray line), aerosols, and other human forcings (light gray line), and natural forcings only (medium gray line).

It is important to note that this figure shows a negligible natural contribution to the long-term trend. According to the hypothesis put forward by the IPCC scientists, natural causes have not been responsible for climate change over the past 170 years and have only played a role in short-term effects. This suggests that without human influence, today's temperature would not be different from that of 1850. However, a major climate change occurred between 1850 and 1940, as shown in figure 79, during which time human CO_2 emissions were less than 10% of the total emissions since 1850.[362] Although many scientists unquestioningly accept the notion that natural climate change has had no effect, the evidence makes it difficult to agree with this claim.

The IPCC report also dismisses the importance of natural internal variability, claiming that its influence on global temperature is minimal over time scales of three decades or more. Instead, the report claims that the long-term trend is primarily driven by human influence. However, downplaying a prominent climate feature that models fail to reproduce can foster unwarranted confidence in model performance. This is evident when models fail to accurately capture temperature trends in the 21st century, leaving modelers uncertain about the reasons for the discrepancy.[363]

Climate models' view of recent climate change is incorrect

Figure 81a compares the latest multi-model average from the 6th Model Intercomparison Project (dashed gray line) with the past temperature record (black line). While the models generally perform well, scientists recognize that most models underestimate the warming observed in the early 20th century and overestimate the warming observed after 1998.[364]

---

[362] ourworldindata.org/co2-emissions

[363] Voosen, P., 2021. Science, 373 (6554) pp.474–475.
doi.org/10.1126/science.373.6554.474

[364] Papalexiou, S.M., et al., 2020. Earth's Future, 8 (10), p.e2020EF001667.
doi.org/10.1029/2020EF001667

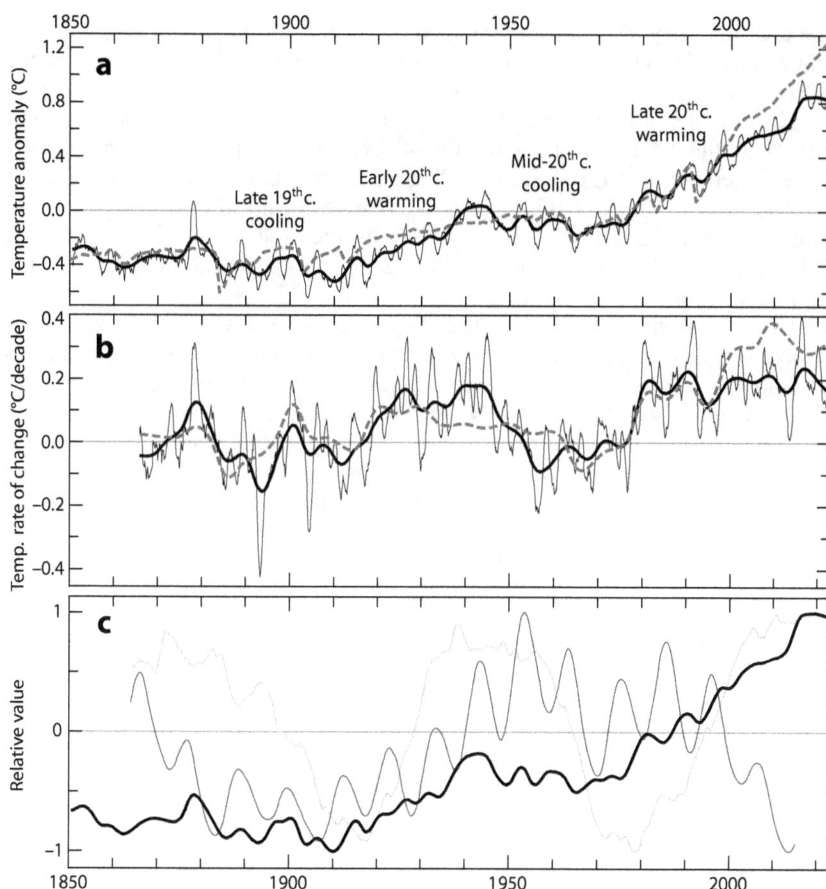

Figure 81. Opposing explanations for recent climate change. a) The black line is the observed climate change, with the thick line representing the smoothed data. The dashed gray line is a similarly smoothed multi-model mean simulation. b) 15-year temperature change rate from (a). c) The black line is as in (a), the medium gray line is the sunspot number, and the light gray line is the Atlantic Multidecadal Oscillation index, both similarly smoothed and expressed as a ratio of their means.[365]

But the problems go deeper. By calculating the rate of warming over a 15-year moving window, we gain insight into the timing and magnitude of temperature changes in figure 81b. This figure compares the rate of warming for the observations (black line) and the multi-model average (dashed gray line). Positive values indicate an overall warming trend, while negative values indicate a cooling trend over the preceding 15-year period.

The cooling observed in the late 19th century is primarily attributed in the models to a stronger response to the 1883 Krakatoa eruption than indicated by the proxies and temperature records. While these records reflect the impact of

---

[365] Temperature data from HadCRUT5, model data for CMIP6 SSP245 from KNMI explorer climexp.knmi.nl/CMIP6/Tglobal/global_tas_mon_ens_ssp245_192_ave.dat sunspot data from SILSO, and AMO data from NOAA.

the eruption, they also show a strong cooling trend that began in 1878 and continued through 1893. Additional cooling was observed in the first decade of the 20th century.

The rate of warming during the early 20th century warming period was comparable to the present, with a rate of 0.18 versus 0.20°C per decade, despite the large difference in CO_2 emissions. This difference is the reason why climate models do not simulate this period. Similarly, the mid-20th-century cooling, which began around 1945, does not begin in the models until the eruption of Mt. Agung in 1963, and even then, they tend to exaggerate the effects of the eruption.

The models successfully simulate the late 20th-century warming period but show the expected increase in the rate of warming that corresponds to the accelerating rise in CO_2 levels. However, the observed rate of warming has not shown a similar acceleration since 1976, despite a large 90 ppm (26%) increase in CO_2 levels. The IPCC reports currently lack an explanation for the absence of this warming acceleration despite continued predictions of accelerated warming and sea level rise. This lack of warming acceleration has led to an increasing failure of models to reproduce temperatures since 1998, an issue discussed in more detail in chapter 49.

An alternative explanation for the recent warming

The Winter Gatekeeper hypothesis provides an alternative view that does not suffer from the problems described above. According to this hypothesis, as shown in figure 81c, the cooling observed in the late 19th century can be attributed to increased heat transport caused by low solar activity and a shift into the negative phase of the multidecadal oscillation. On the other hand, the early 20th century saw a warming initiated by the shift to a positive phase (decrease in heat transport) of the multidecadal oscillation and the influence of the Modern Solar Maximum since the mid-1930s. The cooling in the mid-20th century resulted from the shift to a negative phase in the multidecadal oscillation. This cooling effect was mitigated by a counteracting effect on stratospheric transport due to high solar activity. In the late 20th century, both forcings again contributed to warming. However, in the 21st century, low solar activity has acted as a dampening factor, potentially leading to a decrease in the rate of warming, as the multidecadal oscillation is expected to turn negative in the mid-2020s.

In essence, a significant portion of the observed warming in the 20th century can be attributed to the impact of the Modern Solar Maximum on the Earth's energy budget. It is noteworthy that such an extended period of above-average solar activity has not been documented for more than 600 years (fig. B21, ch. 23). According to the hypothesis, this increased solar activity led to a reduction in the amount of heat transported to the Arctic, resulting in a decrease in cooling during the mid-20th century and an increase in warming during the early and late 20th century.

Rising CO_2 levels have undoubtedly influenced the observed warming trend of recent decades, but the Enhanced CO_2 Effect hypothesis tends to overemphasize its role by neglecting the natural causes considered in the Winter Gatekeeper hypothesis. A more comprehensive understanding of recent climate change should attribute the observed warming to a combination of human emissions and natural variations in poleward heat transport.

Box 27. The polar enigma

Polar amplification refers to the phenomenon whereby temperature changes at high latitudes are more pronounced than the global average. This concept is consistent with the paleoclimate evidence that the Earth has changed its global temperature primarily by affecting regions outside the tropics, especially at high latitudes. We have already examined this phenomenon when discussing the latitudinal temperature gradient (fig. 13, ch. 9) and climate conditions during the Early Eocene (ch. 20).

Polar amplification was first observed in early climate models in the 1970s and is consistently reproduced in current models as a response to increasing greenhouse gases. Identifying the exact causes of this phenomenon has proven challenging. In models, polar amplification is attributed to a combination of radiative forcing patterns and various feedback mechanisms.[366] The most well-established mechanism is the high-latitude surface albedo feedback, driven primarily by changes in sea ice and snow cover. However, the Planck and lapse rate feedbacks are found to be more important in the simulations.

The lapse rate feedback is influenced by temperature changes with altitude (ch. 7). In the tropics, where the upper troposphere is expected to warm due to increased latent heat release, the feedback is negative. However, a more stable atmosphere at high latitudes limits warming to lower altitudes, leading to positive feedback. On the other hand, the Planck feedback results from increased longwave emissions from warmer surfaces and is negative in all regions. However, its effect is more pronounced at warmer lower latitudes. In principle, both feedbacks reduce warming at low latitudes more than at high latitudes.

However, actual observations differ significantly from climate model simulations with respect to polar amplification. In the Southern Hemisphere, polar temperatures in the lower troposphere (fig. B27, gray line) have remained unchanged over the past four decades, even though models predict warming in Antarctica. Similarly, in the Northern Hemisphere, polar temperatures in the lower troposphere (fig. B27, black line) showed a cooling trend from 1979 to the mid-1990s. The warming observed in later years was primarily a result of the 1997 climate shift and the subsequent response of the Arctic, which is discussed in detail in chapter 34.

Climate scientists have yet to explain the large discrepancies between model simulations and actual observations of polar amplification. Recognizing the importance of this issue, the Polar Amplification Model Intercomparison Project was established.[367] The researchers emphasize the complexity of the factors influencing polar temperature trends. They note that although Antarctic amplification has been delayed to date, it is expected to occur in the future in response to continued increases in GHGs.

---

[366] Smith, D.M., et al., 2019. Geosci. Model Dev. 12 (3), pp.1139-1164. doi.org/10.5194/gmd-12-1139-2019

[367] Smith, D.M., et al., 2019. Geosci. Model Dev. 12 (3), pp.1139-1164. doi.org/10.5194/gmd-12-1139-2019

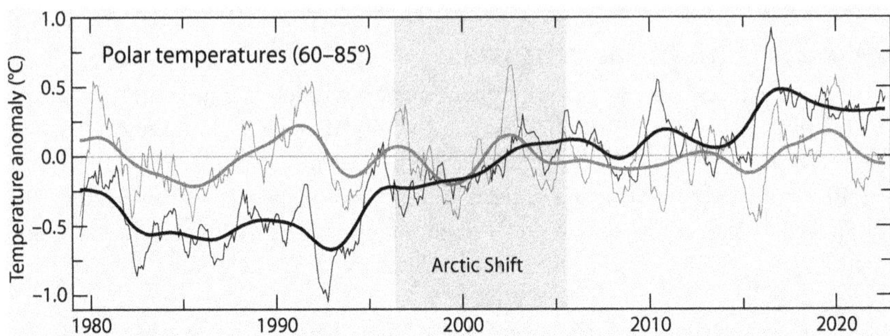

Figure B27. Polar temperatures. Satellite record of temperature changes in the lower troposphere for the 60-85° latitude regions. The black lines correspond to the data for the north polar region and its smoothing, and the gray lines correspond to the south polar region.[368]

However, the explanation offered contradicts the presumed causes of polar amplification. The lapse rate and Planck feedbacks do not primarily act to increase polar temperatures but rather to decrease non-polar temperatures. If their estimated values were inaccurate, the effect would be more significant for global warming as a whole rather than specifically for polar warming.

The Winter Gatekeeper hypothesis provides a coherent explanation for polar temperature trends by linking the strength of the polar vortex to the transport of heat to the poles. The polar vortex is strong in the Southern Hemisphere due to reduced planetary wave activity. As a result, the heat transport to Antarctica remains more stable compared to the large fluctuations observed in the Arctic. Moreover, in the Antarctic atmosphere, which is characterized by low water vapor content and temperature inversions, the increase in CO_2 levels is expected to cause cooling rather than warming by increasing infrared emissions.[369]

In terms of temperatures in the northern polar region, the Arctic experienced a cooling from 1976 to 1997 due to the strong vortex conditions within the low-transport climate regime discussed in Chapter 39 (fig. 61). As noted above, the weakening of vortex conditions following the 1997 climate shift has led to Arctic warming. In chapter 34, we considered how the recent enhanced Arctic warming is due to changes in heat transport rather than Arctic amplification.

In summary

Climate models that assume that recent climate change is primarily driven by changes in CO_2 resulting from human emissions face challenges in accurately reproducing the patterns observed since the end of the Little Ice Age. This difficulty suggests that the main drivers of climate change may be misidentified. Human emissions were relatively low between 1845 and 1940, when much of the climate change of the past 175 years occurred. This casts doubt on the widely accepted view that natural factors have made an insignificant contribution to long-term climate change.

---

[368] Data from the UAH satellite temperature dataset.
[369] van Wijngaarden, W.A. & Happer, W., 2020 arXiv preprint
doi.org/10.48550/arXiv.2006.03098

In contrast, the Winter Gatekeeper hypothesis provides a more compelling explanation for the observed patterns of climate change, which are characterized by alternating 30-year warming and cooling periods. This hypothesis also highlights the importance of the Modern Solar Maximum as a major factor in driving the observed 20th-century warming. It also provides insights into the absence of polar amplification in Antarctica and the appearance of such amplification in the Arctic since the mid-1990s. By taking these factors into account, the Winter Gatekeeper hypothesis provides a more comprehensive understanding of recent climate change dynamics and challenges the dominant assumption based primarily on human CO_2 emissions.

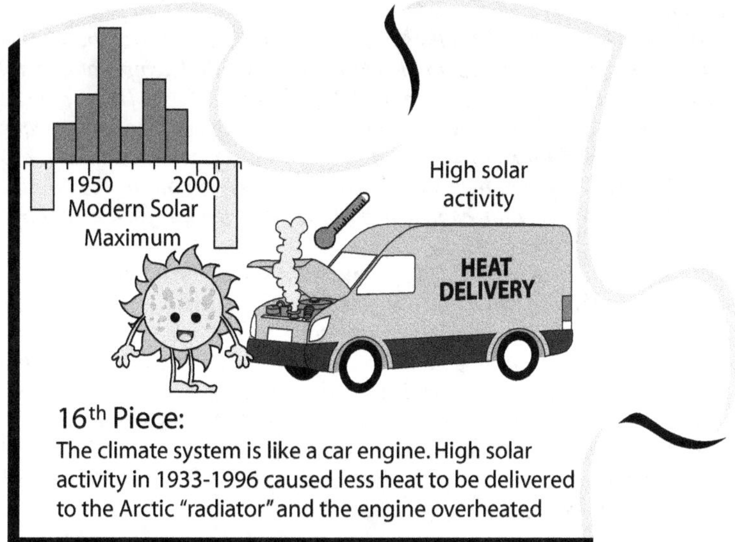

SECTION 13 KEY ISSUES

The Winter Gatekeeper outperforms the Enhanced CO$_2$ Effect hypothesis in explaining past climate puzzles. The isolation of Antarctica by the Circumpolar Current halved Oligocene CO$_2$ levels, leading to a long period of global warming that resulted in the Mid-Miocene Climate Optimum. The new hypothesis also illustrates how small changes in Milankovitch forcing at the top of the atmosphere lead to massive changes in the volume of Earth's ice sheets by showing the effect of current changes in heat transport on Greenland's summer snow cover.

The Winter Gatekeeper hypothesis explains the discrepancy between CO$_2$ and global temperature trends in the Holocene. Global temperatures follow Northern Hemisphere insolation, while CO$_2$ follows Southern Hemisphere temperatures. About 3,000 years ago, an abrupt 350-year cooling event occurred, unrelated to CO$_2$ variations or volcanic eruptions but correlated with two solar grand minimums. This hypothesis explains the evidence pointing to changes in atmospheric circulation and heat transport as the driving factors behind the event.

About half of the post-Little Ice Age warming occurred between the 1840s and 1940s, even though only 10% of human CO$_2$ emissions occurred during that period. Models fail to reproduce this uneven warming, characterized by alternating periods of warming and cooling. This calls into question the Enhanced CO$_2$ Effect hypothesis, which ignores the role of natural factors in long-term climate change. The Winter Gatekeeper hypothesis provides a superior explanation for the warming and its timing over these hundred years. It also explains the observed Arctic amplification since the mid-1990s and accounts for its absence before and in Antarctica.

Section 14. Explaining Climate Change Mechanisms

CHAPTER 47
THE BLIND MEN AND THE ELEPHANT

The Winter Gatekeeper hypothesis introduces a new factor contributing to climate change, suggesting that previous causes may have been misidentified. Contrary to what it may seem, it is not difficult for scientists to have missed a major cause of climate change because those working with the IPCC have focused primarily on attributing climate change to human activities rather than identifying its causes. Finding a missing cause in recent decades would be difficult, but it becomes easier when considering the Little Ice Age, where the main cause remains unidentified. Researchers have found that the climatic equator, where the trade winds from both hemispheres meet, shifted south by 5° latitude during the Little Ice Age. This shift resulted in a significant increase in heat transport from the Southern Hemisphere to the Northern Hemisphere, providing evidence for the existence of an unknown cause capable of driving climate change in this way.

The case for a missing forcing

The Winter Gatekeeper hypothesis introduces a novel climate-forcing mechanism by proposing that changes in poleward heat transport can strongly influence climate. This mechanism affects the radiative flux at the top of the atmosphere, which changes the energy content of the entire climate system. However, for this new forcing to be incorporated into climate change theory, it must fill a gap in the current understanding of climate change. Moreover, the importance of the proposed role of the new driver is directly proportional to the magnitude of the missing forcing it seeks to account for.

Identifying causality in a highly complex system such as the climate is a major challenge, and the scientists involved in the IPCC reports do not actively pursue this endeavor. Rather, their primary focus is on identifying and attributing the impact of human activities on the climate. They assume that the causes of climate change are already well understood. Consequently, if a particular forcing were missing from the established set, any resulting effects would be attributed to one or more of the already identified forcings.

If the Winter Gatekeeper hypothesis is correct, the Modern Solar Maximum should have contributed substantially to the observed 20th-century warming. However, because of the simultaneous influence of anthropogenic forcings, it would not be readily apparent if we were missing a forcing in our understanding of recent climate change. To identify a missing forcing, we need to examine past climatic conditions that challenge the Enhanced CO_2 Effect hypothesis. One such example is the Miocene climatic optimum discussed in chapter 44. During this period, the planet experienced temperatures 5 to 8°C (9-14 °F) higher than today, despite similar CO_2 levels of about 400 ppm. Notably, climate models require CO_2 levels of at least 800 ppm to accurately simulate this past climate, suggesting the existence of an unidentified forcing.

The missing forcing from the Little Ice Age

We don't have to go that far back in time to find another missing forcing; just a few centuries will do. The Little Ice Age presents a similar puzzle, with its primary cause still elusive. Global glacier extent and climate proxies suggest that the dramatic and extensive cooling that began around 1300 and lasted for five centuries cannot be attributed solely to gradual Milankovitch forcing (fig. 35, ch. 21). To complicate matters, CO_2 levels remained stable between 1100 and 1500, volcanic activity was below average for most of this period (fig. 42, ch. 26), and the accepted solar forcing alone is insufficient to explain the observed cooling trend.

Let's explore this elusive missing forcing by focusing on the Intertropical Convergence Zone. Box 2 (ch. 3) defined it as the belt where the warm, moisture-laden trade winds from both hemispheres converge. Here, air rises by convection, creating a band of clouds and storms that encircles the Earth near the equator. It is essentially the climatic equator within the global atmospheric circulation system. Heat is transported from this band toward the poles. The position of this zone varies seasonally. It moves into the summer (warmer) hemisphere to transport more heat to the winter (colder) hemisphere (fig. 17, ch. 11).

Despite its global impact, the Little Ice Age manifested primarily in the Northern Hemisphere. This well-documented cooling period resulted in a notable shift of the Intertropical Convergence Zone toward the comparatively warmer Southern Hemisphere. Scientists have conducted studies on several Pacific islands using a variety of proxies. Their results show that during the Little Ice Age, the climatic equator shifted about 500 kilometers (about 5° of latitude) southward from its current average position.[370]

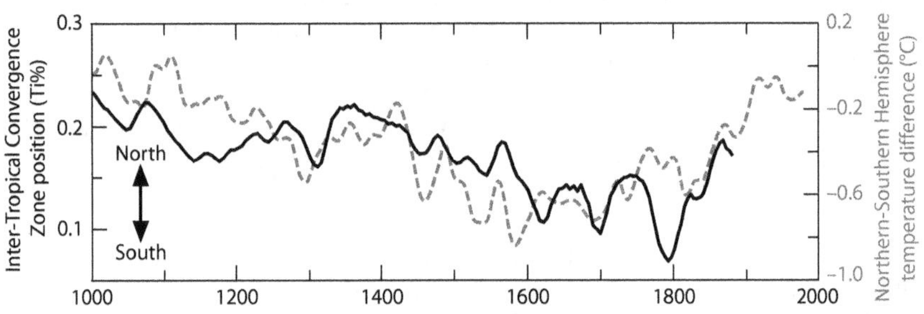

Figure 82. Migration of the climate equator during the Little Ice Age. Position of the Intertropical Convergence Zone (black line) and interhemispheric temperature contrast (dashed gray line) during the last 1000 years.

Figure 82 illustrates the southward migration of the Intertropical Convergence Zone as represented by a precipitation proxy from the Caribbean Sea (black line).[371] In this report, the authors examine the shift in the context of the evolving temperature contrast between the two hemispheres (dashed gray line). In addition, they examine the effect of the changing location of the Intertropi-

[370] Sachs, J.P., et al., 2009. Nat. Geosci. 2 (7), pp.519–525. doi.org/10.1038/NGEO554
[371] Schneider, T., et al., 2014. Nature, 513 (7516), pp.45–53. doi.org/10.1038/nature13636

cal Convergence Zone on the atmospheric energy balance, linking its location to the transport of energy through the atmosphere.

The effect of the position of the Intertropical Convergence Zone on atmospheric heat transport poses an intriguing dilemma that is not addressed in the study. During the Little Ice Age, when the climate equator shifted toward the Southern Hemisphere, thereby directing more heat toward the Northern Hemisphere, an important question arises: if the Northern Hemisphere experienced significant cooling, what happened to that energy? Multiple studies have shown that the ocean cooled during this period, indicating that the energy was not transferred there. It is clear that there was a large energy deficit in the Northern Hemisphere, but the exact cause remains unclear. While changes in ice-snow albedo are often proposed as a contributing factor, it's important to note that albedo operates as a feedback mechanism, meaning that the cooling must occur before the increase in ice-snow albedo can take place. Cooling should also precede the shift in the position of the climate equator.

There is another study that provides valuable insight into the nature of this problem. It quantifies the increase in heat transport across the equator resulting from the 5° southward shift of the Intertropical Convergence Zone during the Little Ice Age at 1.7 petawatts (PW).[372] To put this into perspective, recall from chapter 9 that the current hemispheric transport is about 5-6 PW. Therefore, the observed shift in the position of the zone represents a substantial global reorganization of heat transport. The authors of the study conclude that *"presently, there is no known climate forcing or feedback during that time period that could account for such a large energy perturbation at the hemispheric scale."*

Scientists acknowledge the existence of a large, unidentified forcing that created a significant energy deficit in the Northern Hemisphere during a prolonged period of low solar activity. This phenomenon triggered a profound shift in global heat transport patterns through the atmosphere. The Winter Gatekeeper hypothesis offers a compelling solution to this puzzle, providing a plausible explanation for the missing forcing. It is important to note that attributing this missing forcing solely to the Little Ice Age and assuming that it is unrelated to recent climate change lacks a scientific basis.

A global network of teleconnections

Scientists recognize that decadal climate variability, referred to in this book as ocean oscillations, reflects a global network of teleconnections spanning different ocean basins, tropical and extratropical regions, and ocean and land areas.[373] Despite several decades of extensive research, understanding the underlying causes of this variability remains a formidable challenge. The connections between different basins highlight the global nature of the associated atmospheric mechanism that either drives or responds to oceanic variability. Most scientists believe that this variability is internally generated. This would also require an unknown underlying oceanic mechanism since the atmosphere is not thought to have the necessary memory. This issue will be discussed in the next chapter.

---

[372] Donohoe, A., et al., 2013. J. Clim. 26 (11), pp.3597–3618. doi.org/10.1175/JCLI-D-12-00467.1

[373] Cassou, C., et al., 2018. Bull. Amer. Meteor. Soc. 99 (3), pp.479–490. doi.org/10.1175/BAMS-D-16-0286.1

The IPCC 6th Assessment Report recognizes that natural internal variability involves the redistribution of energy within the climate system.[374] Consequently, studying global changes in poleward heat transport is critical to understanding the influence of this interconnected network on energy redistribution. However, there is a hurdle. When forced to follow the temporal evolution of oceanic oscillations, global climate models confirm that many observed historical climate changes can be partially attributed to these oscillations.[375] However, when these models are allowed to run autonomously, their ability to reproduce these teleconnections is severely degraded, and they exhibit spontaneous ultra-low frequency variations in inter-basin connectivity.[376]

Climatologists face a significant challenge in the absence of global-scale multidecadal variability in state-of-the-art climate models. This limitation hinders their ability to address fundamental questions about its origin and to understand the full extent of its impact on climate change. It also hinders accurate predictions of its potential evolution, which is a serious concern. The influence of multidecadal climate variability is far-reaching, affecting precipitation patterns, temperatures, and the occurrence of extreme weather events that affect the lives of most of the planet's inhabitants. However, these critical aspects are not adequately reflected in the IPCC reports, which emphasize the relatively small impact of natural variability over periods longer than two decades and instead emphasize the dominant role of human influence on long-term climate trends.

Chapter 12 discussed one possible reason for the inability of climate models to reproduce global shifts in poleward heat transport. These models have been developed on the assumption that the Bjerkness compensation hypothesis is correct. However, this hypothesis contradicts substantial evidence that changes in oceanic and atmospheric transport do not compensate for each other. As shown in chapters 17 (fig. 27) and 34 (fig. 54), scientists have observed synchronous increases in atmospheric and oceanic transport to the Arctic during the 21st century. Despite this evidence, the prevailing view, supported by the models, is that the Arctic is warming while the Bjerkness compensation mechanism continues to operate.[377]

The blind men and the elephant

A well-known Indian parable dating back several thousand years tells the story of a group of blind men who venture into a forest and encounter an elephant. Each blind man touches a different part of the elephant's body and then describes it from his limited perspective. As a result, they mistakenly believe they have encountered different animals. The underlying message of this story is that reality can present itself in different forms if a global perspective is not taken into account.

---

[374] Eyring, V., et al., 2021. Climate Change 2021: The Physical Science Basis. 6th AR IPCC. p.517. doi.org/10.1017/9781009157896.005

[375] Ruprich-Robert, Y., et al., 2017. J. Clim. 30 (8), pp.2785–2810. doi.org/10.1175/JCLI-D-16-0127.1

[376] Kravtsov, S., et al., 2018. NPJ Clim. Atmos. Sci. 1 (1), p.34. doi.org/10.1038/s41612-018-0044-6

[377] Burgard, C. & Notz, D., 2017. Geophys. Res. Lett. 44 (9), pp.4263–4271. doi.org/10.1002/2016GL072342

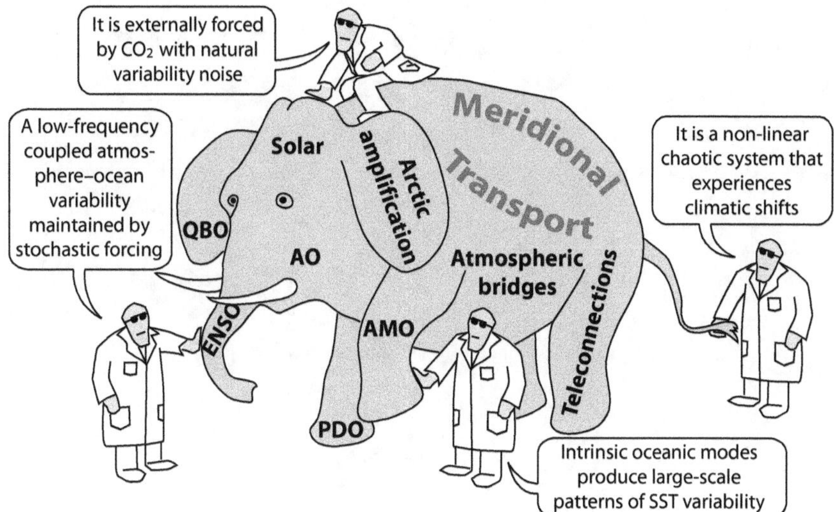

Figure 83. The blind scientists and meridional transport.

Similarly, scientists studying different types of variability, investigating potential causes, exploring teleconnections, and studying atmospheric bridges often fail to recognize that they are studying different facets of a global phenomenon: the fundamental characteristic of global climate – namely, variability in the meridional transport of heat and moisture from the tropics to the poles. The understanding that different factors modulate this variability is the basis of the Winter Gatekeeper hypothesis.

In summary

The atmospheric reorganization that occurred during the Little Ice Age had a profound effect on the Earth's energy balance. The Northern Hemisphere experienced a considerable energy deficit, leading to a substantial increase in heat transport from the Southern Hemisphere. The Winter Gatekeeper hypothesis explains this phenomenon by attributing it to increased poleward heat transport caused by reduced solar activity, resulting in greater energy loss in the Arctic.

Multidecadal ocean oscillations are recognized as a reflection of a global network of teleconnections linking ocean basins to the rest of the globe. These oscillations play an important role in redistributing energy, resulting in variable temperature trends and strongly influencing climate. However, climate models cannot capture global-scale multidecadal variability. As a result, scientists may not realize that this variability is a manifestation of global changes in poleward heat transport, which the current hypothesis does not account for.

CHAPTER 48
UNRESOLVED CLIMATE REGIMES AND THEIR SHIFTS

Regimes and shifts play a critical role in shaping the climate changes we experience during our lifetimes, making them the most significant climatic manifestations for humans. This fact is strongly supported by the substantial ecological response observed in most North Pacific species during the climate shifts that led to their discovery. Surprisingly, these substantial climate changes receive minimal attention in IPCC reports, with their effects dismissed as mere multi-decadal noise overshadowing human impacts. Yet climate regimes profoundly influence various aspects of our climate, including critical factors such as the frequency and intensity of hurricanes. Unfortunately, recognition of their importance is hampered by the unknown origins of these regimes, as scientists are often reluctant to admit their lack of understanding. Despite extensive research to identify an oceanic cause, progress has been limited in recent decades.

The importance of climate regimes and shifts

The climate we experience during our lifetimes is influenced more by multidecadal oscillations than by its long-term trend. These oscillations occur in cycles of about 50-70 years, roughly the span of a human lifetime. As a result, many people perceive noticeable differences in climate from several decades ago, leading to concern about these changes. A historical example, shown in figure 53 (ch. 34), illustrates public concern about Arctic warming in the 1920s. As early as the 1930s, Guy Callendar expressed concern about warming and attributed it to human emissions. The 1970s brought fears of an impending glaciation due to cooling trends. In the 1990s, global warming became a major concern. However, these changes were primarily the effects of multidecadal oscillations interacting with a gradual warming trend that was imperceptible to most people. Multidecadal climate oscillations are, therefore, of great concern to us. While the IPCC reports refer to them as natural internal variability, implying an internal origin, the truth remains that we have yet to identify their causes.

In the 1990s, the field of ecology made a major contribution to climate studies. In 1991, marine ecologists identified an abrupt change in the North Pacific ecosystem that had occurred 15 years earlier, caused by a sudden climate shift.[378] Surprisingly, this change had escaped the attention of climatologists, who were primarily focused on identifying a discernible human impact on the climate. Through detailed investigation, researchers determined that the origin of the change was not the ocean but a shift in global atmospheric circulation

---

[378] Ebbesmeyer, C.C., et al., 1991. Proceedings of the Seventh PACLIM Workshop, April 1990. Interagency Ecological Studies Program Technical Report, 26 pp.115–126. hdl.handle.net/1834/22168

(fig. 48, ch. 31).[379] This shift presented a challenge, however, because climate models do not reproduce such events, leaving their causes shrouded in mystery. In essence, this shift represented a transition between stable states, which have since come to be known as climate regimes.

In ecology, the concept of climatic regimes induced by climate shifts has gained considerable traction, leading to numerous studies in different ecosystems. Surprisingly, climatology has paid relatively little attention to this area of research. For example, the 1997 climate shift receives far less mention in climate studies than the 1976 climate shift, despite being more recent. To illustrate, figure 84 is from a study that focused on the ecological implications of climate regime shifts in the North Pacific.[380] In this study, the authors compiled data from 64 biological time series, consisting primarily of commercially important fish stocks and a smaller representation of invertebrates and zooplankton. They used principal component analysis, a statistical technique that condenses the variation in the data into a few key variables. The second most influential source of variability was attributed to climatic factors, as shown by the black line in figure 84.

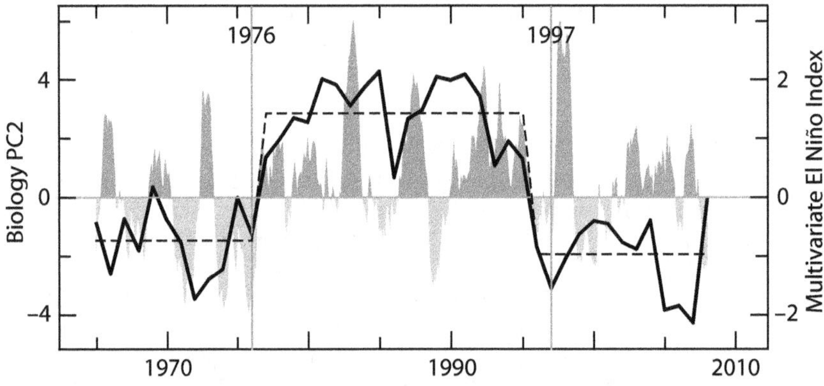

Figure 84. Ecological shifts. The black line represents the second principal mode of ecological variability in the Northeast Pacific. The multivariate El Niño-Southern Oscillation index is shown in dark gray for positive values and in medium gray for negative values.

To link ecological changes in the Pacific Ocean to climatic shifts, an index reflecting the status of the El Niño-Southern Oscillation has been included in the figure.

The large deviation of the analyzed species from previous averages during the two identified climate shifts underscores their importance. It's worth noting that these shifts are not limited to the North Pacific, where they are more obvious, but have a global impact. Surprisingly, despite their importance, climate regimes and shifts resulting from multi-decadal variability are conspicuously absent from the IPCC reports, which focus primarily on the anthropogenic causes of climate change.

---

[379] Graham, N.E., 1994. Clim. Dynam. 10, pp.135–162. doi.org/10.1007/BF00210626

[380] Litzow, M.A. & Mueter, F.J., 2014. Prog. Oceanogr. 120, pp.110–119.
    doi.org/10.1016/j.pocean.2013.08.003

Response of extreme weather events to climate regimes

Climate regimes and shifts are of great importance to humanity, as they play a central role in determining various aspects of our climate and its associated hazards. The media often report the warnings of certain scientists that climate change is exacerbating extreme weather events, increasing their frequency and severity. However, data show that climate regimes and shifts also affect extreme weather events.

Researchers measure their energy levels to gauge the strength of hurricanes (cyclones) in different basins. Data have been available since 1950 for the Northwest Pacific and the Atlantic, the regions most prone to hurricanes. Figure 85 shows the accumulated annual hurricane energy recorded in these regions.[381]

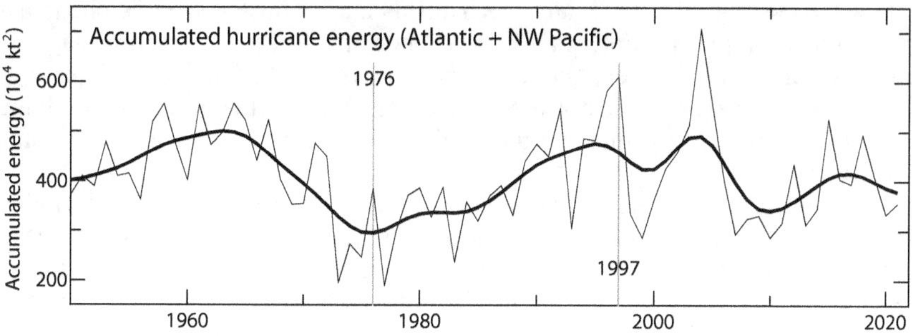

Figure 85. Accumulated cyclone energy in the Atlantic and Northwest Pacific Oceans.

Hurricane energy since 1950 shows no significant trend but rather multidecadal oscillations. However, it shows an increasing trend during the 1976-1997 climate regime and a decreasing trend in the subsequent climate regime established since 1997. Hurricanes are a clear indication of the transport of heat and moisture, primarily facilitated by the zonal (east-west) atmospheric circulation. Consequently, one would expect an increase in both the frequency and intensity of hurricanes as the zonal circulation strengthens at the expense of the meridional (south-north) circulation.

The pattern observed in the accumulated hurricane energy during the 1976-1997 regime is consistent with the information presented throughout the book, indicating a period of reduced heat and moisture poleward transport. As shown in figure 48b (ch. 31), the angular momentum of the atmosphere experienced a sharp increase during the 1976 climate shift. This increase indicates a strengthening of the zonal circulation at the beginning of the period of reduced poleward transport, which lasted until 1997.

The IPCC has become the leading authority in the field of climatology and has a major influence on current and future climate research. As a result, scientists who pursue investigations that fall outside the IPCC's focus often face less successful careers, resulting in fewer studies of those particular aspects. Despite the importance of understanding climate regimes and shifts, elucidating their causes, and developing predictive capabilities, these topics are not given the same level of priority within the IPCC agenda.

---

[381] Accumulated cyclone energy data from NOAA.

Internal versus external forcing

The prevailing view among most scientists is that multidecadal climate variability is caused by internal factors since no known external forcing of matching periodicity and magnitude has yet been identified. However, it is important to recognize that this perspective is an argument from ignorance, and the possibility of an external cause cannot be definitively ruled out. Many scientists contend that the diverse manifestations of this variability suggest a combination of factors at play. While it is clear that the geography of different ocean basins plays a crucial role in its expression, extensive research over several decades has yielded limited progress in uncovering the root cause of these multi-decadal ocean oscillations.

To establish a quasi-periodic pattern, a system must have a mechanism to track the passage of time. Many mechanisms may contribute to this phenomenon. For example, the interplay of two opposing forces with a time lag can produce an oscillation, explaining its prevalence in phenomena as diverse as the business cycle or predator-prey dynamics. A fruitful approach to studying oscillations is investigating the underlying mechanisms responsible for "time counting."

There are three plausible explanations for the presence of quasi-periodic oscillations in the Earth's oceans. The prevailing view is that the ocean serves as the primary source of memory for the system. In contrast, the atmosphere, with its dynamic and chaotic nature, lacks the capacity for long-term memory. Instead, the ocean's memory is thought to be closely tied to the heat stored in the upper layer, known as the mixed layer, especially in dynamically important regions.[382] This thermal inertia gives the ocean the ability to retain a memory of past changes.

An alternative explanation proposed is that the global signal observed in multidecadal variability arises from the synchronization of coupled nonlinear (chaotic) oscillators.[383] This phenomenon is often observed in natural systems. According to this perspective, climate regimes can be seen as synchronization states between different oceanic oscillations. Changes in the strength of their coupling would disrupt this synchronization, leading to the emergence of a new state of synchronization, thereby driving climate change.

A problem arises, however, when the driver of multidecadal oscillations is attributed solely to the ocean. Extensive evidence reviewed throughout the book suggests that the atmosphere plays a prominent role in the observed changes. Unlike the oceans, the atmosphere readily produces global changes that can manifest as shifts in atmospheric angular momentum that affect the Earth's rotation rate. In general, changes in sea level pressure precede changes in sea surface temperature by one to three months. Despite the available evidence, many scientists find it difficult to accept the third possibility: that globally coordinated multidecadal ocean oscillations are induced by atmospheric forcing over the ocean. Yet this explanation remains the most consistent with the available evidence.

---

[382] Monselesan, D.P., et al., 2015. Geophys. Res. Lett. 42 (4), pp.1232–1242. doi.org/10.1002/2014GL062765

[383] Tsonis, A.A. & Swanson, K.L., 2011. Int. J. Bifurcat. Chaos, 21 (12), pp.3549–3556. doi.org/10.1142/S0218127411030714

The problem arises because the lack of memory in the atmosphere introduces a requirement for external forcing. This is a paradigm shift of such magnitude that most scientists are reluctant to consider it without irrefutable evidence. While there is some evidence for external forcing, it remains inconclusive. In particular, the 18.6-year lunar nodal cycle has been observed multiple times in Pacific air and sea surface temperature data. In addition to the well-known Pacific Decadal Oscillation of about 60 years, this ocean exhibits a bidecadal oscillation in sea surface temperature.[384] Scientists have found that the phase and duration of this bidecadal component coincide with the lunar nodal cycle, suggesting that this lunar cycle may influence large-scale heat transfer in the western North Pacific.[385] In their 2007 study, these scientists found phase coincidence between the lunar nodal cycle and some of the most significant El Niño events of the 20th century. Based on this observation, they predicted an increased likelihood of a major El Niño event in 2015 eight years in advance, which ultimately occurred.

A combined solar and lunar influence on climate has been proposed, acting through their effect on the latitudinal temperature gradient.[386] This particular forcing mechanism is consistent with the Winter Gatekeeper hypothesis, which explains how this external forcing affects climate by altering poleward heat transport. If this combined solar-lunar forcing drives the observed multidecadal climate variability, a mechanism similar to the conjecture presented in box 25 (ch. 34) seems plausible to determine its periodicity. This mechanism would provide the necessary memory to account for the atmospheric effect. Because of their different periods, the shifting correlation between lunar and solar effects would result in the observed periodicity, as shown in figure B25 (ch. 34).

However, the Winter Gatekeeper hypothesis does not rely on an external cause for multidecadal ocean oscillations. The source of the transport changes does not alter their climatic impact. If an external cause does exist, it may take decades for it to be widely accepted since models ignore this possibility and most scientists focus their research on exploring an oceanic origin for this climate variability.

In summary

Climate regimes exhibit distinctive characteristics with respect to the intensity of atmospheric circulation, the orientation of heat transport (north-south versus east-west), and their subsequent effects on sea surface temperatures and hurricane frequency. Understanding these climate regimes and their changes is of great importance to us. However, they receive less attention because they have no anthropogenic cause. In fact, their cause remains unknown and various hypotheses have been proposed. The prevailing belief among most scientists is that climate regimes have an oceanic origin, with the upper layer of the ocean providing the necessary memory for their periodicity. Another proposed explanation revolves around the synchronization of coupled chaotic oscillators.

---

[384] Minobe, S., 1999. Geophys. Res. Lett. 26 (7), pp.855–858.
doi.org/10.1029/1999GL900119

[385] McKinnell, S.M. & Crawford, W.R., 2007. J. Geophys. Res. Oceans, 112 (C2).
doi.org/10.1029/2006JC003671

[386] Davis, B.A. & Brewer, S., 2011. Quat. Sci. Rev. 30 (15-16), pp.1861–1874.
doi.org/10.1016/j.quascirev.2011.04.016

Finally, there is the possibility that these regimes are externally triggered by the combined influence of the Moon and Sun on the latitudinal temperature gradient, leading to changes in the global atmospheric circulation. Clearly, we still have much to discover about this crucial phenomenon, and climate models are of limited help because they do not adequately reproduce it.

SECTION 14 KEY ISSUES

The Winter Gatekeeper hypothesis proposes a new cause of climate change, the absence of which is detected by studying the Little Ice Age. The 5° latitudinal shift of the climatic equator during this period represented a massive change in global heat transport that required an unknown factor capable of doing so. The effect was a huge energy transfer from the South to the North Hemisphere, indicating a large deficit of unknown origin in the North. The Winter Gatekeeper hypothesis explains both the energy deficit and the increased heat transport as a result of low solar activity.

Climate regimes and shifts result from global-scale, multidecadal variability in heat transport that is not captured by climate models and scientists struggle to understand. The minimal attention they receive in IPCC reports contrasts with their ecological importance and their critical role in shaping the climate we will experience in our lifetimes. The atmospheric circulation and heat transport characteristics of climate regimes make them determinants of many aspects of climate, including hurricane frequency. Despite decades of effort, the cause of multidecadal variability and its relationship to global heat transport remains unknown.

SECTION 15. A MODELED DISASTER

CHAPTER 49
WHAT'S WRONG WITH MODELS?

Many everyday objects and technologies rely on models to simulate well-understood processes. Climate models, however, deal with one of the most complex phenomena we know, involving many poorly understood processes. These models are complicated and fragile, representing a model climate that is very different from the real climate. They do not even know the temperature of the planet. The many problems associated with climate models suggest that the model climate they produce is only superficially similar to the real climate. In essence, the current state of knowledge makes it impossible for climate modelers to achieve their goals. In addition, there is evidence that these models are missing important climate features, causing their accuracy to deteriorate over time instead of improving.

The model world versus the real world

Computer models are mathematical representations of some aspect of reality. These models are extremely valuable because they help us navigate complexity and provide accurate answers quickly and easily. The complex technologies we encounter in our daily lives rely heavily on models. For example, the design of a new airplane relies heavily on models before construction and flight testing. These models provide the design team with critical information, such as the minimum wing area required for flight. We will return to this example in the next chapter.

We have come to rely on models, but it is critical to keep two important considerations in mind. First, models serve as imperfect representations of reality and exist in a space separate from the actual world. They are creations of our minds, capable of providing answers within their programmed parameters, whether accurate or nonsensical. Second, the construction of reliable models depends on our understanding of the underlying processes. If we do not understand how something works, we cannot develop a robust model that we can trust. Unfortunately, our understanding of climate remains a formidable challenge — one of the most complex problems that science has to deal with. In a 1964 lecture, renowned physicist Richard Feynman famously said, *"I think I can safely say that no one understands quantum mechanics."* The same sentiment can be applied to climate change; no one understands it. Many climate processes lack a solid theoretical foundation, such as how the atmosphere transports heat to the poles through midlatitude storms.[387]

There is no denying that climate models have inaccuracies. The real question is the extent of their errors and their usefulness for different purposes and audiences. In the following chapter, we will try to answer the latter part of this question.

Climate models are incredibly complex and fragile. While there are different types of models, a state-of-the-art general circulation model (such as those par-

---

[387] Barry, L., et al., 2002. Nature, 415 (6873), pp.774–777. doi.org/10.1038/415774a

ticipating in the 6th Intercomparison Project) with a grid resolution of 1x1° (about 100 x 100 km, 60 x 60 miles) and 30 layers consists of a staggering 2 million cells. Typically, these models track seven variables — wind speed in 3 dimensions, pressure, temperature, density, and water vapor content — for each grid cell. They perform calculations that include all known processes that affect these variables and consider how changes in one cell affect neighboring cells. As a result, these models are iterative programs, where the output of one time step serves as the input for the next, making them inherently more prone to instability than other types of computer models.

Despite their high resolution, many physical processes occur at subgrid scales, including those at the molecular level. In addition, certain processes lack equations that can describe them accurately. In such cases, these processes are approximated by numerical parameters. If a model doesn't perform as expected, these parameters are adjusted until the desired result is achieved.[388] A climate model cannot claim to represent reality if it contains the modeler's biases or preconceptions about how climate processes should work.

A climate model may consist of 2 million lines of code. Because of the interconnectedness and iterative nature of their components, these models are exceptionally fragile, unlike the real climate. A recent example of this fragility is when scientists discovered a flaw in the CESM2 model in how it simulated the interaction between moisture and condensation nuclei, which affects cloud formation. It took a team of 10 scientists five months to identify the problem and correct the error in the data.[389] Any change in a model can easily derail it.

A telling example of how climate models represent a model world rather than the real world is their inability to provide an accurate surface temperature for the planet. Figure 86, taken from an article devoted to establishing a connection between climate model projections of global temperature and real-world observations, sheds light on this disparity.[390]

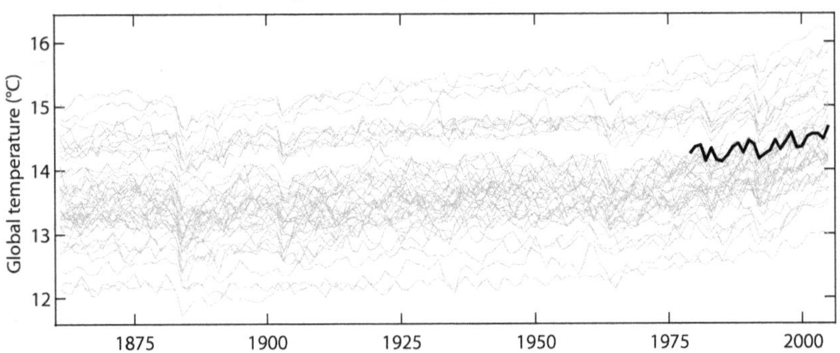

Figure 86. Models do not know the temperature of the planet. Global temperature from historical simulations of 42 models from the 5th Intercomparison Project. The thick black line is from the European reanalysis product.

<hr>

[388] Hourdin, F., et al., 2017. Bull. Am. Met. Soc. 98 (3), pp.589–602. doi.org/10.1175/BAMS-D-15-00135.1

[389] The Wall Street Journal. Feb 06, 2022. Climate scientists encounter limits of computer models and bedeviling policy.

[390] Hawkins, E. & Sutton, R., 2016. Bull. Am. Met. Soc. 97 (6), pp.963–980. doi.org/10.1175/BAMS-D-14-00154.1

Scientists aren't concerned that climate models differ by 3°C (5.4 °F) in simulating the planet's temperature, even though that difference is half the temperature that separates our interglacial period from the Last Glacial Maximum. What matters to them is the consistency of the changes over time within the simulations and the fact that the "cold world" models predict similar changes as the "warm world" models when subjected to the same forcings. It is surprising, to say the least, that a 3°C difference doesn't significantly affect the performance of these models.

To what extent are climate models wrong?

This is a difficult question to answer. There are many papers in the scientific literature pointing out flaws in climate models, but comprehensive lists are not readily available because only the modelers themselves keep track of such problems without publishing them. To provide some insight, I have compiled a collection of notable model failures from climate papers that I have consulted over a period of just four months. It is worth noting that my reading covers only a minimal fraction of what is published on the subject, leaving room for the reader to speculate about the potential scope of this list.

* Models tend to overestimate the cooling and show slower recovery after volcanic eruptions.[391]
* Models fail to accurately represent the stratospheric response to solar changes.[392]
* Models fail to predict severe winter weather resulting from Arctic amplification.[393]
* Models struggle to reproduce the cooling period observed between 1945 and 1975.[394]
* Models fail to simulate climate shifts, such as the one in 1976.[395]
* Models fail to reproduce historical patterns of ocean warming.[396]
* Models fail to capture temperature trends in the tropical troposphere and stratosphere.[397]
* Models failed to predict the divergence of Arctic and midlatitude winter temperature trends.[398]
* Models fail to reproduce Northern Hemisphere snow cover trends.[399]

---

[391] Brohan, P., et al., 2012. Clim. Past, 8 (5), pp.1551-1563.
doi.org/10.5194/cp-8-1551-2012
[392] Misios, S., et al., 2016. Q. J. R. Meteorol. Soc. 142 (695), pp.928–941.
doi.org/10.1002/qj.2695
[393] Cohen, J., et al., 2020. Nat. Clim. Change, 10 (1), pp.20–29.
doi.org/10.1038/s41558-019-0662-y
[394] IPCC AR6 SPM doi.org/10.1017/9781009157896.001
[395] Ibid.
[396] Bronselaer, B. & Zanna, L., 2020. Nature, 584 (7820), pp.227–233.
doi.org/10.1038/s41586-020-2573-5
[397] Mitchell, D.M., et al., 2020. Environ. Res. Lett. 15 (10), p.1040b4.
doi.org/10.1088/1748-9326/ab9af7
[398] Cohen, J., et al., 2020. Nat. Clim. Change, 10 (1), pp.20–29.
doi.org/10.1038/s41558-019-0662-y
[399] Connolly, R., et al., 2019. Geosciences, 9 (3), p.135.
doi.org/10.3390/geosciences9030135

- Most models underestimate warming in the early 20th century and overestimate warming after 1998.[400]
- Models tend to overestimate atmospheric warming.[401]
- Models predict midlatitude ozone trends in the lower stratosphere that are inconsistent with observations since 1998.[402]
- Model predictions are inconsistent with observed changes in the sea surface temperature gradient in the equatorial Pacific Ocean.[403]
- None of the models accurately reproduce the increase in summer high-pressure blocking over Greenland.[404]
- Models lack global-scale multidecadal variability.[405]
- All models show warming in the tropical upper troposphere that is absent from observations.[406]
- Models exhibit a "signal-to-noise paradox" where they predict observed climate variability better than their own variability, indicating an underestimated signal-to-noise ratio.[407]
- Models show a cold bias in the equatorial cold tongue.[408]
- Models do not realistically reproduce the observed annual cycle of albedo.[409]
- Models do not accurately capture the small interannual variability in albedo.[410]
- Models generate a spurious double Intertropical Convergence Zone in the tropical Pacific.[411]
- Models fail to reproduce the interhemispheric albedo symmetry.[412]

---

[400] Papalexiou, S.M., et al., 2020. Earth's Future, 8 (10), p.e2020EF001667. doi.org/10.1029/2020EF001667

[401] Mitchell, D.M., et al., 2020. Environ. Res. Lett. 15 (10), p.1040b4. doi.org/10.1088/1748-9326/ab9af7 McKitrick, R. & Christy, J., 2020. Earth Space Sci. 7(9), p.e2020EA001281. doi.org/10.1029/2020EA001281

[402] Ball, W.T., et al., 2020. Atmos. Chem. Phys., 20, 9737–9752. doi.org/10.5194/acp-20-9737-2020

[403] Seager, R., et al., 2019. Nat. Clim. Change, 9 (7), pp.517-522. doi.org/10.1038/s41558-019-0505-x

[404] Hanna, E., et al., 2018. Cryosphere, 12 (10), pp.3287–3292. doi.org/10.5194/tc-12-3287-2018

[405] Kravtsov, S., et al., 2018. NPJ Clim. Atmos. Sci. 1 (1), p.34. doi.org/10.1038/s41612-018-0044-6

[406] McKitrick, R. & Christy, J., 2018. Earth Space Sci. 5 (9), pp.529-536. doi.org/10.1029/2018EA000401

[407] Scaife, A.A. & Smith, D., 2018. NPJ Clim. Atmos. Sci. 1 (1), p.28. doi.org/10.1038/s41612-018-0038-4

[408] Li, G., et al., 2016. Clim. Dyn. 47, pp.3817–3831. doi.org/10.1007/s00382-016-3043-5

[409] Stephens, G.L., et al., 2015. Rev. Geophys. 53 (1), pp.141–163. doi.org/10.1002/2014RG000449

[410] Ibid.

[411] Si, W., et al., 2021. Geophys. Res. Lett. 48 (23), p.e2021GL094779. doi.org/10.1029/2021GL094779

[412] Stephens, G.L., et al., 2016. Curr. Clim. Change Rep. 2, pp.135-147. doi.org/10.1007/s40641-016-0043-9

- Heat transport is climate-state invariant in models despite large changes in the temperature gradient.[413]
- Models poorly simulate temperature trends in the lower stratosphere and inconsistently reproduce tropical tropopause temperatures and water vapor changes.[414]
- Models hindcast a strengthening of the Brewer-Dobson circulation in the lower stratosphere during the second half of the 20th century, in contrast to observations.[415]
- Models underestimate the Holton-Tan effect, and each model represents the relationship between the Quasi-Biennial Oscillation and the polar vortex differently.[416]
- Models produce ten times smaller interannual changes in ocean latent heat flux than observed.[417]
- Models simulate enhanced warming over Antarctica – Antarctic amplification – whereas no warming has been observed for this continent.[418]

Do climate models reproduce the real climate?

Climate models may give the impression that they simulate the real climate, but appearances can be deceiving. Let's use a popular video game called "The Sims" that my daughter used to play as an analogy. In this life-simulation game, players create virtual characters, place them in houses, and influence their emotions and desires. With each new version and expansion pack, the game offered more features and capabilities for the Sims. While the game was not intended for players to harm their Sims, there were certain circumstances in which such actions were possible. For example, players could have a Sim enter a pool and then remove the ladder, leaving the Sim trapped and eventually drowning. Despite the game's ability to replicate various behaviors, the fact that the Sims couldn't escape a pool without a ladder highlighted a defective simulation in some fundamental aspects.

Similarly, climate models, while seemingly comprehensive, are likely to overlook crucial elements or fail to accurately represent certain phenomena. Just as the Sims' inability to escape a pool without a ladder revealed gaps in the simulation, there are many essential factors and processes that climate models are currently unable to reproduce correctly. In many cases, climate models reproduce certain climate behaviors by adjusting multiple parameters that aren't inherent to the models but are introduced by the modelers. Even when climatologists claim that models accurately capture certain climate characteristics, it

---

[413] Donohoe, A., et al., 2020. J. Clim. 33 (10), pp.4141–4165.
doi.org/10.1175/JCLI-D-19-0797.1

[414] Solomon, S.et al., 2010. Science, 327 (5970), pp.1219–1223.
doi.org/10.1126/science.1182488

[415] Young, P.J., et al., 2012. J. Clim. 25 (5), pp.1759–1772.
doi.org/10.1175/2011JCLI4048.1

[416] Elsbury, D., et al., 2021. Geophys. Res. Lett. 48 (24), p.e2021GL094083.
doi.org/10.1029/2021GL094083

[417] Yu, L. & Weller, R.A., 2007. B. Am. Meteorol. Soc. 88 (4), pp.527–540.
doi.org/10.1175/BAMS-88-4-527

[418] Smith, D.M., et al., 2019. Geosci. Model Dev. 12 (3), pp.1139-1164.
doi.org/10.5194/gmd-12-1139-2019

remains difficult to determine whether they do so for the same underlying reasons as the real climate. If different models give different answers, it is unlikely that the true cause has been identified.

Therefore, when models make predictions about future climate, they are essentially predicting the model's own version of future climate, not the actual future climate. It is crucial for us to keep this distinction in mind. What happens in the models will most likely not happen in the real world.

The climate does not change as the models indicate

The prevailing theory and climate models suggest that the climate has responded almost exclusively to human activity over the past 270 years. This is illustrated in figure B5 (ch. 8). However, despite the steady increase in anthropogenic forcing, the climate hasn't warmed uniformly. Instead, it has experienced multidecadal periods of enhanced warming followed by periods of reduced warming or even cooling, known as hiatuses. These patterns challenge the expectation of a sustained temperature rise.

Given the substantial increase in anthropogenic forcing since 1950 and our continued emissions, models predict an increased rate of warming in the coming decades (fig. 87a). However, actual observations do not match these predictions.

Soon after the 1997 climate shift, the discrepancy between observations and model predictions increases significantly. Within just 25 years, this difference reaches 0.35°C (0.6 °F, fig. 87b). To put this in perspective, the missing warming is about one-third of the observed warming over the last 100 years.

Scientists are well aware of this dilemma, which is made particularly problematic by the failure of models to accurately reproduce the current period.[419] The inherent fragility of climate models is highlighted by the fact that attempts to improve the realism of cloud simulations have increased their sensitivity to rising CO_2 levels. Correcting this problem is a complex task, leading to proposals to exclude models that predict higher levels of warming when calculating multi-model averages. While there may be valid reasons behind such proposals, it can be likened to selectively choosing a preconceived answer, known as cherry-picking.

It is no coincidence, however, that since the 1997 climate shift, which changed the intensity of heat transport to the poles, climate models have indicated an excessive warming trend. Surprisingly, scientists remain largely unaware of the two fundamental problems plaguing climate models. First, the models tend to underestimate the magnitude of the solar forcing due to inadequate incorporation of indirect effects. As a result, the anthropogenic forcing is artificially amplified to compensate, allowing the models to account for observed climate changes. Second, the variability of heat transport, a crucial aspect of the climate system, is inadequately represented in these models. As a result, they overlook two important aspects of the climate, making them inadequate for accurately predicting its future.

---

[419] Voosen, P., 2021. Science, 373 (6554) pp.474–475.
    doi.org/10.1126/science.373.6554.474

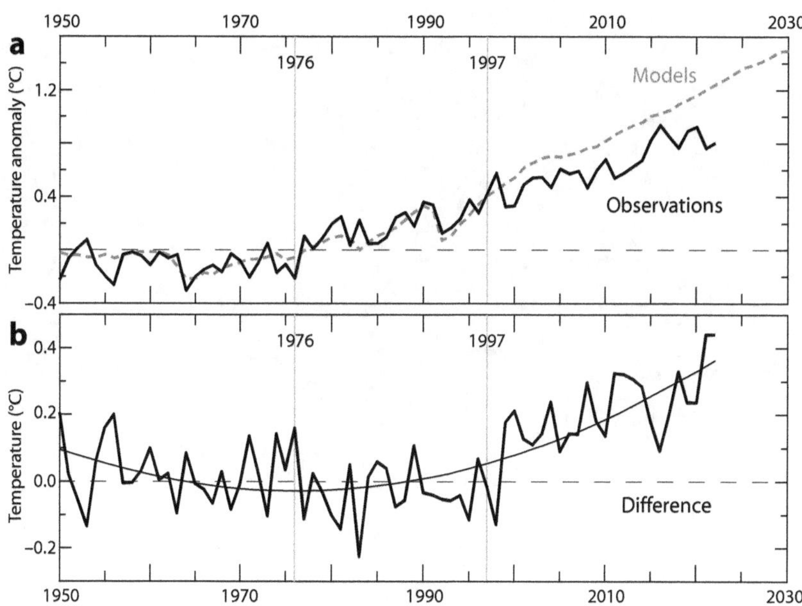

Figure 87. Models run too hot. a) The black line corresponds to the global mean surface temperature. The dashed gray line is the multimodel mean from the 6th Intercomparison Project under an emissions scenario similar to today's.[420] b) Evolution of the difference between models and observations. The temperature difference for the period 1961-1990 is zero because this is the period used as the reference (baseline) for calculating the anomaly.

The Winter Gatekeeper hypothesis offers a potential solution to these problems. However, the likelihood that scientists will recognize these two problems is quite low. As a result, climate models are likely to continue on their trajectory of becoming increasingly complex, fragile, and costly. Significant change may come only when scientists are confronted with the stark reality of their models' inability to accurately predict climate even a few decades into the future.

In summary

Climate models suffer from numerous unresolved problems, and recent adjustments have made their performance worse rather than better. One notable problem is their tendency to overestimate warming after 1998, casting doubt on the accuracy of their future climate projections. The root cause of this failure is likely to be the absence of two crucial climate features: the response to solar variability via indirect effects and the response to variability in poleward heat transport. These fundamental features are central to the Winter Gatekeeper hypothesis, which offers a potential explanation for the models' shortcomings.

---

[420] Data from HadCRUT5 and CMIP6 all members ensemble average under SSP2-4.5 scenario with 1961-1990 baseline.
climexp.knmi.nl/CMIP6/Tglobal/global_tas_mon_ens_ssp245_192_ave.dat The use of a more recent baseline and a reduced multi-model mean in AR6 hides this issue.

CHAPTER 50
CLIMATE MODEL PREDICTIONS ARE NOT USEFUL TO SOCIETY

Modeling is an integral part of science, and it serves a variety of purposes for scientists beyond mere prediction. Climatology relies heavily on models, and even when these models are flawed, they still provide value to scientists. However, complex climate models have drawbacks that limit their usefulness for making accurate predictions. They are prone to the butterfly effect, where small changes in initial conditions lead to very different results. In addition, structural flaws in these models lead to erroneous predictions over time. Due to their complexity, attempting incremental improvements becomes a challenge, as even small changes can have far-reaching effects, to the point where the benefits of such improvements become negative. Consequently, a critical question arises: do uncertain predictions from imperfect models provide any benefit to society? The likely answer is no.

How scientists use models

Modeling is a fundamental aspect of scientific work. In essence, any hypothesis can be viewed as a conceptual model. Numerical models serve many purposes in science, of which prediction is only one. These models provide extensive utility in several areas, including:

* Testing hypotheses
* Suggesting new questions
* Guide data collection
* Elucidate dynamic relationships
* Challenging existing theories
* Identify discrepancies between hypotheses and data
* Educate and train students
* Increase scientific output

In non-experimental fields of science, such as climate science, models play an indispensable role to the extent that a substantial portion of scientific output depends on them. To illustrate this, table 2 presents data on the frequency of the term "model" and its variants in the titles or abstracts of articles published in a leading publication, the Journal of Climate, over four years separated by a decade. The first year of the journal was 1988, and since then, the number of articles published has increased sharply each decade. Since the 1990s, about two-thirds of the articles include references to models in their titles or short abstracts, further evidence of the extensive reliance on models in climate science.

Climate models can be placed in a hierarchy that encompasses a range of complexity. At the simpler end are specific or regional models, while at the other end are more complex models such as general circulation models and Earth system models. These advanced models incorporate biological, geological, or chemical processes into their simulations. Complex models are often involved in model intercomparison projects aimed at establishing a multi-model framework. The most recent of these projects is the 6th edition, with the participation of over 70 models built by 33 modeling groups from 16 countries.

Table 2. Use of model. The number of articles published by the Journal of Climate and the proportion of them containing the word "model" and its variants in the title or abstract for four selected years.

Journal of Climate

Year	Number of articles	Use of model[1]
1988	76	46.0%
1998	184	67.4%
2008	370	67.8%
2018	545	65.5%

[1] Use of word "model*" in title or abstract

It's worth emphasizing that even when a model is obviously flawed, it can still be of great value to scientists. The key is to understand the reasons why the model is flawed and does not accurately represent certain facets of the climate. Constructing flawed models allows scientists to gain insight into areas for improvement while uncovering new knowledge and generating new ideas for climate research. The field of climate science has made remarkable progress through the use of models.

Problems inherent in models affect their predictions

Current state-of-the-art models use time steps of about 30 minutes, so it takes weeks or months in real time on supercomputers to simulate a century of climate evolution. In addition, to compute a single hypothetical evolution of the climate system (a "model run"), an initial condition and boundary conditions are required. The former is a mathematical description of the state of the climate system at the beginning of the simulated period. The latter are the values of all external forcing changes affecting the system, such as solar radiation, greenhouse gases, or aerosol concentrations.

The inclusion of nonlinear mathematical formulas in climate models, coupled with their iterative nature, makes them highly sensitive to initial conditions due to their chaotic properties. This phenomenon is commonly referred to as the butterfly effect. In a fascinating experiment, the Community Earth System Model was subjected to 30 simulations of North American climate over 50 years, starting in 1963.[421] Remarkably, despite initial conditions differing by only an infinitesimal fraction of a degree of temperature, the results in 2012 showed large divergences, many of which would have greatly surprised scientists if they had actually occurred. The authors claim that ensemble averaging reduces natural variability and reveals the warming trend attributed to anthropogenic climate change. However, this claim seems highly unlikely because

---

[421] Deser, C., et al., 2016. J. Clim. 29 (6), pp.2237–2258.
  doi.org/10.1175/JCLI-D-15-0304.1

the models do not reproduce the same kind of natural variability that is observed in the real climate (fig. 57, ch. 36).

The models characterized by mathematical chaos lack the natural variability inherent in the real climate. It is important to note that the mathematical space in which this chaos unfolds is likely to be constrained by different factors than those that constrain the chaos observed in the real climate – factors that are essentially unknown to us. Moreover, it is crucial to understand that averaging outcomes from chaotic processes is different from averaging outcomes from random processes. In the latter case, a true mean can be approximated with a sufficient number of trials. However, chaotic systems cannot be simply averaged to eliminate randomness or variability because they depend entirely on the specific path taken, and the number of possible paths is incalculable. Thus, two different sets may yield completely different averages.[422]

When modeling very complex systems, such as the Earth's climate, it can be assumed that the models are structurally imperfect and that their mathematical description of the climate is misspecified. This introduces a new problem. Even with perfect initial conditions, if the model is structurally imperfect, a large difference in results will emerge over time. This is called the hawkmoth effect.[423] It means that the probability distribution and uncertainty in a forecast of any model will become misleadingly accurate, misleadingly diverse, and erroneous over time.

The expectation that incremental improvements in very complex models will lead to incremental improvements in their representation of reality and in the accuracy of their predictions is probably false. The compound nonlinear effects of small adjustments to the model structure are so large that calibration becomes computationally expensive, and the marginal performance benefit of additional subroutines or processes may be zero or even negative.[424] Adding more detail to a model can make it less accurate and less useful. We are already beginning to see this problem with climate models, as described in the previous chapter in explaining their fragility.

We must remember that a state-of-the-art climate model represents a hypothesis about how the Earth's climate system works. However, it is important to note that even if a model agrees with observations, it cannot be considered correct. It is widely accepted that all models are inherently flawed, as evidenced by the list of model failures outlined in the previous chapter. The process of constructing a model involves numerous simplifications, approximations, and the exclusion of various processes, some of which may remain unknown to us. As a result, the hypothesis generated by a climate model is fundamentally inaccurate. Each of the current climate models is known to produce results that deviate from observational data beyond the bounds of observational uncertainty and error. In the words of a philosopher of science, the notion that any of these models can be empirically adequate cannot be taken seriously, and

---

[422] Hansen, K., 2016. judithcurry.com/2016/10/05/lorenz-validated/

[423] Thompson, E.L. & Smith, L.A., 2019. Economics, 13 (1), p.20190040.
doi.org/10.5018/economics-ejournal.ja.2019-40

[424] Ibid.

agreement between models and observations should not be taken as confirmation of the models' validity.[425]

Climate predictions are often called projections to indicate their dependence on specific forcing scenarios, such as greenhouse gases and aerosols. When several models in an ensemble produce consistent results, this is considered a robust result. For example, if all models predict a global average temperature increase of more than 4°C (7 °F) by the end of the century under a particular emissions scenario, the result is considered robust. However, it is important to understand that robustness alone does not justify an increase in confidence. This is because models are not completely independent entities.[426] Many models share code borrowed or inherited from previous models, and they all share common errors, technological limitations, and knowledge constraints. These common errors are widely known, and the lack of model independence has been demonstrated. Relying on the model agreement based on common errors can lead to overconfidence in projections.

Climate model projections are not useful to society

Earlier, I argued that climate models are of great value to scientists even when they contain clear errors. This is because their primary goal is not to make accurate predictions but to improve understanding and refine the models. When it comes to society, however, inaccurate predictions can have detrimental effects. The large degree of uncertainty associated with climate projections, often greater than acknowledged, means that relying on such predictions can leave society in a worse position than if there were no predictions at all. In the absence of model-based predictions, societies have traditionally relied on historical evidence of past changes as a guide.

To illustrate the importance of this, let's take sea level rise projections as an example. A widely referenced 2014 article presented a comprehensive set of probability distributions that incorporated inputs from expert community assessments, expert elicitation, and process modeling. These projections were based on data from a global network of tide gauges.[427] For example, the study predicted that under a high emissions scenario, San Francisco (USA) could experience a sea level rise of 0.6 to 1.0 meters (2-3.3 feet) by 2100. As a result, the California Ocean Protection Council, which is responsible for providing sea level rise projections to various agencies for planning purposes, set benchmark targets to prepare for a 1-meter rise by 2050.

Underscoring the importance of proactive measures, the Federal Emergency Management Agency states that $1 invested in pre-disaster preparedness can prevent up to $6 in subsequent public and private losses. Taking these projections seriously, the Governor of California signed a bill in 2021 that officially includes sea level rise as a critical issue to be addressed by the California Coastal Commission. This legislation establishes a mechanism to provide up to

---

[425] Parker, W.S., 2009. Suppl. Proc. Aristot. Soc. 83 (1) pp.233–249. doi.org/10.1111/j.1467-8349.2009.00180.x

[426] Frigg, R., et al., 2015. Philos. Compass, 10 (12), pp.965–977. doi.org/10.1111/phc3.12297

[427] Kopp, R.E., et al., 2014. Earth's future, 2 (8), pp.383–406. doi.org/10.1002/2014EF000239

$100 million annually in grants to local and regional governments to help them prepare for the challenges posed by rising sea levels.

Figure 88 shows the historical sea level rise at San Francisco since 1900. The graph shows a consistent long-term trend of +1.96 mm per year, unaffected by rising atmospheric CO_2 levels or accelerated melting of polar ice caps and mountain glaciers.

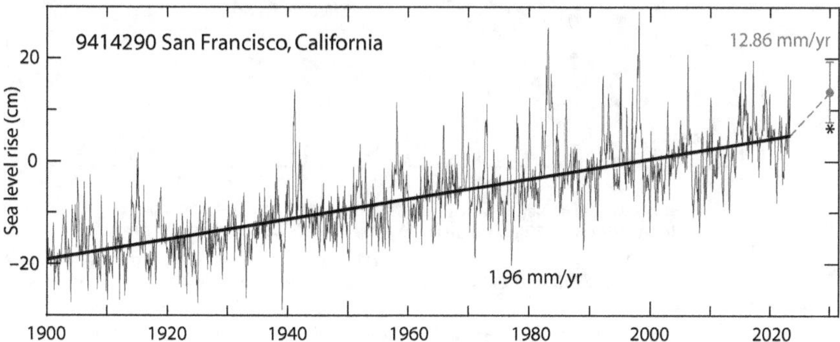

Figure 88. Sea level rise at San Francisco since 1900. The thick line is the +1.96 mm/yr trend for the past 123 years. An asterisk marks the continuation of this trend for the next seven years. The gray dot and bars show the median and very likely (90%) range predicted by a 2014 study. Reaching this median requires a 6-fold increase in the rate of sea level rise.

Since the above paper includes predictions for sea level rise in 2030, we can assess the progress of the prediction for San Francisco so far. According to the paper, it is very likely (90% probability) that San Francisco will experience a rise of 7 to 19 cm (2.8 to 7.5 inches) between 2000 and 2030 for the emissions scenario that is closest to the emissions produced. So far, however, the actual sea level rise has been only 4.5 cm (1.8 inches), even though more than 75% of the projected period has passed. For San Francisco to reach the study's mean projection by 2030, the rate of sea level rise would have to accelerate from the historical rate of +1.96 mm per year over a century to six times that rate over the next seven years.

It is reasonable to conclude that the world is unlikely to experience the sea level rise predicted by experts and models, given that the methodology used for the 2030 projection is the same as that used for the 2100 projection. It is in society's best interest to rely more on empirical evidence than on flawed models and inaccurate expert predictions. The money that California and other regions are spending to prepare for a sea-level rise that will not occur is more than just wasteful because it has an opportunity cost. Such expenditures distract from potential investments and divert resources from addressing other pressing societal needs.

Trusting model predictions over evidence is a mistake

The current state of affairs has led society to be alarmed by predictions made by models that have already been proven wrong by the time they are published, but this often goes unnoticed. A recent example of this phenomenon is shown in figure 89. In June 2023, news headlines around the world highlighted a scientific study that warned of the possibility of ice-free summers in the Arctic by the 2030s, regardless of our efforts to reduce emissions.

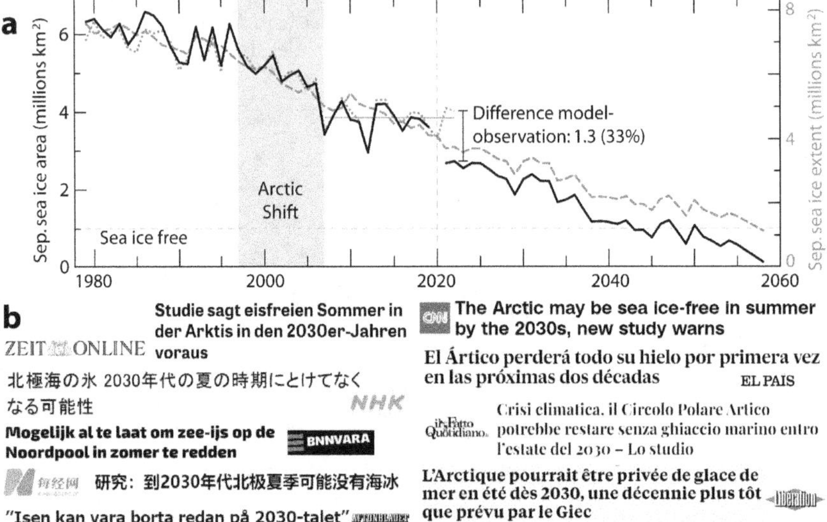

Figure 89. Arctic sea ice projections and their implications. a) Results of a modeling study. The black line before 2020 is the observed change in the September sea ice area, and after 2020 is the sea ice area projected in the study. The dashed gray line is the mean Arctic sea ice area from the 6th Coupled-Model Intercomparison Project. The dotted light gray line is the September sea ice extent, a related measure of sea ice, and the horizontal line shows the lack of trend over the past 16 years. b) Examples of media headlines following the June 6, 2023 press release.

The article presents projections based on observations of an ice-free Arctic even under a low emissions scenario.[428] However, it should be noted that the data in the article only cover observations through 2019, although data for 2020-22 were available at the time of publication. In addition, the model projections in the study begin in 2021. Figure 89 shows the results of the study under an intermediate emissions scenario similar to the current situation. However, a significant problem arises when considering the acceptance and publication of the paper, as the model projections for 2021 and 2022 differ greatly from the observed data, with a staggering difference of 1.3 million km² (0.5 million sq miles) or 33% lower. This obvious problem, which undermines the study as a whole, raises questions about how the paper was accepted for publication.

Furthermore, it is important to emphasize that there has been no significant trend in Arctic summer sea ice extent over the past 16 years. This casts serious doubt on the entire premise of the study. Regardless of the emissions scenario, it is highly unlikely that the Arctic will become sea ice-free by the 2030s or 2040s since there has been no decline over the past decade and a half.

---

[428] Kim, Y.H., et al., 2023. Nat. Commun. 14 (1), p.3139.
  doi.org/10.1038/s41467-023-38511-8

How could such a blatantly flawed article successfully pass the peer review process? Moreover, who determines its suitability for widespread dissemination in a global media landscape that seems incapable of questioning or scrutinizing these predictions? The refuting data derived from observations are readily available to anyone with an Internet connection and can be easily located with a simple search engine query. The current method of communicating predictions from uncertain climate models to society is undeniably inadequate, and it is truly surprising that no authoritative scientific voice has addressed this issue and voiced its disapproval.

Uncertain climate model predictions that increase the anxiety of vulnerable individuals provide little discernible benefit to society as a whole. This is especially true for young people, who may lack the experience and healthy skepticism necessary to critically evaluate the information presented to them by authorities. Such predictions can also mislead policymakers and lead them to make the wrong decisions. Unfortunately, some scientists have put personal gain from amplifying climate alarmism ahead of maintaining scientific rigor. This situation hinders the development of a constructive, trusting relationship between society and its scientific community.

Climate model projections are used to justify phasing out the use of fossil fuels. This energy transition requires a profound transformation of the global economy. Even assuming that the models are correct, it implies great risk. To return to the example in the previous chapter of trusting the models used to design a new airplane, the risk is similar to building a new airplane based solely on computer models: If the new airplane had never been flight-tested, would we allow people to buy tickets and board it for its first flight? No matter how much we trust these computer models, we would never allow it, yet we are willing to take the global economy on a test flight based on climate models that we are sure are flawed, hoping they are not too flawed.

In summary

Climate models are very useful to scientists and make an invaluable contribution to our understanding of climatology. However, climate model predictions are not useful to society because of their uncertainty and high probability of being unacceptably wrong. Two examples are presented. In San Francisco, as in many other places, the sea level has risen linearly over the past century, unresponsive to rising CO_2 levels, temperature, or melting ice sheets. However, models and experts predict a large increase in the global rate of sea level rise, which is expected to affect San Francisco. These predictions have led policymakers to spend large sums of money on disaster preparedness. Since the prediction is for the period 2000-2030, we already know that the expected acceleration will not occur and that the prediction will be wrong by a wide margin. When considering the loss of summer sea ice in the Arctic, none of the models considered the possibility of a lack of decline over the past 16 years. As a result, some of the widely reported media predictions of an ice-free Arctic in the 2030s are implausible. These examples show that climate model predictions are worse than useless, causing undue anxiety and misallocation of resources.

SECTION 15 KEY ISSUES

Climate models include many poorly understood processes, miss important features, and are very complex and fragile. They don't even know the temperature of the planet. They produce a model climate that, despite a superficial similarity, is fundamentally different from the real climate. The current state of knowledge makes it impossible for climate modelers to achieve their goals. Recent changes cause models to overestimate warming after 1998, casting doubt on their future climate projections.

While modeling is an integral part of science, the shortcomings of climate models limit their usefulness for making accurate predictions. Their sensitivity to initial conditions and structural flaws make their projections highly uncertain, even when different models agree, because they are not truly independent. Analysis of sea level and sea ice predictions shows that climate model predictions can have negative impacts on society.

SECTION 16. FUTURE CLIMATE

CHAPTER 51
TWO CONTRASTING FUTURES

Frequent statements from world leaders often paint a future of "climate hell" if we fail to phase out fossil fuels. Yet despite 30 years of effort, our dependence on them has actually increased by 60%. This contradiction between the urgent need to act and the impossibility of doing so is causing climate anxiety and depression among vulnerable people and leading to the emergence of climate radicalism. However, this pessimistic climate future is based solely on uncertain and flawed models and requires warming rates far beyond anything observed to date. In contrast, current warming rates show no signs of acceleration and may even be declining due to natural variability.

The Winter Gatekeeper hypothesis, driven primarily by natural factors, offers a contrasting perspective on the future of our climate. To assess its differences with the Enhanced CO_2 Effect hypothesis, they are compared using an intermediate emissions scenario through 2050. Potential changes in anthropogenic aerosols, solar activity, and multidecadal ocean oscillations are considered to provide a conservative projection of climate change by this time. Given the large divergence in predictions between the two hypotheses, we should expect one of them to be proven wrong within the next two decades.

An extraordinary popular delusion?

The Secretary-General of the United Nations is a position filled by a compromise candidate, usually a politician or career diplomat. They are elected from a middle-power country on a regional rotation basis. Nevertheless, the position wields considerable influence, providing the world's most visible pulpit for making speeches, drawing attention to global issues, and occasionally playing a crucial role in conflict mediation.

The current UN Secretary-General is António Guterres, a former president of Portugal and of the Socialist International. Among major world leaders, Guterres takes an extreme position on climate change. In several recent speeches, he has expressed concern that climate change is out of control, that countries must phase out coal and other fossil fuels to avoid a climate "catastrophe," and that humanity is on a "highway to climate hell." He states that we have passed global warming into an "era of global boiling."[429]

Statements like these, coming from one of the world's most prominent leaders, indicate overconfidence about the future of our planet's climate if we don't achieve a fundamental transformation of the global energy system and economy. António Guterres goes beyond the assessments of the IPCC reports to present a more pessimistic view of our future climate. Unfortunately, this view is widely shared by the global media.

In 1992, the importance of reducing our dependence on fossil fuels was recognized with the adoption of the United Nations Framework Convention on Climate Change, which aims to stabilize atmospheric concentrations of GHGs.

---

[429] news.un.org/en/story/2023/07/1139162

Over the next 30 years, the proportion of the world's primary energy derived from fossil fuels fell from 87% to 82%. However, the amount of energy derived from fossil fuels has increased enormously, from 300 to 500 exajoules (quintillion joules), an increase of 60%!

It should be clear to everyone that it is not possible to substantially reduce the use of fossil fuels in the coming decades, which is why it is not happening. We currently lack a viable alternative energy source that can meet the growing demands of our expanding population while replacing a significant portion of our existing fossil fuel-based energy. Any attempt to reduce the use of fossil fuels by curtailing energy consumption would have a profound impact on the global standard of living and lead to social unrest. The urgent need to rapidly reduce our dependence on fossil fuels to "save the planet" – as expressed by many world leaders – is being met with the impossibility of doing so. As a result, many people are experiencing climate anxiety, despair, and depression.[430] This has led to the emergence of activist groups advocating radical measures, including attacks on masterpieces in art museums.

Given our limited understanding of the Earth's climate and the inherent problems with climate models, our certainty about climate conditions several decades into the future is quite low. The climate models used in the 6th Intercomparison Project predict an average temperature increase of 2°C (3.6 °F) by 2100 compared to today's temperatures (mean 2015-2022) in the scenario that closely matches current emission levels (fig. 90a). However, this projected increase is subject to considerable uncertainty, ranging from +1°C to +3°C at the 90% confidence level. In addition, achieving this scenario would require drastic reductions in CO_2 emissions in the second half of the century.

The projection of this intermediate scenario faces a problem in that it relies on a sustained temperature increase of 0.25°C per decade to reach its predicted average value. However, this rate of warming is significantly higher than what has been observed to date. Over the past 40 years, despite rapid increases in CO_2 emissions, the warming rate has been 0.2°C per decade and has actually declined over the past seven years (fig. 81b, ch. 46). Lower tropospheric temperature data from the University of Alabama in Huntsville, derived from satellite measurements, show a lower warming rate of 0.14°C per decade since 1979. Satellite data are less affected by the urban heat island effect, which occurs when surface temperatures are measured in areas affected by human activity. The difference in warming rates between the two methods cannot be attributed to less warming in the lower troposphere, as this would cause a significant increase in the lapse rate (the decrease in temperature with altitude). Such an increase would act as negative feedback to counteract the enhanced greenhouse effect.

In addition, the influence of multidecadal ocean oscillations on global temperature and its rate of change is substantial (ch. 19). Consequently, as these oscillations enter their cooling phase, the rate of global warming could decrease. There is no evidence or historical precedent to suggest that the average rate of warming can increase to as much as 0.25°C per decade. In contrast,

---

[430] Hickman, C., et al., 2021. Lancet Planet. Health 5 (12), pp.e863–e873. doi.org/10.1016/S2542-5196(21)00278-3

there is evidence that the rate of warming turned negative in the 1960s and early 1970s, even in the presence of rising CO_2 levels.

Given our understanding of the Earth's warming rates and the recognition that the models overestimate temperature increases (fig. 87, ch. 49), it seems unlikely that the planet will experience a 2°C rise by 2100. Moreover, even if warming were to reach 1°C above the present temperatures, which is within the lower uncertainty range projected by the models, would this really constitute a climate "catastrophe"? It seems that part of humanity has succumbed to an extraordinary popular delusion.[431]

Choosing a likely scenario

This book examines two contrasting hypotheses about the primary factors driving climate change: the widely supported hypothesis centered on the enhanced effect of CO_2 changes and a new hypothesis focused on natural variations in heat transport to the Arctic pole during winter. While these hypotheses are not mutually exclusive, they lead to divergent predictions of future climate change. Therefore, if future climate change is consistent with one hypothesis, it will falsify the other.

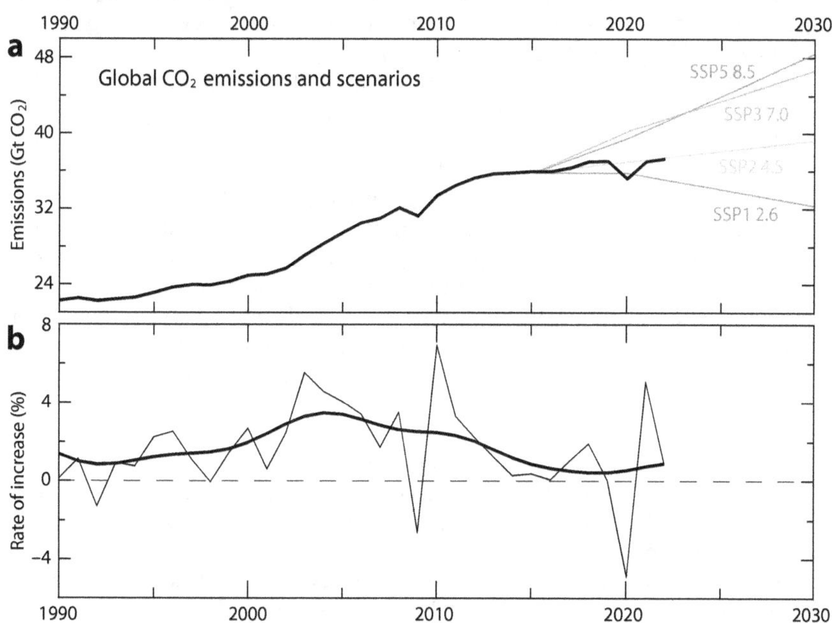

Figure 90. Anthropogenic CO_2 emissions. a) Recent CO_2 emissions (thick black line) and emissions for different scenarios in the 6th Assessment Report (thin gray lines). b) Annual rate of increase in CO_2 emissions (thin line) and a smoothing of the data (thick line).[432]

---

[431] The expression is taken from Charles Mackay's 1841 book on mass delusions "Extraordinary Popular Delusions and the Madness of Crowds."

[432] CO_2 emissions data from Gilfillan, D. & Marland, G., 2021. Earth Syst. Sci. Data, 13(4), pp.1667–1680. doi.org/10.5194/essd-13-1667-2021 and the Energy Institute Statistical Review of World Energy, 72 ed. www.energyinst.org/statistical-review

Before we can predict future climate, we must anticipate changes in the factors that drive it. This is why scientists refer to climate projections, which are predictions that depend on specific forcing scenarios. Each scenario includes a range of possible changes in CO_2 levels, aerosol concentrations, and solar activity. Because volcanic eruptions are unpredictable, even if a projection is accurate, the climate outcome in the event of a major eruption could be colder than expected.

The rate at which atmospheric CO_2 increases depends primarily on changes in human emissions. These emissions have contributed to a marked acceleration in the growth of CO_2 levels, from 0.85 ppm per year in the 1960s to 2.45 ppm per year in the last decade (2013-2022). To inform the IPCC's 6th Assessment Report, scientists have developed a new set of scenarios, some of which are similar to those in the previous report. Figure 90a shows global CO_2 emissions and four scenarios, ranging from the optimistic SSP1 2.6 to the pessimistic SSP5 8.5. The second number in the name of each scenario indicates the projected increase in radiative forcing by the year 2100 in W/m^2. The SSP2 4.5 scenario assumes a large reduction in emissions starting in the 2040s, although it is currently closest to current emissions. It is worth noting that our emissions show a declining trend in their rate of change since the early 2000s, which was not expected (Fig. 90b).

Predicting future solar activity is less important for the Enhanced CO_2 Effect hypothesis, as historical changes in solar forcing are not expected to have contributed significantly to climate change under this hypothesis. However, it is of great importance for the Winter Gatekeeper hypothesis, which poses a challenge because solar activity has proven difficult to predict accurately. In response to this problem, I developed a simple solar model in 2018. This model simplifies the analysis by assuming an average length of 11 years for each solar cycle and focuses only on the total number of sunspots in a cycle. The model does not attempt to predict the maximum number of sunspots in a cycle or their exact number in a given year. Instead, it focuses on comparing the activity of one cycle to the rest, which should provide sufficient information to define a future climate change scenario.

The model accounts for the influence on solar activity of five long solar cycles with durations ranging from 50 to 2500 years, as revealed by the ^{14}C and sunspot records. A similar low-frequency modulation model accurately predicted the extended cycles 24-25 minimum years before they occurred.[433] The model, shown in figure 91, has already demonstrated success in predicting higher solar activity for cycle 25 compared to cycle 24. It also predicts an increase in solar activity over the next 35 years, leading to the establishment of a new solar grand maximum in the 21st century.

Trying to predict climate change eight decades into the future is of limited value. By 2100, most of today's population will be dead, and advances in science and economic development are likely to reveal the incompleteness of our current knowledge, leading to a society very different from our current expectations. It is more practical to focus our climate projections on the next 25

---

[433] Clilverd, M.A., et al., 2006. Space Weather, 4 (9) S09005.
doi.org/10.1029/2005SW000207

years. This time frame allows us to test our hypotheses about future climate change and provides a more meaningful and relevant basis for analysis.

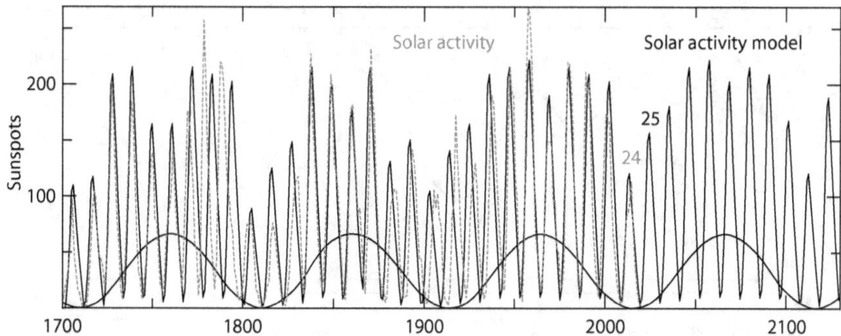

Figure 91. Solar activity model. Annual sunspots from 1700-2018 (dashed gray line) are compared with the output of a solar activity model for the interval 1700-2130 (black line). Four centennial oscillations are shown. The current solar cycle is 25.

Changes in climate drivers over the next 25 years

The scenario used in this analysis to project future climate evolution through 2050 takes an intermediate approach, incorporating the greenhouse gas and aerosol forcings from SSP2 4.5 and solar activity conditions similar to those of the past three decades. This conservative scenario assumes no significant changes in our economy, energy system, or the behavior of natural climate drivers. We can now explore how this scenario affects known climate drivers, taking into account some of their expected changes.

The Winter Gatekeeper hypothesis emphasizes two natural climate drivers on a multi-decadal time frame: solar activity and multi-decadal ocean oscillations. Solar activity is projected to increase from the present to the end of the period considered (fig. 92a, black line). The Atlantic Multidecadal Oscillation is representative of global ocean oscillations that strongly influence poleward heat transport. If the periodicity observed during the 20th century continues, it is expected to enter its cold phase within the next 15 years (fig. 92a, dashed gray line).

The Enhanced CO_2 Effect hypothesis emphasizes two major human drivers of climate change: GHGs and industrial aerosols. Because our emissions are expected to continue, CO_2 levels will continue to rise. However, there is a possibility that the rate of increase in atmospheric CO_2 levels will slow slightly due to the observed decline in emission rates shown in figure 90b. By 2050, CO_2 levels could reach about 475-480 ppm (fig. 92b, black line). This projection is lower than the 507 ppm projected in the SSP2 4.5 scenario.[434]

Aerosols play a cooling role by increasing the atmospheric albedo, thereby reducing the amount of solar energy that reaches the Earth's surface. The forcing effect of industrial aerosols stopped increasing in the 1990s and has been decreasing over the past decade. This continuing decline in aerosol levels is

---

[434] Meinshausen, M., et al., 2020. Geosci. Model Dev. 13 (8), pp.3571–3605.
doi.org/10.5194/gmd-13-3571-2020

contributing to warming and is expected to continue. The aerosol projection shown in figure 92b (dashed gray line) is from NASA.[435]

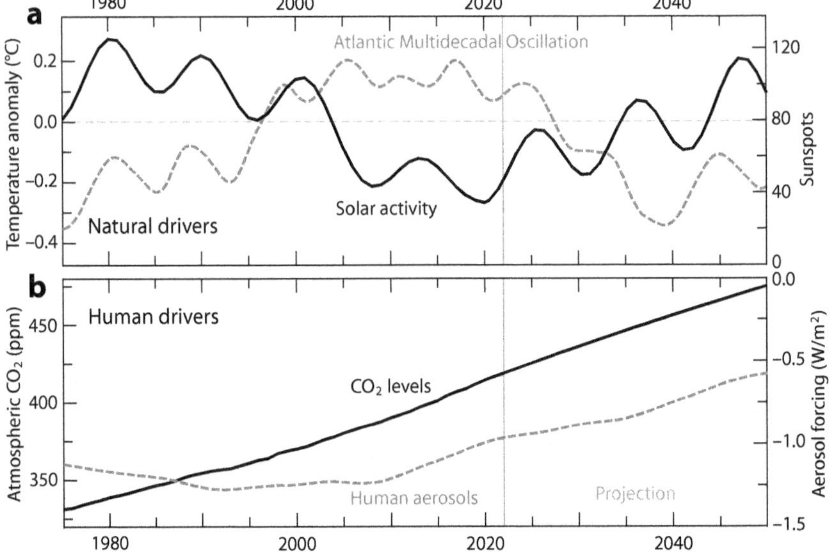

Figure 92. Some drivers of climate change. a) Natural drivers. The smoothed data for solar activity (black line) up to 2022 are projected to 2050 according to the solar model in figure 90. The smoothed data for the Atlantic Multidecadal Oscillation index (dashed gray line) are projected according to past behavior. b) The increase in annual CO_2 levels (black line) is projected to 2050. The human aerosol forcing (dashed gray line) is projected to 2050, according to the NASA climate model.

Two hypotheses lead to two different future climates

As anthropogenic climate forcing increases, models based on the Enhanced CO_2 Effect hypothesis project a continued and rapid temperature rise. According to these models, global temperatures are expected to exceed the 1961-1990 average by about 2°C (3.6 °F) by 2050 (fig. 93a, dashed gray line).[436] However, achieving this prediction would require a sustained warming rate of about 0.3°C per decade, which is 50% higher than observed in the past. It is highly unlikely that such a level of warming will occur within the next 25 years, even under the intermediate scenario.

According to the Winter Gatekeeper hypothesis, a projected phase shift in the Atlantic Multidecadal Oscillation, coinciding with below-average solar activity, is expected to lead to a moderate cooling effect until 2040 (fig. 93a, black line). However, as solar activity is expected to continue to increase thereafter, the warming trend is likely to resume. By 2050, the global mean surface

---

[435] Miller, R.L., et al., 2021. J. Adv. Model. Earth Syst. 13 (1), p.e2019MS002034. doi.org/10.1029/2019MS002034

[436] Data from CMIP6 all members ensemble average under SSP2-4.5 scenario with 1961-1990 baseline. climexp.knmi.nl/CMIP6/Tglobal/global_tas_mon_ens_ssp245_192_ave.dat

temperature could potentially be one degree Celsius below the model's average temperature projections.

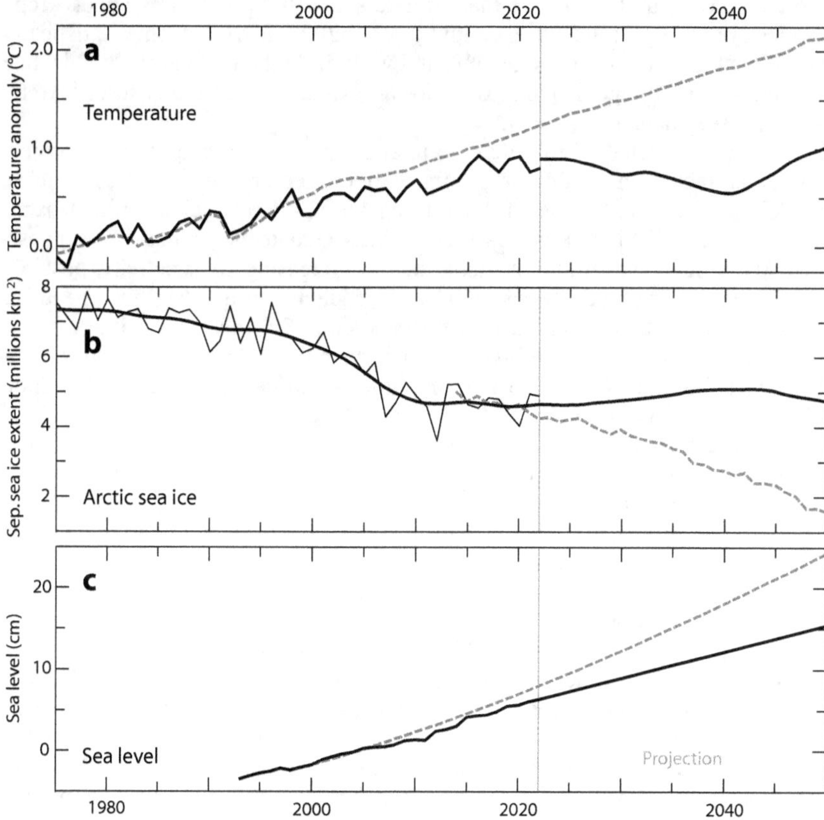

Figure 93. Contrasting climate projections to 2050 from two competing hypotheses. a) Temperature data from the HadCRUT5 dataset (black line to 2022). Temperature projections from models (dashed gray line) and the new hypothesis (black line from 2022 to 2050). b) September Arctic sea ice extent data (black line to 2022). Sea ice extent prediction from models (dashed gray line) and the new hypothesis (black line from 2022 to 2050). c) Sea level data from NASA (black line to 2022). Predicted sea level rise from models (dashed gray line) and the new hypothesis (black line from 2022 to 2050).

After the 1997 climate shift, there was a marked acceleration in the rate of Arctic sea ice loss. Scientists noticed this trend about a decade later and became increasingly concerned about the prospect of an ice-free Arctic.[437] However, researchers were surprised by the recovery of sea ice in 2013 when it became clear that there had been no net loss since 2007. Using models, they cal-

---

[437] Stroeve, J.C., et al., 2005. Geophys. Res. Lett. 32 (4).
doi.org/10.1029/2004GL021810

culated a 34% chance of a 7-year hiatus (pause).[438] However, the hiatus has now extended to 17 years, and the probability has dropped to 10%. In other words, there is a 90% chance that climate scientists' predictions about Arctic sea ice are wrong. If the pause continues until 2027, it will become statistically significant ($p<0.05$, or less than 5%) and will refute the hypothesis of a decline caused by anthropogenic emissions. For an explanation of the observed Arctic changes, see chapters 34 and 42.

With the expected steady increase in anthropogenic forcing, models project a gradual decline in Arctic sea ice (fig. 93b, dashed gray line). By 2050, these models predict a reduction of 2.4 million km^2 (925,000 square miles) from current levels.[439] In stark contrast, the Winter Gatekeeper hypothesis suggests that Arctic sea ice will remain stable due to increasing solar activity, possibly even experiencing modest growth until subsequent warming triggers a negative trend similar to that observed in the 1980s (fig. 93b, thick black line). If this prediction is correct, it would render future changes in Arctic sea ice inexplicable under the Enhanced CO_2 Effect hypothesis, similar to the current situation observed in Antarctica.

Sea levels have been rising for the past two centuries, and the trend is very likely to continue for the next 25 years.[440] In previous IPCC reports, projections of sea level rise were relatively conservative until the 5th Assessment Report, which introduced projections of a significant acceleration, although none has yet been observed. In the previous chapter, we discussed an article that combined models from the 5th Assessment Report with expert judgment to generate sea level projections.[441] According to this study, the intermediate scenario projected sea level rise for 2000-2050 is estimated to be 26 ± 8 cm (10 ± 3 inches; fig. 93c, dashed gray line). NASA data indicate that the sea level in 2022 is already 8 cm higher than in 2000. Based on the Winter Gatekeeper hypothesis, which assumes a continuation of the current trend, an additional 9 cm of sea level rise is projected for the period between 2022 and 2050, resulting in a total sea level rise of 17 cm (6.7 inches; fig. 93c, black line) in the first half of the 21st century. This prediction falls below the lower bound of the 90% confidence interval of the combined model-expert prediction.

The differences in predictions between the competing hypotheses are striking, as shown in figure 93. By 2050, actual climate conditions may provide insight into the relative influence of the anthropogenic drivers proposed by the Enhanced CO_2 Effect hypothesis versus the natural drivers proposed by the Winter Gatekeeper hypothesis. If the outcome falls somewhere between the two sets of predictions, it could indicate a comparable contribution of anthropogenic and natural climate drivers in shaping climate change.

In figure 93, the most recent year of available data is 2022 (indicated by the vertical gray line). Despite the relatively short period (less than a decade) since

---

[438] Swart, N.C., et al., 2015. Nat. Clim. Change, 5 (2), pp.86–89. doi.org/10.1038/nclimate2483

[439] SIMIP Community, 2020. Geophys. Res. Lett. 47 (10), p.e2019GL086749. doi.org/10.1029/2019GL086749

[440] Jevrejeva, S., et al., 2008. Geophys. Res. Lett. 35 (8). doi.org/10.1029/2008GL033611

[441] Kopp, R.E., et al., 2014. Earth's future, 2 (8), pp.383–406. doi.org/10.1002/2014EF000239

the model results, represented by the dashed gray lines, were obtained, they show clear trends that diverge from the observed data, suggesting a pessimistic bias. If the Winter Gatekeeper hypothesis is correct, we would expect this divergence between model predictions and observations to increase over time. But even if this new hypothesis is incorrect, it is important to recognize that the claim by the UN Secretary-General and others that we are on a "highway to climate hell" is not evidence-based. It is an exaggerated claim based on uncertain predictions derived from flawed models, with damaging consequences for people's mental health.

In summary

According to the Enhanced CO_2 Effect hypothesis, if CO_2 emissions continue at current levels, we can expect a temperature increase of 2°C (3.6 °F), a decrease in Arctic summer sea ice extent of 2.4 million km^2 (0.9 million sq miles), and a sea level rise of 18 cm (7 inches) above present levels by 2050. Conversely, the Winter Gatekeeper hypothesis, which assumes increased solar activity and a shift of the Atlantic Multidecadal Oscillation into its cold phase, predicts minimal temperature change, no significant loss of Arctic sea ice, and a sea level rise of only 9 cm (3.5 inches). The stark contrast between these predictions provides an opportunity to challenge and possibly falsify one of these hypotheses within the next two decades. If the results fall somewhere in between, it would suggest that both natural and human factors contribute similarly to climate change. In either case, the narratives of climate catastrophe seem grossly exaggerated and may indicate that an extraordinary popular delusion has taken hold in our society.

SECTION 16 KEY ISSUES

Despite 30 years of warnings of impending "climate hell" if we do not phase out fossil fuels, our dependence on them has increased by 60%. This contradiction is leading to climate anxiety and depression among vulnerable populations and the emergence of climate radicalism. Yet warming, sea level rise, and Arctic sea ice loss have not accelerated in recent decades. According to the Enhanced CO_2 Effect hypothesis, if CO_2 emissions continue at current levels, we can expect a temperature increase of 2°C (3.6 °F), a reduction in Arctic summer sea ice extent of 2.4 million km^2 (0.9 million sq miles), and a sea level rise of 18 cm (7 inches) above the current levels by 2050. The Winter Gatekeeper hypothesis, which takes natural factors into account, predicts minimal temperature change, no significant loss of Arctic sea ice, and a sea level rise of only 9 cm (3.5 inches) by that date. The stark contrast between these predictions provides an opportunity to reject one of the hypotheses in two decades.

CHAPTER 52
CONCLUSION

The climate is incredibly complex, and this book reflects that complexity. If you've made it to this final chapter, you've put in a lot of effort and deserve congratulations. My purpose in writing this book was not to confuse you but to shed light on the fact that there are no simple, definitive answers to why the climate is changing. This message has great value, especially in an age of false certainties.

The book provides much information about what we currently know about climate change and even more about what we do not know because that is the reality of the issue. We ignore more than we know about climate change. It is a work in progress, not a settled matter.

Climate change is a complex phenomenon driven by energy changes. The first part of the book focuses on our understanding of the vertical flux of energy within the climate system and our poor knowledge of changes in albedo (reflectance) and horizontal energy transport. It is often overlooked that the greenhouse effect varies across the globe, being strong over the tropics but weak over the polar regions in winter. This inherent variability makes changes in horizontal heat transport non-neutral to radiative fluxes at the top of the atmosphere, thus potentially contributing to climate change. When heat is moved to a location where it can be more easily radiated to space, the outgoing radiation increases, leading to a reduction in the energy content of the system.

The challenge, however, lies in our limited understanding of heat transport due to the current lack of accurate measurement methods. In the first part of the book, we provide an in-depth review of our current knowledge of meridional heat transport, essentially the movement of heat toward the poles. Interestingly, this knowledge gap may be a reason for the general overlooking of the role of the El Niño-Southern Oscillation in the transport system and the failure to recognize multidecadal oceanic oscillations as manifestations of global transport changes.

The second part of the book deals with natural climate variability. This fascinating subject has not received sufficient attention because it is generally believed not to have contributed to recent climate change. Our knowledge of the factors behind past changes remains remarkably poor. We can't explain the times when palm trees grew at the poles and frogs lived in Antarctica. We don't know why the Earth cooled for 50 million years. The drop in CO_2 levels could not have been the cause because most of it occurred during the Oligocene, at the beginning of a long warming period. Many abrupt climate events during the Holocene, such as the Little Ice Age, are puzzling. In essence, we do not know how to explain climate change if changes in CO_2 cannot explain it. While volcanic eruptions have short-term effects, we overlook the indirect effects of changes in solar activity despite ample evidence of their existence. Our inability to understand past climate change stems from our failure to recognize changes in heat transport as a driver of climate change.

In the third part, we critique the existing consensus hypothesis, which unfortunately falls short in several key respects. It fails to explain the prominent climate features we experience during our lifetimes, including the prevalence of multidecadal climate regimes interspersed with sudden shifts. It also fails to account for many past climate changes, ignoring crucial factors such as changes in heat transport and indirect solar forcing. We need new hypotheses that can address these limitations. In this context, I present my contribution – the Winter Gatekeeper hypothesis – which seeks to provide a more comprehensive and accurate understanding of climate change.

The purpose of this book is not to convince you that my hypothesis is correct but rather to demonstrate the inadequacy of our consensus hypothesis, which is primarily based on CO_2 changes. At best, the consensus hypothesis can only partially explain the observed changes in our climate. The evidence presented in this book points strongly in that direction. I provide you with the evidence and give you the tools you need to make your own informed judgment. Unlike the Enhanced CO_2 Effect, my hypothesis offers a viable explanation for the evidence presented. However, I can't be sure that it is correct. It took decades to establish the validity of Darwin's theory of evolution by natural selection or Milankovitch's theory of glaciations due to orbital changes. But it is important to realize that we need a better hypothesis than the one we currently have.

The fourth part discusses why the Winter Gatekeeper hypothesis is superior to our current consensus hypothesis. We shed light on the fallacy that climate models are the final arbiters of climate change science. The real judgment lies in the divergent predictions of the two hypotheses, especially when looking at a relatively short period, such as the next 25 years. While the changes we see over the next 25 years will not confirm the validity of either hypothesis, they should reveal which one is flawed.

If you've read this book, you've gained a great deal of insight into the nuances of climate change – much more than many public figures who confidently discuss the topic. If you weren't skeptical before, you should now question the proclaimed crisis and the proposed solutions. Skepticism is at the heart of scientific inquiry, and even as a scientist myself, I adhere to one of the most important principles of science: *"Nullius in verba,"* or take no person's word for it. Scientists have a responsibility to present the evidence while acknowledging that their answers are tentative and subject to possible error or incompleteness.

Supporting the policies you believe are best for your society is the essence of democracy. To ensure that your decisions are truly yours, it's important to be vigilant against deception. Skepticism is your only armor against this. Cultivate your skepticism and embrace doubt, for it is better *"to live with doubt and uncertainty and not knowing than to have answers which might be wrong."*[442]

---

[442] Feynman, R., 1981. In: "Feynman: The Pleasure of Finding Things Out" BBC Horizon, Series 18, episode 9 (23 November 1981) vimeo.com/340695809

The Winter Gatekeeper

The greenhouse effect is strong in the tropics and weak at the poles in winter. As a result, increased heat transport to the poles makes the planet cooler because they act as cooling radiators. Increased heat transport in winter makes the Earth spin faster.

The climate exhibits decades-long heat transport regimes separated by abrupt shifts. These regimes manifest as oceanic oscillations that reflect different transport intensities, resulting in different surface temperature trends.

Heat transport and energy loss from the Arctic is determined by the strength of the polar vortex, which is undermined by atmospheric planetary waves. These waves are modulated by multiple factors, including the Quasi-Biennial Oscillation, multidecadal oceanic oscillations, El Niño, volcanic eruptions, and solar activity. They act as the gatekeepers of winter heat transport.

Solar activity affects El Niño, the Earth's rotation rate, and heat transport. Its climate effects are manifested in Arctic temperatures and the frequency of cold winters in the Northern Hemisphere. Its short-term effects are often obscured by the other gatekeepers, but the energy change caused by solar activity anomalies is cumulative. Over several decades of solar divergence, it becomes important, and when it persists for a century or two, it causes the most significant climate changes in thousands of years.

High solar activity in the period 1933-1996 was responsible for the retention of more energy in the climate system and global warming.

GLOSSARY

^{14}**C**: Unstable isotope of carbon with an atomic weight of 14 and a half-life of about 5,700 years. It is produced by the action of high-energy cosmic and solar radiation on atmospheric nitrogen. It is used for radiocarbon dating back about 40,000 years and as a proxy for past solar activity. Its production is affected by solar magnetic activity and geomagnetic variations.

– A –

Abrupt climate change: Climate change characterized by a sustained change in one or more climate variables at a rate greater than that observed 80% of the time, resulting in a different climate state that may last for decades or longer.

Abrupt climatic event: A period of centuries that exhibits significantly altered climate variables on a global or hemispheric scale as a result of abrupt climate change, constituting a different climate state.

Albedo: Is the fraction (percentage) of solar radiation that is reflected by a surface. Atmospheric albedo due to cloud cover is the largest contributor to the Earth's albedo. Surface albedo is highest for ice and generally lowest for ocean.

Angular momentum: A vector quantity determined by the rotational momentum of a rotating body or system, which is equal to the product of the angular velocity of the body or system and its moment of inertia with respect to the axis of rotation. The direction of the vector is the axis of rotation.

Anomaly: Refers to a temperature scale, usually in Kelvin or Celsius degrees, where the zero value has been placed at the average temperature over a period of time, usually 30 years. The name is unfortunate because it suggests that temperature changes are anomalous.

Anthropogenic: Caused by past and present human activities.

Aphelion: The point in an orbit that is farthest from the sun. For the Earth, it currently occurs around July 5.

Arctic Oscillation: Also known as the Northern Hemisphere Annular Mode, the Arctic Oscillation is a mode of climate variability that affects the winds that circulate counterclockwise around the Arctic. A positive Arctic Oscillation is characterized by strong winds, a ring-like jet stream, low surface pressure in the Arctic, and cold air masses confined to the polar regions. A negative Arctic Oscillation is characterized by weaker winds, a meandering jet stream, high surface pressure in the Arctic, and cold air masses penetrating non-polar latitudes. The Arctic Oscillation index is calculated by comparing the 20-90°N 1000 mBar geopotential height field with its main mode of variability for the period 1979-2000.

Atlantic Meridional Overturning Circulation: A system of surface and deep ocean currents in the Atlantic responsible for the transport of heat, salt, carbon, and nutrients. Surface currents transport heat and moisture northward from the tropics, while deep cold currents transport salt southward. Regions of overturning at either end link the two subsystems.

Atlantic Multidecadal Oscillation (AMO): A recurrent mode of climate variability in the North Atlantic associated with changes in sea surface temperature, changes in North American, European, and North African precipitation, and the intensity of North Atlantic hurricanes. It is characterized by alternating phases of 20-40 years with an amplitude of about 0.6°C in sea surface temperature.

Atmospheric wave: Atmospheric waves are motions of air in the Earth's atmosphere that have different spatial (meters to thousands of kilometers) and temporal (minutes to weeks) scales. They are periodic perturbations of some atmospheric variable (pressure, temperature, or wind speed) that may propagate or remain in their place of origin. The important waves for this book are planetary waves, a type of Rossby wave.

Atmospheric window: In the infrared, the frequency range of 8.5-13.5 µm that allows about 17% of the long-wave radiation emitted from the surface to pass through the atmosphere unimpeded. About 12% of the solar energy received at the surface is lost to space through this atmospheric window. Other important windows exist in the visible and radio frequencies.

Austral: Of or relating to the Southern Hemisphere.

<h2 style="text-align:center">– B –</h2>

Bjerknes compensation: The 1964 proposition by Jacob Bjerknes that variability in latitudinal heat transport by the ocean is largely compensated for by variability of the opposite sign in latitudinal heat transport by the atmosphere. Although not formally demonstrated due to difficulties in measuring ocean heat transport, it is generally accepted.

Boreal: Of or relating to the Northern Hemisphere.

Bray solar cycle: A solar activity periodicity of about 2500 years, first described by Roger Bray in 1968 and linked to a climate periodicity of the same period and phase.

Brewer-Dobson circulation: A global atmospheric circulation pattern in which tropical tropospheric air rises into the stratosphere and then moves poleward as it descends. It plays a key role in transporting mass (including ozone) and heat in the stratosphere from the equator to each pole.

<h2 style="text-align:center">– C –</h2>

Climate: The general pattern of weather conditions for an area. Climate is defined statistically in terms of the mean and variability of relevant climatological variables over time scales ranging from months to thousands or millions of years.

Climate change: A change in climate identified by statistically significant changes in its climatological variables that persist over an extended period of time, typically decades or longer. By this definition, climate is always changing.

Climate regime: A climatic state characterized by little change in one or more climate variables over a period of time.

Climate sensitivity: See equilibrium climate sensitivity.

Climate shift: A small rapid change in climate from one climate regime to another.

Climate system: An interactive system consisting of five major components: the atmosphere (air), the hydrosphere (water), the cryosphere (frozen water), the land surface and the biosphere (living things).

Conduction: Conduction is the transfer of heat between particles by collisions. The flow of energy is spontaneous from a hotter to a colder body, and its rate depends on the temperature gradient and the properties of the conducting medium.

Convection: The transfer of a property of the atmosphere or ocean, such as heat, moisture, or salinity, by predominantly vertical mass motions of water or air. In meteorology and oceanography, it is the vertical equivalent of the predominantly horizontal advection.

Coupled Model Intercomparison Project (CMIP): A collaborative framework for improving knowledge of global coupled ocean-atmosphere general circulation models. Organized in 1995 by the Coupled Modeling Working Group of the World Climate Research Programme. The most recently completed phase of the project (2014-2020) is Phase 6.

Cryosphere: Part of the Earth's surface where water is in solid form, including frozen ground (permafrost). It accounts for about 7% of the Earth's surface.

– D –

Dansgaard-Oeschger event: An abrupt glacial climate event centered in the North Atlantic-Nordic Seas region, characterized by abrupt warming measured in Greenland ice cores at 7-13°C over a period of seven decades, followed by a slower return to glacial conditions over several centuries to a few millennia. Their effect is hemispheric, and they are coupled with isotope changes in Antarctica to produce a global climate feature that is well registered in global methane levels.

Dark Ages Cold Period: A climatic interval after the Roman Warm Period and before the Medieval Warm Period, characterized by cooling. It is usually dated to about 400-900 AD.

– E –

Early Holocene: The first part after the division of the Holocene into three periods of similar length. It previously had a variable span depending on the area and proxy studied, but in 2018 the International Union of Geological Sciences established its correspondence to the Greenlandian stage between 11,700-8,326 years before 2000.

Early 20th-century warming: The period of global warming between 1910 and 1945 that was of comparable magnitude (0.5°C versus 0.6°C) to the late 20th-century warming between 1975 and 2000, despite a much smaller increase in atmospheric CO_2 levels.

Earth system model: A model that incorporates biogeochemistry and the carbon cycle, and from an emissions pathway generates a model of the resulting atmospheric CO_2 levels.

Easterlies: Prevailing pattern of surface winds from east to west. Known as trade winds in the Hadley cell and polar easterlies in the Polar cell.

Eddy: A fluid flow with a different direction than the general flow. They are responsible for most of the energy and angular momentum transfer within the fluid. The size and number of eddies is a measure of turbulence. Examples of atmospheric eddies include hurricanes, cyclones and anticyclones, and Rossby waves. Oceanic eddies are responsible for upwelling and downwelling.

Eddy solar cycle: An approximately 1000-year periodicity of solar activity named after John A. Eddy, who described it in 1976.

El Niño-Southern Oscillation: Irregular, periodic 2-5-year oscillation in sea surface temperatures and prevailing wind strength over the tropical eastern Pacific Ocean that affects weather over much of the world.

El Niño: The warm phase of the El Niño-Southern Oscillation associated with warm surface waters in the central-eastern Pacific Ocean and a weakening or reversal of the easterly trade winds.

Enhanced CO$_2$ Effect hypothesis: The hypothesis that the amount of CO$_2$ in the Earth's atmosphere is the main factor controlling the temperature of the Earth's surface and that changes in CO$_2$ levels have caused most of the major climate changes in the past and are responsible for the current global warming.

Entropy: A measure of the unavailability of a system's energy to do work. It also expresses the irreversibility of a process due to the dispersion of matter or energy.

Equable climate problem: Refers to the inability of climate models to reproduce past hothouse climates of the Earth (e.g., early Eocene, Cretaceous), characterized by a reduced temperature difference between the equator and the poles, warm polar regions with reduced seasonality, and ice-free conditions at both poles, without resorting to unrealistic greenhouse gas concentrations or altered physical parameters.

Equilibrium climate sensitivity: The warming caused by a doubling of atmospheric CO$_2$ after the oceans have had time to equilibrate.

– F –

Feedback: A feedback occurs when some of the output of a system is added to or subtracted from the input, altering the result. Amplifying feedbacks are positive and damping feedbacks are negative. Systems dominated by negative feedbacks are inherently stable, and systems dominated by positive feedbacks are unstable.

Ferrel Cell: Part of the atmospheric circulation pattern proposed by William Ferrel in 1856 to explain prevailing wind patterns between 35° and 60° latitude in both hemispheres. A portion of the rising air at 60° diverges at high altitude westward and toward the equator, where it meets the opposite circulation of the Hadley Cell at 30° latitude. There, it subsides and strengthens the ridges of high pressure below it. The air then flows eastward and northward near the surface. The Ferrel Cell is driven by the presence of the Hadley and Polar Cells because its air rises in a colder region and falls in a warmer one, being driven mechanically rather than thermally. This makes it a weaker cell with more mixed winds, and its characteristic surface winds are called prevailing westerlies. The Ferrel cell is not a very good representation of reality, as strong westerlies are usually found at 10 km altitude.

Forcing: Any process or disturbance that drives climate change, usually by changing the radiative flux at the top of the atmosphere.

$$-\text{G}-$$

General circulation model: Numerical models that represent physical processes in the atmosphere, ocean, cryosphere, and land surface.

Geopotential height: It is the actual height of a pressure surface above mean sea level and is related to the density of the air below. A low geopotential height indicates the presence of cold, dense air masses below, while a high geopotential height indicates the opposite. It is measured in meters relative to a given pressure. On weather maps, height contours connect points of equal geopotential height.

Glacial cycle: The alternation of glacial and interglacial periods during the Pleistocene according to Milankovitch orbital frequencies.

Glacial inception: The transition from an interglacial to a glacial period. The establishment of glacial conditions after an interglacial period is a very slow process that can take 15,000 years or more. It is generally accepted that glacial inception begins when non-polar ice sheets begin to grow and the sea level begins to fall significantly.

Glacial period: A period of time within an ice age when the surface temperature is several degrees colder than today and the polar and mountain ice sheets are much more extensive, covering large parts of the Northern Hemisphere.

Glacial termination: A period of about 5-10 thousand years during which the transition from a glacial to an interglacial period occurs. They are usually dated at the midpoint, defined as the time when sea level rise reaches 50% of its change.

Greenhouse effect: It is the difference between the temperature at which a planet must emit infrared radiation to balance the absorbed solar radiation and the temperature at its surface. It is caused primarily by greenhouse gases in the atmosphere that absorb and emit infrared radiation. Because of the greenhouse effect, the Earth's surface is 33°C warmer than it would be with an atmosphere transparent to infrared radiation or with no atmosphere at all.

Greenhouse gas (GHG): A gas that absorbs and emits energy in the thermal infrared portion of the spectrum. The major GHGs in the Earth's atmosphere are water vapor (H_2O_v), carbon dioxide (CO_2), methane (CH_4), nitrous oxide (N_2O), ozone (O_3), chlorofluorocarbons (CFCs), and hydrofluorocarbons (HFCs). In climatology, the term may refer only to non-condensing greenhouse gases, excluding water vapor.

Greenhouse theory: Theory that describes how the balance between absorbed solar radiation and emitted infrared radiation determines the surface temperature of a planet with an atmosphere containing greenhouse gases. Because of the presence of greenhouse gases, most of the infrared radiation emitted into space comes from the atmosphere rather than the surface, and the surface temperature becomes warmer. Changes in the amount of greenhouse gases cause an imbalance between absorbed and emitted energy due to a change in the amount of infrared radiation emitted. The balance is restored by a change in surface and atmospheric temperature, causing a change in climate.

– H –

HadCRUT: A global surface temperature dataset produced by the Hadley Centre of the UK Met Office and the Climatic Research Unit of the University of East Anglia. The current version is HadCRUT5.

Hadley Cell: Part of the atmospheric circulation pattern proposed by George Hadley in 1735 to explain prevailing wind patterns near the equator (trade winds). High solar radiation in the equatorial band causes warm air to rise. At high altitudes, the warm air moves poleward and is deflected eastward by the Coriolis force. At 30° latitude, the air sinks and closes the loop by moving equatorward and westward at the surface, creating the trade winds (easterlies).

Hiatus: In climatology, any period during the instrumental era of temperature measurement (since 1850) when little or no warming has occurred. Hiatuses appear to be the low phase of an approximately 65-year periodicity. The first hiatus occurred between 1879 and 1909. The second hiatus occurred between 1944 and 1974. A third hiatus, popularly known as "the pause," began in 1998 and lasted until 2014.

Holocene: The current interglacial and geologic epoch. The International Union of Geological Sciences has stratigraphically defined the base of the Holocene as 11,700 years before 2000.

Holocene Climatic Optimum: A period within the Holocene when the highest global average temperatures were reached. Although its timing varied between different regions, it can be considered to have occurred globally between approximately 9600 and 5500 years before present.

Holton-Tan effect: A phenomenon in which the strength of the northern stratospheric winter polar vortex synchronizes with the equatorial quasi-biennial oscillation. The vortex becomes stronger and colder when the quasi-biennial oscillation is in its westerly phase and weaker and warmer when it is in its easterly phase.

– I –

Ice age: Any geological period in Earth's history characterized by the presence of large continental ice sheets. We are currently in the Quaternary Ice Age, as ice sheets cover both Greenland and Antarctica. Within an ice age, colder glacial periods or stadials alternate with warmer interglacial periods or interstadials. Historically and popularly, the term ice age is used as a synonym for glacial periods, leading to confusion.

Indo-Pacific Warm Pool: A large area ($>30 \times 10^6$ km^2) in the tropical western Pacific and eastern Indian Oceans, comprising about 7% of the Earth's surface, that is permanently above 28 °C (82 °F). The high temperature causes deep convection, producing clouds as high as 15 km and significant atmospheric circulation effects. It is an important part of the global climate system.

Infrared radiation: Radiation with a wavelength between 0.75-1000 μm. The infrared relevant to climate is the thermal infrared, between 3-15 μm.

Insolation: The amount of solar energy received per unit of surface area for the period under consideration.

Intergovernmental Panel on Climate Change (IPCC): The United Nations body charged with producing reports that evaluate the published science on climate change.

Intertropical Convergence Zone (ITCZ): Is the climatic equator of the planet, the area around the Earth where the northeast and southeast trade winds converge, creating what sailors call the calms. It is formed by high tropical insolation that drives the convection of warm, moist air, forming the ascending branch of the Hadley cell. As the air rises, it cools, forming a band of clouds and thunderstorms that encircles the globe near the equator. The location of the ITCZ varies with the seasons, moving north from January to July and south from July to January, following the band of maximum solar flux. Tropical monsoons are linked to the position of the ITCZ, and long-term changes in its position due to changes in insolation resulting from precessional and obliquity changes have a very important effect on paleoclimate evolution.

$$- \textbf{L} -$$

La Niña: The cold phase of the El Niño-Southern Oscillation associated with cold surface waters in the central-eastern Pacific Ocean and a strengthening of the easterly trade winds.

Lapse rate: The rate of change in atmospheric temperature with increasing height. It is positive when the temperature decreases with increasing height and negative when it increases.

Last Glacial Maximum: The time during the last glacial period when ice sheets were at their greatest extent. It is defined based on the sea level being 125 meters below present levels between 26,500 and 19,000 years ago.

Late Cenozoic Ice Age: The current ice age that began 33.9 million years ago at the Eocene-Oligocene boundary with the onset of Antarctic glaciation. It spans the second half of the Cenozoic Era or "Age of Mammals."

Late Holocene: The last part after the division of the Holocene into three periods of similar length. It previously had a variable span depending on the area and proxy studied, but in 2018 the International Union of Geological Sciences established its correspondence to the Meghalayan stage from 4,250 years before 2000 to the present.

Late 20th-century warming: The period of global warming between 1975 and 2000 that was of comparable magnitude (0.6°C versus 0.5°C) to the early 20th-century warming between 1910 and 1945, despite a much larger increase in atmospheric CO_2 levels.

Latitudinal insolation gradient: The gradient, determined by the angle of incidence of solar radiation, in the amount of energy received from the Sun at the Earth's surface in a given period of time (e.g., kWh/m^2 day) that varies with latitude. This gradient acts on the climate system through differential solar heating, which determines the Earth's latitudinal temperature gradient that drives atmospheric and oceanic circulation and creates the different climate zones. The latitudinal insolation gradient changes with the seasons and, on longer timescales, with changes in obliquity and precession.

Latitudinal temperature gradient: The surface temperature gradient, determined mainly by differential solar heating, that changes with latitude and by the efficiency of heat transport from the tropics to the poles. The latitudinal temperature gradient drives atmospheric and oceanic circulation and creates the different climate zones. The latitudinal temperature gradient changes with the seasons and on longer timescales with changes in obliquity and precession. However, unlike the latitudinal insolation gradient, it also changes when there is a latitudinal change in surface temperatures, such as with recent Arctic warming. The latitudinal temperature gradient is a central property of the Earth's climate system.

Length of day (LOD): A measure of variations in the length of the day determined by the difference between the astronomical length of the day and 86,400 International System seconds.

Little Ice Age: A climatic interval following the Medieval Warm Period characterized by cooling and the expansion of mountain glaciers. There is no agreement on the duration of the Little Ice Age. In this book, the Little Ice Age is considered to span the period 1300-1845.

Low-gradient paradox: The physical paradox posed by equable climates with warm poles requiring enhanced meridional heat fluxes to maintain mild high latitude temperatures while keeping low latitudes from becoming excessively warm, and the turbulence theory tenet that the meridional heat flux is proportional to the meridional temperature gradient.

Lunisolar: Caused by the Sun and the Moon.

<h2 style="text-align:center">– M –</h2>

Medieval Warm Period: A climatic interval following the Dark Ages Cold Period and preceding the Little Ice Age, characterized by warming and contraction of mountain glaciers. It is usually dated to about 950-1250 AD.

Meridional transport: The north-south transport of heat, moisture, clouds, chemicals, and angular momentum along the Earth's meridians.

Meridional wind circulation: The north-south component of the atmospheric circulation.

Middle Holocene: The second part after the division of the Holocene into three periods of similar length. It previously had a variable span depending on the area and proxy studied, but in 2018 the International Union of Geological Sciences established its correspondence to the Northgrippian stage between 8,326 and 4,250 years before 2000.

Milankovitch theory: The theory proposed by Milutin Milanković in 1920 to explain the alternation of interglacial and glacial periods during the Pleistocene as a result of long-period changes in the Earth's orbit caused by the gravitational pull of the Sun, Moon, and planets. In 1976, Pleistocene climate proxies were shown to follow the orbital frequencies proposed by Milankovitch.

Modern global warming: The period of warming since the end of the Little Ice Age between 1845 and the present.

Modern Solar Maximum: The period 1935-2000, which is the longest period of above-average decadal solar activity in the 275-year-long sunspot record.

– N –

Neoglacial: The period of the Holocene between about 5200-400 years before present, characterized by an increase in global glacier advance and a decrease in global temperature. It is thought to have been driven by the decrease in Earth's obliquity and the decrease in northern summer insolation due to precession.

Neoglaciation: The increasing trend in global glacier advance following the Holocene Climatic Optimum identified and named by François Matthes in the 1940s.

North Atlantic Oscillation: A north-south dipole of the atmospheric pressure mode of variability over the North Atlantic that exhibits pronounced climatic teleconnections. One center of the dipole is over Greenland, and the other center of opposite sign is in the central North Atlantic between 35-40°N. The North Atlantic Oscillation index is constructed from the pressure difference between the Icelandic Low and the Azores High. The oscillation alternates between a positive mode with a strong Icelandic Low and Azores High and a negative mode with a weak Icelandic Low and Azores High. Strong positive phases of the North Atlantic Oscillation show above-average temperatures in the eastern United States and northern Europe and below-average temperatures in Greenland and often in southern Europe and the Middle East. They are also associated with above-average winter precipitation over northern Europe and Scandinavia and below-average winter precipitation over southern and central Europe. Opposite patterns of temperature and precipitation anomalies are typically observed during strong negative phases of the North Atlantic Oscillation.

– O –

Obliquity: The angle between the Earth's orbital plane (ecliptic) and the equatorial plane, also called axial tilt. It can vary between 22.1° and 24.5° and is currently 23°26′ (23.44°) and decreasing. It is the main Milankovitch parameter for orbital forcing of climate, responsible for the spacing and occurrence of interglacials.

Oceanic Niño Index: NOAA's El Niño-Southern Oscillation index based on sea surface temperature in the Niño 3.4 region (5°N-5°S, 120-170°W).

Ozone layer: The part of the stratosphere that contains about 90% of the Earth's ozone. Most of the ozone is between 20 and 35 kilometers (12-22 miles) high. It plays a critical role in protecting life on land by absorbing the most energetic and damaging wavelengths of ultraviolet light.

– P –

Pacific Decadal Oscillation (PDO): A mode of climatic variability in the North Pacific with extensive teleconnections. It is defined as the dominant pattern of sea surface temperature anomalies in the North Pacific basin. It is strongly influenced by the El Niño-Southern Oscillation and represents a long-term envelope of the El Niño-Southern Oscillation variability. Its phases can last decades and when positive, present negative sea surface temperature anomalies in the central and western North Pacific and positive sea surface temperature anomalies in the eastern North Pacific and vice versa. A weak mirror image of these anomalies occurs over the South Pacific.

Pause: See hiatus.

Perihelion: The point in an orbit at which the Sun is closest. For Earth, it currently occurs around January 4.

Petrological tracer: A mineral sediment whose origin can be traced to geological formations within a given region.

Planetary wave: A type of Rossby wave with very long wavelengths (thousands of kilometers) that can propagate vertically into the stratosphere if they are large enough and stratospheric conditions allow.

Polar Cell: Part of the atmospheric circulation pattern. Very cold air at high altitude in the polar regions sinks, creating an area of high pressure. It then moves equatorward and westward at the surface (polar easterlies) toward the 60° parallel, where it meets opposite, warmer, more humid winds from the Ferrel Cell. The air rises and diverges, with some of it moving poleward and eastward at high altitude to close the loop.

Polar front: The weather front boundary between the polar cell and the Ferrel cell around 60° latitude, near the polar regions, in both hemispheres. At this boundary, a sharp temperature gradient occurs between these two air masses, each with a very different temperature.

Polar vortex: A large region of low-pressure cold air that rotates cyclonically (clockwise in the Southern Hemisphere, counterclockwise in the Northern Hemisphere) around both poles, manifesting in both the troposphere and stratosphere. The stratospheric polar vortex is an autumn-spring phenomenon, while the tropospheric polar vortex usually persists, albeit weakened, throughout the summer.

Precession: In a rotating body or system, precession is the relatively slow (compared to the rate of rotation) change in the orientation of the axis of rotation. Earth's axial precession is responsible for the slow shift of the equinoxes (and seasons) along its orbit, with very important climatic implications, and is one of the Milankovitch orbital forcings. The Earth's orbit around the Sun also has an axis of rotation that exhibits precession (apsidal precession), which modifies the frequencies of the precession of the equinoxes.

Proxy (climate): Preserved physical characteristics of the past that allow the reconstruction of past climatic conditions.

– Q –

Quasi-Biennial Oscillation (QBO): Is a quasi-periodic oscillation in the strong stratospheric winds that circle the planet high above the equator, descending about 1 km per month. The new belt that develops above the old one has the opposite orientation. At a given altitude (measured at 30 hPa), westerlies and easterlies alternate every about 14 months. The amplitude of the easterly phase (QBOe, negative values of wind speed) is about twice as strong as that of the westerly phase (QBOw, positive values of wind speed) and lasts a little longer, but low-speed easterly winds (-5-0 m/s) behave climatically like westerly winds. The QBO has important effects on the climate of the Northern Hemisphere, especially in winter, by influencing the strength of the polar vortex and the jet stream.

Quaternary: The current and most recent of the three periods of the Cenozoic Era, spanning the last 2.59 million years and divided into two epochs: the Pleistocene (2.59 million to 11,700 years ago) and the Holocene (11,700 years ago to present).

– R –

Radiative forcing: The net change in the energy balance of the Earth system due to an imposed perturbation.

Reanalysis: A scientific method for developing a comprehensive record of how weather and climate change over time. It combines past model-based short-range weather predictions with observations through data assimilation to produce a synthesized estimate of the state of the climate system. The process mimics the production of daily weather forecasts for climate applications.

Roman Warm Period: A very long climate interval after the 2.8 kiloyear event and before the Dark Ages Cold Period, characterized by warming and contraction of mountain glaciers. Some authors date it to 2550-1650 years ago (550 BC-350 AD), while others limit it to 250 BC-350 AD. Historical and climatic evidence suggests that the Roman Warm Period may have been as warm or warmer than today.

Rossby wave: A type of inertial wave generated in rotating planets due to differences in the Coriolis effect with latitude. Atmospheric Rossby waves are giant meanders in high-altitude winds with wavelengths of several hundred kilometers. Oceanic Rossby waves are much smaller and are generally associated with the thermocline.

– S –

Seasonality: The difference between the seasons. In paleoclimatology, this difference has changed over time due to changes in precessionally linked insolation. Currently, winters in the Northern Hemisphere are warmer and summers are cooler than during the early Holocene, showing a decrease in seasonality over time.

Stadium Wave hypothesis: The hypothesis, proposed by Marcia Glaze Wyatt in 2012, of a multidecadal climate signal propagating across the Northern Hemisphere in a network sequence of synchronized ocean, atmosphere, and sea ice indices. All indices vary on the same timescale of approximately 64 years peak-to-peak throughout the 20th century, with one index leading the next in a consistently ordered lead-lag fashion.

– T –

Thermocline: Thin layer in a large body of fluid that separates a zone of increased temperature mixing from a zone of decreased temperature mixing, resulting in a more rapid rate of temperature change than above and below.

Thermodynamics: The branch of physics concerned with heat, work, and temperature and their relationship to energy, entropy, and the physical properties of matter and radiation.

Top of the atmosphere: The height at which the energy exchange between space and Earth is assumed to occur for energy budget calculations. It must be below the height at which satellites measure the Earth's outgoing radiation. Most studies use an altitude of 100 km (62 miles).

Torque: A measure of the force that causes an object to rotate and gain angular acceleration. It is equal to the product of the magnitude of the force and the distance from its point of application to the axis of rotation.

Total solar irradiance: The total amount of solar radiation in W/m^2 received outside the Earth's atmosphere on a surface perpendicular to the incoming radiation and at the Earth's mean distance from the Sun. It can only be reliably measured by satellites, and the record goes back only to 1978. The solar cycle variation of total solar irradiance is on the order of 0.1%.

Treeline: The edge of habitat, at high altitudes or high latitudes, beyond which trees cannot grow.

– W –

Westerlies: Dominant pattern of surface winds from west to east. In the Ferrel cell, they are known as prevailing westerlies.

Winter Gatekeeper hypothesis: A hypothesis that proposes long-term changes in the amount of heat and moisture transported poleward as a major cause of climate change. The main effect of this mechanism on decadal to centennial time scales is due to changes in the amount of heat transported to the Arctic during winter. Solar variations are an important modulator or "gatekeeper" of this transport.

– Z –

Zonal wind circulation: The longitudinal (east-west) component of the atmospheric circulation.

INDEX